GREAT AMERICAN DISEASES

SECOND EDITION

GREAT AMERICAN DISEASES

Their Effects on the Course of North American History

IAN R. TIZARD
JEFFREY M.B. MUSSER

ELSEVIER

ACADEMIC PRESS
An imprint of Elsevier

Academic Press is an imprint of Elsevier
125 London Wall, London EC2Y 5AS, United Kingdom
525 B Street, Suite 1650, San Diego, CA 92101, United States
50 Hampshire Street, 5th Floor, Cambridge, MA 02139, United States

For accessibility purposes, images in electronic versions of this book are accompanied by alt-text descriptions provided by Elsevier. For more information, see https://www.elsevier.com/about/accessibility.

Publisher's note: Elsevier takes a neutral position with respect to territorial disputes or jurisdictional claims in its published content, including in maps and institutional affiliations.

Notices
Knowledge and best practice in this field are constantly changing. As new research and experience broaden our understanding, changes in research methods, professional practices, or medical treatment may become necessary.

Practitioners and researchers must always rely on their own experience and knowledge in evaluating and using any information, methods, compounds, or experiments described herein. In using such information or methods they should be mindful of their own safety and the safety of others, including parties for whom they have a professional responsibility.

To the fullest extent of the law, neither the Publisher nor the authors, contributors, or editors, assume any liability for any injury and/or damage to persons or property as a matter of products liability, negligence or otherwise, or from any use or operation of any methods, products, instructions, or ideas contained in the material herein.

ISBN: 978-0-443-31404-9

For information on all Academic Press publications visit our website at
https://www.elsevier.com/books-and-journals

Publisher: Peter Linsley
Editorial Project Manager: Ellie Barnett
Production Project Manager: Prasanna Kalyanaraman
Cover Designer: Christian J. Bilbow

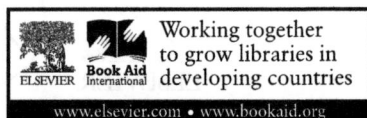

Working together
to grow libraries in
developing countries

www.elsevier.com • www.bookaid.org

Typeset by TNQ Tech

Contents

Acknowledgments

As always, this book would not have been possible without the support of our wives, Claire and Karen.

Introduction: How infectious diseases have shaped American history

Ian R. Tizard, Jeffrey M.B. Musser

It is now more than 105 years since the great influenza pandemic of 1918 killed over 600,000 Americans. It has been over 60 years since Jonas Salk's polio vaccine began to eliminate that scourge from the continent. Infectious diseases that had killed thousands and caused widespread despair and death in the past have been largely brought under control. The emergence of the coronavirus that causes COVID-19 came as an ugly reminder that infectious diseases have not been eliminated. They have been controlled temporarily, and only in developed countries. They will inevitably continue to return and kill because they are an integral part of life on Earth. In this microbial world, humans and other living creatures are merely a resource to be exploited by microbes.

The history of the Americas has been defined by major infectious disease epidemics, beginning with the massive destruction of Native American societies by European smallpox, to diseases that determined the course of the War of Independence, and to the major engineering projects designed to provide citizens with clean water and efficient sewage removal. Infectious diseases have defined and in a large part controlled our history. The Americas are not unique in this respect. Other countries and continents have also been profoundly affected by infectious disease outbreaks.

That said, there has been a perception that those epidemics belonged to history. For much of the past 50 years, infectious diseases were no longer considered a threat. Many believed that if more diseases arrived, they could be handled. Certainly, the history of the 20th century suggests a progressive and inexorable improvement in our health resulting from a decline in infectious diseases. On the face of it, humans appeared to have won; we beat the microbes.

This, of course, was an illusion. First, we must not extrapolate the situation in the developed world to less developed regions. Second, we can see cracks appearing in our own medical defenses, ranging from bacterial antibiotic resistance to vaccine hesitancy by large segments of the population. As the disease threat appears to be insignificant, so the imperative to vaccinate is reduced. Conversely, the results of global overpopulation, encroachment on wilderness areas, and climate change have increased the potential for disease transmission and provided bacteria and viruses (as well as their insect vectors) with opportunities to spread into previously disease-free environments.

In rich countries such as the United States, life expectancy, which had hovered around 30 to 40 years for hundreds of years, had reached 50 by 1900 and 80 by 2010. We long outlive the lifespan of our grandparents' generation. This increase has been predominantly due to the control of infectious diseases. Antibiotics, water chlorination, pasteurization, adequate sewage treatment, and, above all, vaccination have profoundly changed the human condition. The great diseases described in this book were largely considered to belong to the distant past.

Notwithstanding the significant increase in life expectancy and an effective doubling of our lifespan, what we have seen over the past 75 years has been no more than a temporary lull in hostilities. We have generated a pause in the severity and prevalence of infectious diseases. Victory has not been total. Infectious diseases still prevail in less developed countries. Increasing human populations and global travel provide infectious agents with new opportunities. COVID-19 spread from China to the United States in 14 hours!

Another feature of our current expectations of prolonged good health is an intolerance of massive casualties. Recent disease outbreaks caused by West Nile virus, SARS-1, or Zika viruses killed only a few thousand individuals, a far cry from the massive pandemics of the past. Nevertheless, they resulted in a panic that subsided once the threat was removed. This "panic-and-forget" pattern was a feature of American responses to infectious diseases until COVID-19 arrived. It is appropriate to remind ourselves of that overused adage that "those who do not remember history are doomed to repeat it." If we are to maintain our relative resistance to infectious diseases, we need to remember another cliché: The price of freedom (from disease) is also eternal vigilance.

The diseases that we have chosen to explore in this text are those we consider the most historically significant. After all, many of the outbreaks we describe have been covered previously with entire texts dedicated to individual diseases or epidemics. Nevertheless, we believe that a holistic approach serves to place the North American infectious disease history in perspective.

This book follows, in part, the curriculum of our introductory microbiology course at Texas A&M University titled "Great Diseases of the World." That course was conceived as an introductory course in microbiology.

Rather than teach the dry topics of biochemical reactions and microbial diseases, we chose to focus on the "cool bits" of microbiology and the lurid history of many infectious diseases. Therefore this book is not designed to be a comprehensive list of all the major infectious and parasitic diseases that have affected Americans over the years. Rather, it is a look at the most important and, in our opinion, the most interesting. We hope that you will agree.

This book does not need to, nor is it intended to be, read in a linear manner from the first to the last page. The reader or instructor is free to pick and choose the disease of interest and topics within each chapter. Most aspects of biology and history are not linear but convoluted and often circular. However, the order of the disease chapters in this book is chronological based on when they were first identified and began to cause problems in North and Central America.

One consistent feature of the history of medicine is an ongoing debate regarding specific diagnoses prior to the modern era. Until about 150 years ago, diagnosis was a subjective art. Disease diagnosis was a matter of physician opinion rather than science. Molecular biology has changed all that. The DNA from the teeth in human skeletal remains has enabled us to identify microbial DNA in bodies more than 25,000 years old. Most of the diseases described in this text have been around for many thousands of years. There have been few surprises, although the origins of the sexually transmitted disease syphilis still remain uncertain.

In January 2020, COVID-19 emerged from Wuhan, China. Within a remarkably short time, it spread worldwide. The disease caused by this coronavirus, COVID-19, appears to have peaked, but the virus is likely to persist. We have sought to place COVID-19 in its historical context. The pandemic has many features in common with epidemics of the past. Technology may change, but human behavior does not. Indeed, COVID-19, unlike the other diseases in this text, has become highly politicized. The rise of the internet and modern communications technology have enabled disinformation to spread. Anti-science, especially anti-vaccine, sentiment has resulted in more fatalities than necessary in the United States. That is not a topic for discussion in this text. The numbers speak for themselves. Likewise, neither history nor science will stop at COVID. More infectious diseases and the microbes that cause them are lurking. The next pandemic is just waiting for its opportunity.

CHAPTER ONE

Infectious diseases and their causes

Contents

Infectious diseases have had, and continue to have, profound effects on the course of American history. Historically they killed millions, caused untold suffering, and played a key role in the colonization and development of both the United States and Canada. And they have not ceased. In 2020 the SARS-CoV-2 coronavirus, the cause of COVID-19, escaped from China and spread around the world. It has killed millions of people and caused enormous social disruption.

In addition to social factors, the impact of infectious disease outbreaks, epidemics, and pandemics is determined largely by two key factors. One is the ability of an organism to cause serious disease—its virulence. The other is the ability of humans to combat these infections—in other words, their immunity. The outcome of such diseases is determined by the balance between virulence and immunity. The decline in the apparent importance of infectious diseases over the past 100 years has been a direct result of improved sanitation practices as well as our growing ability to prevent and

Great American Diseases
ISBN: 978-0-443-31404-9
https://doi.org/10.1016/B978-0-443-31404-9.00001-X

treat these infections through the use of vaccines and drugs such as antibiotics. In the absence of vaccines and drugs, organisms may spread unchecked, especially in large, dense populations [1]. That is the case in our ongoing outbreak of COVID-19. It was certainly the case in the past before modern medicine achieved its successes and it will remain the case in the future.

While the idea that some diseases were caused by invisible microorganisms was floated from time to time, this was mere speculation until Antonie van Leuwenhoek in the Netherlands learned how to make lenses capable of magnifying up to 275 times or greater. He used these in single-lens microscopes to examine a diverse mixture of biological specimens. As a result, he was the first to observe bacteria and protozoa [2]. When he examined the gunk from between his teeth, he saw all sorts of small organisms that he called *kleine diertjens*—Dutch for "small animals." This word was translated into English as "animalcules." These creatures were swimming around and he drew and described their shapes. He also examined his own feces and found a parasitic protozoan that we now call *Giardia*. Van Leuwenhoek reported this finding of oral bacteria in a letter to the Royal Society in London in September 1683.

The significance of these microscopic animalcules was not readily apparent at the time, and while other thoughtful individuals such as Cotton Mather in Boston made statements speculating that invisible organisms were the cause of disease, it was not until the science of microbiology developed in the mid-19th century that specific organisms were recognized as causes of specific diseases. In 1838, Gottfried Ehrenberg in Germany called these animalcules bacteria, from the Greek *bakteria*, meaning "a little stick" because they were rod shaped.

Robert Koch in Germany was the first to link a specific bacterium with a disease when he identified the anthrax bacillus [3]. Koch was a physician working in practice in Wöllstein, in eastern Germany. As the district medical officer, he was confronted by an ongoing outbreak of anthrax, a disease that had killed hundreds of people and thousands of cattle. On examining the blood of dead animals under a very primitive microscope, Koch observed rod-shaped structures. He also showed that blood from a sheep that had just died of anthrax, when injected into a mouse, would kill the mouse within 24 hours. Then he found the rod-shaped structures in the blood and tissues of the mouse! Spleen tissue from the dead mouse, when injected into another mouse, would give the same result—rapid death. Koch determined that the rod-shaped structures were bacteria. He eventually developed a method of growing the bacteria in culture and was thus the first to link a specific bacterium to a specific disease. His paper demonstrating that anthrax was caused by the bacterium *Bacillus anthracis* was published in 1876.

As a result of his growing reputation, Koch moved to Berlin and joined the Imperial Health College. Here he made more significant advances. He found that he could "fix" bacteria to a glass microscope slide by drying them from a liquid solution, he could stain them with aniline dyes, he worked on improving microscopes, and he was the first to publish photographs of bacteria. Others in his laboratory also made significant advances. For example, the plate technique of growing bacteria in pure culture was developed in his laboratory. Two of his assistants, Walter and Fannie Hess, discovered that agar, a gel obtained from seaweed, could be used to make a nutrient medium on which bacteria could be cultured. Another of his assistants, Julius Petri, designed a shallow dish (now called a Petri dish) to hold the agar-based medium and permit bacteria to grow without contamination. Subsequently, Robert Koch went on to describe other disease-causing bacteria such as *Mycobacterium tuberculosis* in 1882 and *Vibrio cholerae* in 1883. Koch, as the founder of the science of bacteriology, received the Nobel Prize for Medicine in 1905.

From the point of view of history, it is important to emphasize the dates of Koch's discoveries. Any disease diagnosis prior to the 1870s must be considered speculative. Physicians could diagnose some very obvious diseases such as smallpox or yellow fever, but in many other disease outbreaks reported prior to the 1870s, their causes must remain speculative. This is especially true of diseases with nonspecific clinical signs such as fever or diarrhea.

As a result of Koch's discoveries, bacteriology rapidly developed as a discipline, and competing scientists sought to be the first to identify the cause of specific diseases. It was not until the end of the 19th century that protozoan parasites, also visible under the microscope with suitable stains, were detected and associated with specific diseases such as malaria and Texas Fever. Fungi were also identified as a cause of some diseases around the same time.

Bacteria are not the only organisms that can cause disease. Viruses can invade and destroy cells and cause many serious infections. Viruses are small molecular complexes that cannot be seen under a light microscope. It was not until 1938 that the first virus was observed by electron microscopy. Prior to that time, however, their existence had been inferred by filtering body fluids to remove bacteria and demonstrating that the filtered material was still infectious. Thus the first report of a "filterable virus" was in 1892 when it was found to be the cause of a disease of tobacco plants.

A world of microbes

While we are largely unaware of it, we live in a world dominated by microbes, especially bacteria. Half the Earth's biomass consists of microorganisms. (Plants account for 35% of the biomass and animals only about 15%.) Microbes live everywhere, including inside and on our bodies. All our body surfaces carry a dense microbial population collectively called the microbiota. The densest population of bacteria on this planet lives within our large intestine. The soil we stand on has a huge microbial population, as do the oceans. With every breath, we inhale about a million bacteria. We have about the same number of bacteria in our body as we have human cells. Almost all of these organisms have evolved to make use of the sources of energy in their environment. A few of them, very much in a minority, have evolved to obtain their energy from the human body. It is these select few that can invade and cause disease under appropriate circumstances.

Bacteria

As noted earlier, enormous populations of bacteria live on our body surfaces. They live on the skin where they are responsible for body odor. They live in our respiratory tract. They live in our mouths where they are responsible for, among other things, tooth decay and bad breath. They live in the genital tract. The greatest population of bacteria lives within our intestinal tract. They share our food with us. They send chemical signals to the body, ensuring that our defenses stay in tip-top shape, and when we die, they make our bodies decompose.

Properties and classification

The early microbiologists learned to recognize bacteria under the microscope and grow them in nutrient-rich culture media. Thus the first method of classifying what they saw was simply by their shape [4]. The round grape-like bacteria they called cocci, and the rod-shaped bacteria they called bacilli (Fig. 1.1). The cocci were subdivided into those that formed chains, like the streptococci, and those that formed clusters, like the staphylococci. Those that formed pairs they called diplococci. Some bacilli are consistently curved into a comma shape and vibrate as they move, so they are called vibrios.

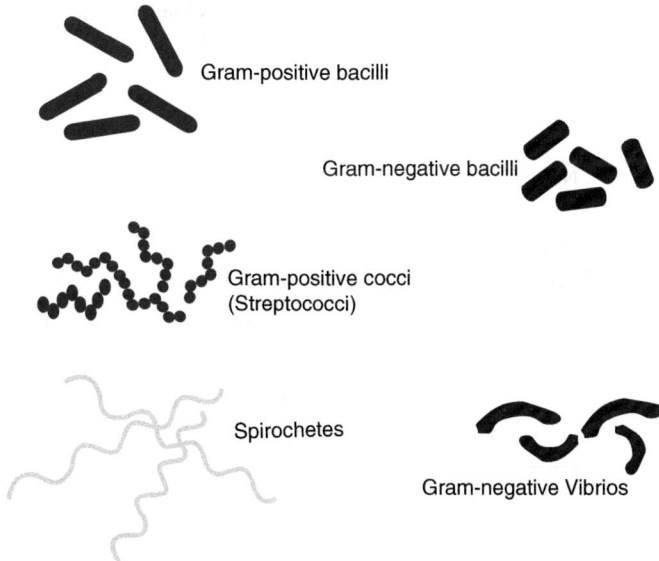

Gram-positive bacilli

Gram-negative bacilli

Gram-positive cocci
(Streptococci)

Spirochetes

Gram-negative Vibrios

Fig. 1.1 *The major bacteria and their shapes. Gram-positive bacteria stain strongly with the dye methylene blue when stained by the Gram method. Gram-negative bacteria do not retain the blue dye so that it can be washed out and they can then be stained by a red counterstain.*

Some bacteria possess long filamentous processes called flagella (Fig. 1.2). When flagella are waved around, they enable the bacteria to move; in other words, some bacteria are "motile." The precise arrangement and numbers of these flagella are also an aid to bacterial classification.

The next step in the discovery process involved the staining of bacteria by chemical dyes. The most useful and important staining technique turned out to be one developed by Hans Christian Gram, a Danish microbiologist in 1884. He was examining stained tissue sections from the lungs of patients autopsied after dying from pneumonia and noticed that his dye stained the bacteria very strongly but not the lung tissue. Upon investigation, Gram found that he could stain some bacteria with a dye called crystal violet. More importantly, he found that in some bacteria, after staining, the crystal violet was easily washed out by alcohol, while in others it persisted. Those bacteria that stain strongly with crystal violet are classified as gram positive, whereas those that readily lose the dye are classified as gram negative. Gram-positive bacteria stain blue, whereas gram-negative bacteria can be counter-stained by, for example, the red dye safranin. These differences between gram-positive and gram-negative bacteria are significant because they reflect

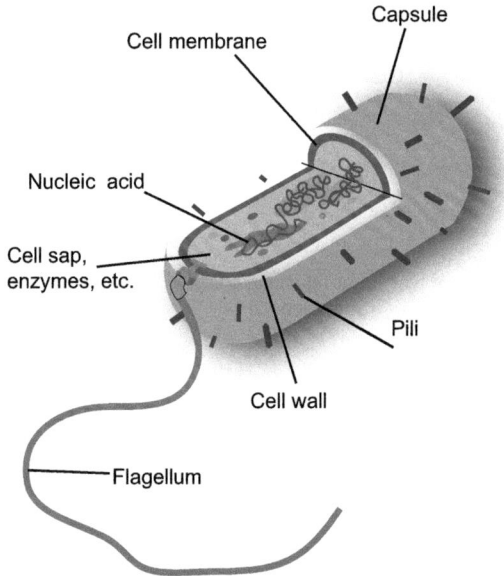

Fig. 1.2 *Pili are cell surface structures by which bacteria can attach to cells.*

major structural and biochemical differences between their cell walls. The cell walls of gram-positive bacteria are significantly thicker than those of gram-negative organisms. Some bacteria, such as the treponemes that cause syphilis, do not stain strongly with Gram stain, thus confirming their distant relationship to other bacteria.

A second staining procedure was developed by the German chemist and pathologist Paul Ehrlich and modified by Franz Ziehl and Friedrich Neelsen in 1882. This involves the use of a dye called carbol fuchsin to stain the bacteria. Once stained, certain lipid-rich bacteria can retain the stain after washing with hydrochloric acid in alcohol because it binds to lipids called mycolic acids. In those bacteria that lack mycolic acids, the carbol fuchsin is readily washed out and the bacteria are decolorized. Those that retain the dye are stained red and are classified as acid-fast bacteria. Acid-fastness is characteristic of the group of bacteria called Mycobacteria. These are the cause of diseases such as tuberculosis and leprosy. The lipids in the cell walls of these bacteria tend to make them impermeable to Gram stain. These lipids also protect them from destruction once they invade the body.

In addition to their physical properties, bacteria may be classified according to their metabolic needs. For example, some require lots of oxygen and are called aerobes, whereas others only grow in the absence of oxygen and are called anaerobes. It should come as no surprise therefore that an aerobe such as *M. tuberculosis* prefers to grow in well-oxygenated tissues like the lung. On the other hand, *Clostridium tetani*, a strict anaerobe that causes tetanus, prefers to grow in dead tissue, such as in deep wounds where oxygen tension is low. Some bacteria, such as *V. cholerae*, prefer a low oxygen tension and are therefore considered "microaerophilic."

Bacteria, with relatively simple metabolic systems, can readily use sugars as an energy source. They differ, however, in which sugars they can use. This too can be used to classify them. For example, two bacteria that cause diarrhea are *Escherichia coli* and *Salmonella enterica*. They look exactly the same, and both are gram-negative bacilli, but *Salmonella* cannot use lactose as a food source, whereas *E. coli* can.

It has long been convention to give bacteria a Latin name using two words (binomial). The first name denotes the genus while the second denotes the species. For example, the organism that causes syphilis is called *Treponema pallidum*, while the bacterium that causes the related disease yaws is called *Treponema pertenue*. Other examples are the organisms of the genus *Brucella*. All cause undulant fever, but *Brucella abortus* infects cattle; *Brucella melitensis* infects goats; *Brucella suis*, as its name implies, infects pigs; and *Brucella canis* infects dogs. All can cause disease in humans (Table 1.1).

These bacterial names may also change as more information accumulates about the organism and its properties. For example, when Alexandre Yersin first identified the bacillus that caused the plague, he called it *Bacillus pestis*. Several years later it became clear that the organism did not really belong in the genus *Bacillus*, so it was renamed *Pasteurella pestis*. Finally, in 1970, another reassessment moved it to another genus, and it is now called *Yersinia pestis*.

As more is learned about bacteria, this binomial system appears to be increasingly inadequate. For example, over 2500 distinct varieties of *S. enterica* have been identified. These varieties are designated serovars or subspecies because they can be distinguished by the use of specific antibodies made in immunized animals. These antibodies are found in blood serum, hence the term serovar.

The normal microbiota

As mentioned earlier, animal bodies are not sterile. We have enormous populations of bacteria living in our intestines, in our upper airways, and on our

Table 1.1 The characters in this book.

Name of organism	Name of disease	Agent
Bacteria		
Treponema pallidum	Syphilis	Spirochete
Borrelia burgdorferi	Lyme disease	Spirochete
Rickettsia prowazekii	Epidemic typhus	Intracellular coccobacillus
Rickettsia rickettsii	Rocky Mountain spotted fever	Intracellular coccobacillus
Salmonella enterica serotype typhi	Typhoid fever	Gram-negative bacillus
Vibrio cholerae	Cholera	Gram-negative vibrio
Mycobacterium tuberculosis	Tuberculosis	Acid-fast bacillus
Yersinia pestis	Plague	Gram-negative bacillus
Legionella pneumophila	Legionnaires disease	Gram-negative bacillus
Viruses		
Variola major	Smallpox	Orthopoxvirus
Measles virus	Measles	Morbillivirus
Yellow fever virus	Yellow fever	Flavivirus
Poliovirus	Poliomyelitis	Enterovirus
Influenza A virus	Influenza	Orthomyxovirus
Human immunodeficiency virus	AIDS	Retrovirus
SARS-CoV-2	COVID-19	Betacoronavirus
Protozoa		
Plasmodium falciparum, P. vivax, P. ovale, P. malariae	Malaria	Hemoprotozoan parasite

skin. These bacteria benefit us in several ways. Thus they help us digest food. They generate metabolites that serve as essential nutrients. They release otherwise unavailable energy from foods, and they stimulate the development of our immune system. If the microbiota is disrupted by, for example, excessive antibiotic use, the resulting "dysbiosis" can lead to changes in the immune system that may provoke allergies. The intestinal microbiota also serves a protective function insofar as it is well adapted to the intestinal environment. They can therefore outcompete and exclude poorly adapted organisms that should not be there. The intestinal microbiota are considered to be commensals, with mutually beneficial interactions between them and their host [5].

How bacteria cause disease

Bacteria that cause disease are said to be pathogenic. Obviously, there are degrees of pathogenicity, and this is measured by virulence. Thus a highly virulent pathogen is one that causes severe disease. Less virulent bacteria can cause mild disease, whereas avirulent bacteria do not cause disease. In all the major bacterial diseases described in this book, the causal agents were highly virulent.

There are many reasons why a bacterium will act as a virulent pathogen. These may be directly related to the organism's mode of growth or the lack of resistance in its host. Bacteria that invade the body, in effect, seek to feed on it. If they are successful, they may grow rapidly, and the disease will progress. For example, *Y. pestis*, the cause of the plague, possesses enzymes and toxins that kill the white blood cells defending the body. These toxins also destroy blood vessel walls so that blood leaks into tissues such as the lungs. The organism multiplies explosively and death occurs within days. Other bacteria, such as *M. tuberculosis*, grow much more slowly. The host immune system keeps most such infections under control so that the majority of individuals infected with *M. tuberculosis* appear healthy. Only if the host's immunity fails does disease develop.

A few bacteria can secrete potent protein toxins that kill their hosts. For example, strict anaerobes such as *B. anthracis*, the cause of anthrax, and *C. tetani*, the cause of tetanus, benefit by killing their hosts. Inside the dead animal, there is an anaerobic environment where they can thrive. On the other hand, most bacteria do not benefit from killing their animal host. If they lose their host, they too may die. Thus an organism such as *V. cholerae*, the cause of cholera, simply causes severe diarrhea that spreads the bacterium throughout the environment. *M. tuberculosis* causes lung infection. The resulting coughing by the victim results in the spread of tuberculosis. Many bacteria, especially gram-negative ones, possess a cell-surface coat of lipopolysaccharide molecules. This lipopolysaccharide coat is recognized by the host and triggers a defensive response. The response triggers behaviors including fever, malaise, and all the other uncomfortable signs we associate with sickness. Thus these lipopolysaccharides are also called endotoxins.

Resistance to bacterial infections is inherited. When a pathogenic bacterium invades a population there is great variation between individuals in their defensive responses. Some individuals are highly resistant, whereas others are very susceptible. Most of the population lies somewhere in between. Over time, repeated infections will kill the most susceptible, and as a result, the population as a whole will become more resistant. If, however, a completely new infectious disease appears to which nobody has resistance, then a lethal pandemic can ensue. This is what happened during the Black Death due to *Y. pestis*, when the population of Europe had never previously experienced the disease. Nobody was resistant and, as a result, mortality was very high. In general, high death rates in an pandemic reflect the invasion of a completely new microorganism. This was also the case with COVID-19.

Resistance to an infection changes in a predictable way during a disease epidemic. Thus, at the onset of the epidemic, the most virulent organisms will kill their victims, and the most susceptible victims will die. However, the organisms also die when their victims die, especially in viral diseases. As a result of the preferential elimination of the most virulent agents and the death of most susceptible victims, the severity of the epidemic will gradually decline. This was seen well in the smallpox epidemics in the Americas following the European invasions and in the 1918 influenza pandemic. The longer the infection persisted, the less severe it became.

Antibiotics

There were no really effective drugs that worked against bacterial diseases until the beginning of the 20th century. At that time, the German chemist Paul Ehrlich developed sulfonamides and salvarsan, chemicals that could selectively kill bacteria without harming the patient. The situation was further transformed when Alexander Fleming discovered penicillin, the first antibiotic. Antibiotics are chemicals produced by fungi and bacteria. Within their natural environment, organisms use these antibiotics to remove competing microbes and so enhance their own chances of success. Following the discovery of penicillin, a massive search was conducted for other antibiotics. Many new and different antibiotics were discovered. As a result, most significant bacterial diseases were rapidly brought under control, and in some circles, bacterial infections were no longer regarded as a threat, especially in developed countries. But, just as some human victims may survive infections, so too can some bacteria survive antibiotics. If used inappropriately in small doses, or for too brief a time, those resistant bacteria preferentially survive—a classic example of survival of the fittest. As a result of the growth of antibiotic resistance, their effectiveness has dropped significantly, and bacterial diseases have become a significant threat once again [6].

Viruses

Viruses are totally unrelated to bacteria. The two are not synonymous. Unlike bacteria, which consist of complex single cells, viruses are simple molecular constructs. They consist of one or two nucleic acid chains packaged within a protein coat—the capsid. In some viruses the capsid may also be covered by a lipid layer—the envelope. Viruses vary in both size and shape. Some, such as poliovirus, are very small indeed. Others, such as the poxviruses, are 10 to 20 times larger and correspondingly more complex (Fig. 1.3). They come in various shapes, but the most common are spheres or icosahedrons. Some may have spikes over their surface like the COVID-19 coronavirus. Others may be long curved rods, whereas others are bullet shaped.

Viruses do not use metabolic pathways; they do not move, and they depend totally on living host cells for their replication. Outside living cells, they are, in effect, inert particles. They can, however, bind, enter, and infect prokaryote cells such as bacteria, as well as eukaryotes such as animal and plant cells.

Viruses are so small that they cannot be seen under a conventional light microscope. All the photographs of viruses you see were made using an electron microscope. (The wavelength of visible light is too long to interact with viruses. Thus a short wavelength source such as electrons must be used.

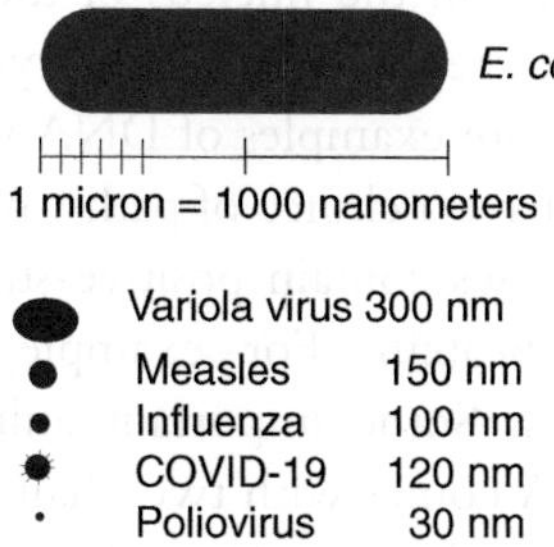

E. coli 2 microns

1 micron = 1000 nanometers

Variola virus 300 nm
Measles   150 nm
Influenza   100 nm
COVID-19   120 nm
Poliovirus   30 nm

Fig. 1.3 *The relative size of a typical bacterium (*Escherichia coli*) and some important viruses.* Bacteria can be observed under a conventional light microscope but viruses cannot.

Electrons have a wavelength up to 100,000 times shorter than the photons of visible light.) Therefore, the first viruses were not observed until 1938. Before then, their existence had to be inferred by passing fluids through ultrafilters that excluded bacteria.

In 1884, Charles Chamberland, working in Louis Pasteur's laboratory, developed a porcelain filter. The pores in the filter were small enough to stop any bacteria from getting through. Chamberlain envisioned his filter being used to purify drinking water in the home. The first virus, a plant virus called tobacco mosaic virus, was detected in 1892. It was found by passing a tissue suspension from an infected plant through a Chamberland filter and transmitting disease with the filtrate. This "filterable virus" was clearly smaller than a bacterium and could not be seen under the microscopes of that time. The first animal viral pathogen to be identified, foot-and-mouth disease virus from cattle, was found by Friedrich Loeffler and Paul Frosch using a similar filtering method in 1898.

Properties and classification

Viruses are classified on the basis of their nucleic acid content [7]. Some contain DNA and others contain RNA. Unlike other organisms, viruses consist of few proteins and hence require very few genes. The simplest viruses, such as yellow fever virus or poliovirus, have less than 10 genes. On the other hand, pox viruses such as smallpox or cowpox may contain more than 100 genes. When they enter cells, viruses force the cells to make new viruses. You may recall that the genetic information in cells is normally encoded in the DNA chain. This information is then transcribed into RNA, and the RNA is translated into the proteins needed for cell growth. Thus, when a cell is infected with a DNA-containing virus, the viral DNA is transcribed to RNA and the RNA then encodes the viral proteins. Most DNA viruses replicate within the nucleus of the host cell although poxviruses are an exception; they replicate in the cell cytoplasm. Poxviruses, herpesviruses, and parvoviruses are examples of DNA viruses.

Viruses that contain RNA have a choice of pathways by which proteins can be made. The simplest viruses contain positive-stranded RNA that is simply translated into viral protein. For example, coronaviruses are positive-stranded RNA viruses. Some important animal viruses contain negative-stranded RNA. (RNA comes with two chains that have complementary sequences and bind to each other. One, the positive strand, is used for translation; the other, the negative strand, serves as a template for making more positive-stranded RNA.) Thus a negative-stranded RNA virus must first produce a positive strand before proteins can be made. Examples of negative-stranded RNA viruses include measles, rabies virus, and influenza viruses. A third way that an RNA virus can make its proteins is by reversely transcribing its RNA to generate DNA. This DNA is then transcribed to RNA and translated to protein. Viruses that undergo reverse transcription are called retroviruses. The most important example of a retrovirus is human immunodeficiency virus (more commonly referred to as HIV), the cause of AIDS.

Some DNA viruses may have a very large genome and consist of many different proteins. One such example is smallpox virus (variola). Its double-stranded DNA encodes many different enzymes and nucleoproteins. An important feature of DNA viruses, in general, is that they replicate their DNA faithfully. Any accidental mutations that occur during replication are identified and repaired. As a result, smallpox virus does not change significantly as it passes through humans. Immunity to smallpox lasts for a very long time. The smallpox vaccine has not changed for many years. It therefore became possible to eradicate the disease simply by vaccinating everyone once.

The influenza viruses are RNA viruses. Their genome consists of eight single-stranded segments of negative-stranded RNA. As a result, they must be copied into positive strands before they can be translated into protein. These 8 RNA segments encode 10 to 14 proteins depending on the strain. In order to make positive-stranded RNA, the virus uses enzymes called RNA polymerases. However, these enzymes are prone to make mistakes. The changed (or mutated) RNA sequences result in changes in the amino acid sequences of the resulting viral protein. As a result, over a few years, influenza viral proteins may change to such an extent that the immune system can no longer recognize them and immune individuals become susceptible again. Thus influenza undergoes "antigenic drift." As a result, if an individual needs to be protected, they must receive a current influenza vaccine every year. Under some circumstances the influenza gene segments from two different flu viruses may recombine to create a completely new virus. This "antigenic shift" generates a virus to which no one is immune, and if it is transmitted between humans, a pandemic may result.

HIV is a retrovirus. Retroviruses use reverse transcriptase to reversely transcribe their RNA to DNA. This enzyme is highly inaccurate, so the proteins that make up HIV are constantly changing. When an individual is infected with HIV, within a few weeks, there will be large numbers of virus variants circulating in their bloodstream. The immune system can recognize and respond to most of these variants—but not all. As a result, it is constantly playing catch-up. Importantly, how can we select a virus strain to make a vaccine? We can protect against one, but not against hundreds of variants. This is why there is currently no vaccine available against AIDS.

Because viruses are so simple and are metabolically inactive, it is very difficult to produce effective antiviral drugs. Antibiotics have no effect on them. Some drugs are available to control influenza, and many are available to fight HIV. For most of the rest, we must rely on vaccination, good nursing, and bed rest.

How viruses cause disease

Viruses, when they enter the body, first attach to their target cells. They usually bind to existing cell surface receptors. Once they bind, the cell takes them in using a process called endocytosis. Once inside the cell, the viruses shed their protein coat. They then use their genes to compel the cell to produce copies of the viral genome. These genes are translated into new viral proteins. The nucleic acids and the proteins are then assembled into new virions. Once assembled, newly formed viral particles may bud off from the infected cell taking some cell membrane with them to form a structure called an envelope. Alternatively, they may simply rupture the cell as they make their escape. A few viruses, such as the herpesviruses, may hide inside cells and only break out when the host is immunosuppressed. Some viruses cause their target cells to become cancerous. Once they escape from cells, viruses proceed to invade other nearby normal cells until the host can mount an immune response that blocks further viral replication and spread. The clinical effects of a viral infection depend upon the cells attacked. Thus influenza and human coronaviruses mainly attack and destroy cells in the respiratory system, resulting in respiratory disease. Poliovirus destroys selected cells in the nervous system, causing paralysis, whereas yellow fever virus attacks liver cells, resulting in their destruction and the development of jaundice.

Protozoan parasites

The bacteria described earlier are prokaryotic organisms. That is, their cell DNA is not contained within a nucleus. However, there are disease-causing microorganisms that are eukaryotes. They consist of cells with nuclei containing their DNA, and these cells divide by mitosis. These are classified as Protista. Many of these organisms have two sexes and use sexual reproduction. The Protista include algae and protozoa. Many protozoans cause significant diseases in humans and other animals. The most important of these are the Plasmodia that cause malaria. (Because they are found in the bloodstream, they are called hemoprotozoa.) The Plasmodia are transmitted by mosquitos, enter liver and blood cells, and then rupture so that the cell contents are released into the victim's bloodstream. For hundreds of years, malaria was a major killer in North America, especially in the southern United States. It remains today one of the greatest killers of humans in many tropical countries.

Other parasites

In addition to protozoa, other parasitic creatures include insects such as mosquitoes and flies, as well as arachnids such as ticks. Equally important but much less obvious are parasitic worms. These, like many enteric pathogens, have been largely eliminated in developed countries as a result of effective sewage disposal and the availability of clean food and water. This, unfortunately, is not the case in many developing countries. Tapeworms and most roundworms are generally no more than an inconvenience and embarrassment now. However, some, most notably the blood-sucking roundworm, the American hookworm (*Necator americana*), had serious debilitating effects in the moister areas of the American South.

Zoonoses

Where do infectious diseases come from? Some human infections are very old indeed and probably affected our anthropoid ancestors. An example of such "heirloom" diseases includes the Treponema bacteria that cause diseases such as yaws (a nasty skin infection). *Salmonella typhi*, the cause of typhoid fever, is also restricted to humans and likely has a very ancient origin. Other examples of such heirlooms are the ectoparasites that live on the skin, such as body lice, and endoparasites that live inside the body, such as pinworms. However, other infections are acquired from other animals and have been called "souvenir" species. Diseases that are transmitted from animals to humans are called "zoonotic" diseases [8]. They can be acquired from animal bites, from eating raw meat, or from occasional contact with animal feces, especially by drinking contaminated water. Examples include rabies, COVID-19, and salmonellosis. When animals were first domesticated and began to live in close proximity to humans, then some of their infections readily jumped to their handlers. Examples include brucellosis, influenza, and measles. In a broad biochemical sense, the differences between mammals are relatively trivial, and it may take very few changes in the genetics of a microbe for it to switch hosts. It was long considered that humans acquired tuberculosis from their cattle, but recent evidence suggests that, in fact, things worked the other way around (Chapter 4).

We are currently in the midst of another transition period, and the recent interactions between humans and wildlife have led to the emergence of important diseases such as Lyme disease, yellow fever, and HIV/AIDS as well as COVID-19, Ebola, SARS, and Zika fever. Not all recent diseases are of animal origin. A good example of a disease arising from a manmade environment is *Legionella*, the cause of Legionnaires disease. *Legionella* grows in air conditioner cooling water!

Prior to the coming of the Europeans, Native Americans would have suffered from heirloom diseases that were carried to the continent by their first ancestors. Examples include yaws and tuberculosis as well as intestinal parasites. They would also have suffered from souvenir diseases picked up from the animals they hunted. They would, however, never have encountered the diseases associated with domestic animals or the massive "crowd diseases" that developed in European cities. As a result, they were very vulnerable when smallpox and measles were introduced by European invaders.

Secondary and opportunistic infections

Many of the organisms described in this book are considered to be primary pathogens. That is, by themselves, without assistance, they can invade the human body and cause disease. These are obviously very serious diseases, but they do not always represent the entire problem. Some organisms may be secondary pathogens. These are bacteria and viruses that may not be able to attack healthy, normal individuals but take advantage of their increased vulnerability caused by other infections. An example of this occurs in influenza. Influenza virus is a primary pathogen that invades the respiratory tract and destroys the cells lining the airways. The raw sites exposed in this way can then be colonized by secondary bacterial invaders. These include organisms such as *Haemophilus influenzae* or pneumococcus. As a result, some influenza victims develop severe bacterial pneumonia that eventually kills them. It is important to distinguish between primary and secondary infections. In the great influenza pandemic of 1918, much time, effort, and money was spent combating these secondary bacterial invaders rather than the primary pathogen, influenza A virus. Similar problems arose with COVID-19 [9].

Another group of infectious diseases may be caused by opportunistic organisms. These are infectious agents that make the most of an opportunity provided by immunosuppressive infections. The most important examples of these occur in individuals infected with the human immunodeficiency virus. As a result of destruction of the patient's immune system, the body's defenses are destroyed and otherwise nonpathogenic organisms such as *Mycobacterium intracellulare* or fungi such as *Pneumocystis jirovecii* can invade and kill the patient.

Pandemics and epidemics

The term endemic refers to an infectious disease prevalent in a human population. (The term enzootic refers to a disease prevalent in animals.) An epidemic refers to a situation where the number of cases of an infectious disease increases over the normal background level. Small epidemics may be called outbreaks. The scale of the disease epidemic is generally fairly large, but this will vary with the disease and its significance. There is no objective line that divides a large epidemic from a small pandemic. A pandemic, on the other hand, is used to describe a disease outbreak that affects an entire continent or multiple countries worldwide. Examples of historic pandemics include the Black Death in the 14th century that affected both Asia and Europe, the first cholera pandemic that stretched from Japan to India to Russia in the 1830s, and the Spanish flu of 1918 that occurred worldwide. The 2020 outbreak of COVID-19 coronavirus began as an outbreak in Wuhan, China, then became a local epidemic before spreading worldwide to become a pandemic.

References

[1] Dobson Andrew P, Carper E Robin. Infectious diseases and human population history. BioScience 1996;46(2):115−26. 00063568. https://doi.org/10.2307/1312814.
[2] Folkes M. Leeuwenhoek's curious microscopes lately presented to the Royal Society by Martin Folkes, 1723. From the Royal Society Catalogue Reference Number CLP/2/17.
[3] Blevins MS. Bronze, Robert Koch and the "golden age" of bacteriology. Int J Infect Dis 2010;14(9):744−51.
[4] Schleifer Karl Heinz. Classification of bacteria and archaea: past, present and future. Syst Appl Microbiol 2009;32(8):533−42. 07232020. https://doi.org/10.1016/j.syapm.2009.09.002.

[5]   Thursby Elizabeth, Juge Nathalie. Introduction to the human gut microbiota. Biochem J 2017;474(11):1823–36. 0264-6021. https://doi.org/10.1042/bcj20160510.

[6]   Hutchings M, Truman A, Wilkinson B. Antibiotics: past, present and future. Curr Opin Microbiol 2019;51:72–80. 18790364. https://doi.org/10.1016/j.mib.2019.10.008. Available from: http://www.elsevier.com/locate/mib.

[7]   Lwoff André, Tournier Paul. The classification of viruses. Annu Rev Microbiol 1966; 20(1):45–74. 0066-4227. https://doi.org/10.1146/annurev.mi.20.100166.000401.

[8]   Han BA, Kramer AM, Drake JM. Global patterns of zoonotic disease in mammals. Trends Parasitol 2016;32(7):565–77. 14715007. https://doi.org/10.1016/j.pt.2016. 04.007. Available from: www.elsevier.com/locate/pt.

[9]   Vaillancourt Mylene, Jorth Peter. The unrecognized threat of secondary bacterial infections with COVID-19. mBio 2020;11(4):2161-2129. https://doi.org/10.1128/mbio. 01806-20.

CHAPTER TWO

Immunity and recovery from infectious diseases

Contents

If animals were totally defenseless against invasion by bacteria and viruses, we would simply be eaten alive. We must therefore fight back and, in effect, wage constant battles against the microbial invaders. Most times, humans win those battles and the invaders are rapidly and effectively destroyed. Sometimes the battles are much more even, the outcome may hang in the balance, and victories are not assured. On other occasions, we lose those battles and hence the war. Many of the diseases described in this book reflect these battles, victories, and defeats. These defeats and the resulting deaths were a result of the failure of the immune system to overcome the virulence of the pathogen. Thus, if we are to understand severe infections, epidemics, pandemics, and their outcomes, we must examine the mechanisms by which the human body defends itself.

Immune defenses

The defense of the body against a diversity of hostile microbes is a complex task. There are many different, potentially pathogenic, bacteria

Great American Diseases
ISBN: 978-0-443-31404-9
https://doi.org/10.1016/B978-0-443-31404-9.00002-1

and viruses, some of extreme virulence. They attack in different ways by many different routes. Defenses that are effective against one type of invader may be ineffective against others. Some pathogens fight back and destroy the body's defenses. Unexpected pathogens, such as SARS-CoV-2, are continually emerging as we interact with other animals and invade new environments. Over the past 100 years, we have seen the balance of the fight shift against the microbes. Vaccines and antibiotics have changed the outcome in many cases. But as the COVID-19 outbreak has shown, conflicts continue to occur. The microbes are still out there and will take advantage of any weakness in our defenses. The return of measles after we thought that it was eliminated in the United States emphasizes the fact that our victories have not removed the threats—with the notable exception of smallpox. Social responses such as failure to vaccinate have permitted infectious diseases to reemerge. The outbreak of COVID-19 demonstrates our continuing vulnerability to new agents when scientific and societal responses fail to match the threat. Once social barriers fail, then each individual must rely on their body's own defenses. As in any defensive battle, a single barrier against invasion is of limited benefit. If microbial invaders are to be completely excluded from our bodies, then multiple defensive barriers are required. For example, the avoidance of sick people and social distancing can be considered to be a component of the behavioral immune system. If done appropriately, this will effectively prevent infectious disease transmission [1,2].

Physical barriers

The body uses physical barriers to simply exclude invaders. The most obvious of these is the skin, a thick, elastic, self-sealing barrier that very few microbes can penetrate without assistance. Thus invasion through the skin is largely restricted to cuts or injections by the bites of insects or ticks. Yellow fever is a good example of a virus injected by biting mosquitos. The skin is not, however, the weakest link in our defenses. The respiratory tract presents a much greater defensive challenge. The airways extend deep into the body to deliver oxygen to our lungs and provide pathways for opportunistic microbial invasion. However, there are physical defenses in the respiratory tract. For example, coughing ejects particulate material from the airways, whereas sneezing ejects material from the upper respiratory tract. The walls of our airways are coated with sticky mucus that will capture any inhaled microbes that collide with the walls. Thus air is largely cleaned on inhalation, and relatively few microbes reach the lungs when we breathe normally. Nevertheless, pathogenic bacteria such as *Mycobacterium tuberculosis* or viruses such as smallpox, coronavirus, and influenza can get into the body through the airways. Therefore it is no coincidence that respiratory tract infections are the leading contributor to disease burden in many countries.

The other weak links in the body's surface defenses are in the gastrointestinal tract. As a result, diarrheal diseases are the second leading cause of death in developing countries. The cells that line the intestine are bound tightly together and present a continuous barrier to invasion. However, this barrier is only one cell thick, and many pathogens have evolved ways to penetrate it. Bacteria such as Salmonellae and *Escherichia coli* release toxins that can kill the barrier cells. *Vibrio cholera* simply poisons the intestinal cells with its toxin. Viruses such as polio or norovirus can directly invade and kill the intestinal surface cells.

Thus physical barriers cannot alone provide the degree of security or reliability required to exclude microbial pathogens. They are important but not sufficient.

Innate immunity

The second line of defense against microbial invasion is the innate immune system. This is a rapidly acting, reflexive set of defenses triggered by exposure to two stimuli: the presence of invading microbes and the detection of cell and tissue damage. Cell surface receptors called pattern recognition receptors can identify molecules located on bacterial cell walls as well as the characteristic nucleic acids of viruses. When microbial molecules bind to these pattern recognition receptors, they trigger responses that lead to the production and release of a mixture of proteins called cytokines. These cytokines control the immune responses and cause both inflammation and the feeling of sickness. Some, such as the interferons, also have potent antiviral activity.

Inflammation is the body's immediate response to microbial invasion and tissue damage. It is characterized by four main signs, redness, swelling, heat, and pain; these are known as the cardinal signs of inflammation. Defensive cells are attracted to the site of invasion. Local blood flow increases so that antimicrobial and antiviral molecules are concentrated at the invasion site. The infected region gets red and swollen as defensive cells and molecules converge on the invaders. This inflammatory response can effectively capture and kill the invaders and repair tissue damage in short order—if it works.

Acute inflammation is probably sufficient to detect and destroy microbes of moderate virulence—most of the time. It can be regarded as the initial skirmish against the invaders, and hopefully the defenders win. Inflammation, however, like all battles, is a nasty business. The outcome is not assured, and collateral damage may be extensive. The fight may sometimes end in a draw where the host can stop the spread of the invaders but not eliminate them completely. This is what happens in tuberculosis.

The cytokines, those proteins generated in response to microbial invasion, do not just act locally. They get into the bloodstream and circulate around the body. Some cytokines reach the brain, where they cause the set of behaviors that we call sickness. For example, they reset the body's thermostat. The brain is led to believe that the body is too cold and responds by triggering heat retention reflexes. The individual begins to get chills and as a result, shivers, sometimes violently. Their metabolism rises and the victim develops a fever. Fevers are protective in that high body temperatures enhance the activities of the cells engaged in both innate and adaptive immune responses. Likewise, bone marrow stem cells respond to a fever by replenishing the body's supply of white blood cells. These white blood cells, called leukocytes, assist by phagocytizing (eating) and destroying the pathogen and by producing antibodies. Additionally, these higher temperatures inhibit the growth of some bacterial invaders. The best example of this temperature effect is seen in the venereal disease syphilis.

In 1927 an Austrian doctor, Julius Wagner-Jauregg, won the Nobel Prize for Medicine for curing insanity!! Patients with late-stage syphilis develop "general paralysis of the insane" as a result of severe brain damage. Wagner-Jauregg observed that syphilis patients who also had malaria developed high fevers and improved clinically. It was also known that high temperatures killed the syphilis bacterium *Treponema pallidum*. Therefore he infected his neurosyphilitic patients with malaria. This malaria therapy worked! The malaria parasites triggered an innate immune response, the patients developed a very high fever, and the high body temperature killed *T. pallidum*, the cause of neurosyphilis (Box 2.1).

In addition to causing a fever, the mixture of cytokines produced during inflammation also induces the subjective feelings of sickness: malaise, fatigue, loss of appetite, and muscle and joint pains. These are all consequences of systemic cytokine release and reflect a change in the body's priorities as it fights off the invaders. Cytokines promote the release of sleep-inducing molecules so that lethargy and fatigue are commonly associated with sickness. This response reduces other energy demands and conserves the resources available

BOX 2.1 Dropping body temperatures!

As described in the text, one consistent response of the body to infection is a rise in body temperature—fever. It is believed that this increased temperature results in improved T-cell responses and hence resistance to infection. In order to determine normal body temperature, it has been usual to simply measure the temperatures of several thousand healthy individuals and determine the average. However, there are degrees of "healthy," and it is perhaps inevitable that some apparently normal individuals may in fact be suffering from a minor infection with a mild fever. Recent analysis of "normal" body temperatures suggests that males born in the early 19th century had temperatures 0.59°C higher than men today. In females, body temperature has dropped by 0.32°C since the 1890s. It has been suggested that in the 1800s, many individuals would have been suffering from inapparent infections such as syphilis, tuberculosis, and periodontal disease. Over the past century and a half, these infections have greatly diminished, resulting in decreased inflammation, improved health, longevity, and as a result, a lower body temperature.

(From M. Protsiv, C. Ley, J. Lankester, et al. Decreasing human body temperature in the United States since the Industrial Revolution. eLife 2020. https://doi.org/10.7554/eLife.49555.)

for defense and repair. Likewise, some cytokines suppress the hunger centers in the brain and so are responsible for the loss of appetite associated with sickness. The benefits of this are unclear, but it may permit the patient to be more selective about their food and redirect energy elsewhere.

Cytokines also act on muscle to release amino acids. Although this eventually results in weight loss, the newly available amino acids can help feed the immune system. When humans are subjected to long-term, low-grade inflammation, they lose weight and red blood cells, becoming very thin and anemic. Chronic cytokine release is responsible for the major weight loss seen in patients with slowly progressing bacterial diseases such as tuberculosis—hence the old name for this disease, "consumption."

Pathogenic bacteria, such as *Staphylococcus aureus*, *E. coli*, and *M. tuberculosis*, require iron for growth. Humans also require iron for vital functions such as oxygen transport and energy production. As a result, bacteria and their hosts compete for the same metal. Sick patients produce iron-binding proteins that can hide the iron from the bacteria. The result of this competition may determine the outcome of an infection, but the lack of iron results in a failure to produce red blood cells, so anemia is often a feature of chronic infections, especially tuberculosis.

When patients die from infectious diseases, death is not always due to a direct effect of the pathogen or its toxins but may be a result of an excessive innate immune response. For example, after massive tissue damage, large amounts of cell debris escape, and these may trigger the release of more cytokines. This "cytokine storm" can cause severe illness or death as a result of "shock." Many viruses, such as influenza, coronaviruses, and yellow fever, as well as the malaria parasite, can also trigger these storms. Septic shock caused by severe bacterial infections accounts for about 9% of human deaths in the United States. Excessive cytokine release also causes kidney, liver, and lung injury. Tissue damage, severe systemic inflammation, and the release of damaged cell fragments raise the levels of molecules that trigger blood coagulation within the bloodstream. This too is often lethal.

Although innate immune responses are rapid and work against many pathogens, they do not present a long-term solution to the defense of the body. Inflammation is a damaging process that is painful and uncomfortable. Inflammation alone cannot keep us healthy and safe. Fighting endless battles with invading microbes will eventually exhaust the host and result in septic shock. What is really needed is an effective automatic defense system that works quietly in the background to destroy microbial invaders. This is the task of the adaptive immune system.

Adaptive immunity

The body contains enormous numbers of small, round cells called lymphocytes. These lymphocytes carry receptors on their surface that can bind foreign molecules. These foreign molecules (mainly proteins) are called antigens. Invading microorganisms, both bacteria and viruses, consist of many diverse foreign antigens. During adaptive immune responses, the lymphocytes detect and then respond to these microbial antigens. This lymphocyte response ends in the destruction of the invaders. More importantly, the body retains a memory of this response—hence the term "adaptive"—so that if the organisms try a second time, the lymphocytes will be ready, and as a result, the invaders are killed much more efficiently. This memory response explains why, for instance, many viruses only cause disease once in an individual's lifetime. The second time around, the body is ready for them. Likewise, it explains how vaccines work. Vaccines mimic an infection, so they activate the adaptive immune system and generate memory cells [3]. It is these memory lymphocytes that persist in our bodies for months or years and confer immunity to reinfection (Fig. 2.1).

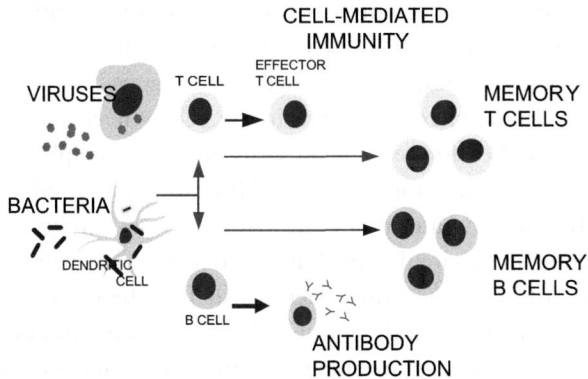

Fig. 2.1 *The pathways of the adaptive immune response. Antigen is captured, processed, and presented to T cells and B cells. If it is presented to B cells, then antibodies are produced. Antibodies play the major role in immunity to bacteria. If the antigen is presented to T cells, then a cell-mediated immune response will develop. Cell-mediated immune responses play the major role in immunity to viruses. Both antibody- and cell-mediated immune responses generate large numbers of memory cells.*

The first steps in an adaptive immune response involve catching a sample of the invaders, breaking it up, and presenting its antigen fragments to the lymphocytes. This has to be done correctly because it is essential that our lymphocytes do not recognize or respond to normal body components. A specialized population of cells called dendritic cells is tasked with catching and processing invading organisms such as bacteria. Dendritic cells, so named due to their tree-like, branching projections (Greek, *dendro*, a tree), catch and present the antigen fragments to the lymphocytes. During the first encounter with a microbe, an individual may only have a few lymphocytes with the correct antigen receptors, and as a result, immunity is slow to develop and sickness may result. However, when foreign antigens bind their receptors, lymphocytes respond by dividing rapidly and generating two new cell populations. One population, called effector cells, are responsible for the immediate destruction of the invader. The second population is the memory cells that are stored for future use. Should the invader come again, it will therefore be met by many more lymphocytes, resulting in a vastly enhanced response, and it will be eliminated before it can do further damage.

Immunity to bacteria and viruses

Adaptive immune responses must differentiate between bacteria and viruses [4]. Bacteria are relatively large organisms that, in general, live outside cells.

They can be destroyed through the use of proteins called antibodies that are produced by responding lymphocytes. (Do not confuse antibodies with antibiotics.) Antibodies bind to the bacterial antigens and mark them for destruction. Viruses, in contrast, live inside cells where antibodies cannot reach. The body therefore eliminates virus infections simply by using lymphocytes to kill the virus–infected cells. This process is called cell–mediated immunity. It is important to point out that antibodies can neutralize any viruses found free in the circulation. They bind to the virus particles and block their invasion of host cells. Some bacteria, such as the cause of tuberculosis (M. tuberculosis), may also hide within cells. As a result, antibodies are ineffective, and cell–mediated immune responses are therefore necessary to kill Mycobacteria and generate resistance to tuberculosis. Should the body make a mistake and only mount an antibody response to tuberculosis, the individual will not be protected and so may die from the disease.

Although all lymphocytes look alike, they actually consist of two major populations. One population matures within the bone marrow and as a result are called B cells. B cells are responsible for the antibody-mediated immune responses. B cells can sense thousands of different antigens and then produce antibodies against them. In order to do this, each B cell carries thousands of identical copies of its specific antigen receptor on its surface. When the B cells are activated, they release their receptors into the bloodstream, where they act as antibodies.

The second lymphocyte population matures within the thymus, an organ found within the chest in infants and young children. These cells, called T cells, regulate the immune responses and also mediate cell-mediated immunity. Thus there are also two types of T cells. One type acts as helper cells. They are needed for the body to produce optimal immune responses. Their characteristic feature is a protein on their surface called CD4. (There are hundreds of different proteins found on the surface of cells. They are denoted by a specific cluster of differentiation [CD] number.) The second major population of T cells is known as effector cells because they act to generate cell-mediated immunity. Their characteristic feature is a protein on their surface called CD8. Both $CD4^+$ and $CD8^+$ T cells circulate in the bloodstream. This is not a purely academic concept because the targets of the AIDS virus are $CD4^+$ T cells. When these cells are destroyed, patients lose their ability to mount cell-mediated immune responses. As a result, they become susceptible to many different viral and fungal diseases and die from many diverse infections.

Population immunity

When a pathogenic microorganism first enters a population that has not encountered it previously and has no preexisting immunity, the organism will be able to grow and spread freely for a while. As a result, the initial cases of disease may be severe or lethal. However, humans and other animals have diverse genetics and adaptive immune systems, and some individuals will be able to mount a strong adaptive response and so resist the infection. As a result, no matter how lethal an infectious disease outbreak is, there will be some survivors. Even the most lethal diseases do not wipe out complete populations, although they may come close. Because of our genetic differences, there will always be survivors who can breed, pass on their protective genes, and maintain the species [5].

Vaccination

Sometime after the 12th century, the Chinese noted that persons who recovered from smallpox did not get the disease a second time. As a result, they devised the process of variolation (Chapter 8) in order to cause a mild natural infection that triggered immunity. (Smallpox is also called variola.) This was still a hazardous procedure until the technique was modified by Edward Jenner in 1796 by substituting cowpox (vaccinia), a much less virulent virus, for the variola. As a result, the new procedure was called vaccination. The antigens of the two viruses are sufficiently similar that the protective immune response against cowpox also protected against smallpox. The mechanisms by which smallpox vaccination worked were completely mysterious. However, in 1880, Louis Pasteur demonstrated that immunity was not restricted to smallpox. He found that a similar process could protect animals against anthrax, which is a bacterial disease, and against rabies, which is caused by a virus. Vaccination jump-starts the adaptive immune responses. Vaccines, in effect, generate large populations of memory cells so that vaccinated individuals respond rapidly and effectively when they subsequently encounter the invader. Once the general principles of vaccination were established, it was employed against many different diseases such as typhoid, plague, and cholera. As a result, the major infectious diseases were brought under control, and the significance of infectious diseases was greatly reduced in developed countries. However, there remain diseases, such as syphilis, AIDS, and tuberculosis, where effective vaccines are not yet available.

Despite the incredible successes of vaccines in totally eliminating diseases such as smallpox and rinderpest (cattle plague), the number of companies manufacturing vaccines in the United States is now relatively few. For many vaccines, such as measles, mumps, rubella, diphtheria, pertussis, tetanus, and polio, there is only a single manufacturer. For influenza vaccines, there are only two producers, so significant vaccine shortages resulted when one of these companies had to close down temporarily. The reasons for these shortages reflect the high cost of developing a new vaccine, stringent government inspections and regulations, and product liability lawsuits that result in low returns on the investment and poor vaccine profitability. As a result, it is likely that life-threatening vaccine shortages will become increasingly common. Vaccination resistance by individuals refusing to vaccinate themselves or their children is an additional problem adversely affecting the prevention of infectious diseases. The return of measles in 2018 and 2024, and the resistance to vaccination against COVID-19 in 2021, may just be harbingers of future infectious disease crises.

Herd immunity

When an infectious disease spreads through a population, the speed of spread and the number of individuals infected will depend upon their level of adaptive immunity. If nobody is immune, then there are few impediments to disease spread and almost everyone will get infected. This is the situation that occurred when smallpox was introduced to America by the first Europeans and when the plague first arrived in Europe to cause the Black Death. Conversely, if everyone is immune, then the disease cannot spread, or cases will be limited. This was the situation with polio prior to the 20th century when the infection was so common that exposure in infancy resulted in intestinal colonization by the virus and lifelong immunity. When community hygiene improved and children did not encounter the polio virus until adolescence, their lack of immunity at that age rendered them susceptible to disease. Vaccination will, of course, protect a portion of the population, and the number of nonimmune, susceptible individuals will be reduced. It is not necessary for 100% of a population to be completely protected in order to prevent an epidemic from occurring. Depending upon the infectivity of the agent, an appropriate level of "herd immunity" may be sufficient to prevent explosive disease outbreaks.

The main purpose of vaccinating humans is to protect each individual from suffering and death. It is expected that clinical disease will be minimized. It is also expected that vaccinated individuals will transmit fewer pathogens. In any such population, the benefits of vaccines result from the collective impact of the procedure on all individuals and the collective decline in pathogen shedding. This decline in shedding, together with collective immunity, contributes to herd immunity.

When vaccines are used to control disease in a population rather than in individuals, herd immunity must be considered. Herd immunity refers to the resistance of an entire group of people to a disease as a result of the presence of many immune individuals [6]. Herd immunity reduces the probability of an infected person encountering a susceptible one so that the spread of disease is slowed or prevented. The public health community seeks to ensure that as many persons as possible are vaccinated in order to maximize herd immunity. Measles outbreaks in the United States have mainly occurred in populations with reduced vaccination coverage (Chapter 3). One goal of the COVID-19 vaccination program in the United States is to reach an effective level of herd immunity.

The spread of an infectious disease is, of course, dependent upon the close proximity of susceptible individuals. Solitary individuals are much less likely to encounter other infected individuals [7]. This is the situation with small migrating tribes of hunter-gatherers in large wilderness areas. On the other hand, humans living in towns and cities encounter numerous other individuals. In such situations, density-dependent (or crowd) diseases can establish themselves. Thus the practice of "social distancing" during the COVID-19 pandemic was effective because it reduced exposure to the virus from infected people [1].

The most important factor that influences the spread of an infection and the development of herd immunity is the basic reproductive number of the disease, termed R. R is the expected number of secondary cases resulting from each primary case, in other words, the probability of transmission of an infectious agent.

R is not a constant. It varies by time and place, and it is an estimate based on many assumptions. An R of 1 indicates that each individual primary case generates one secondary case, and the prevalence of the disease will remain unchanged. A disease with an R of less than 1 means that one case will, on average, generate fewer than one secondary case. As a result, the infection will decline and eventually die out. Conversely, an organism with an R greater than 1 means that each primary case will generate a larger number of secondary cases, and as a result, the number of such cases will increase. The higher the R number, the more difficult it is to prevent an infectious disease. R depends on the effective contact rate between individuals over time, the size of the population, and the duration of infectivity. Thus R will vary as a result of population density (the number of contacts), environmental effects, cultural effects, any quarantine/social distancing, seasonal effects, duration of infection, the degree of immunity, and the presence of newly susceptible individuals. In an epidemic, if the pool of susceptible individuals drops as a result of death and immunity, then R will drop until the organism runs out of "fuel."

A variant of R, called R_0 and pronounced "R naught," assumes that everyone in a population is completely susceptible to the infection. In a "real-world" situation, R is of greater practical value, although it may have to be calculated retrospectively.

Vaccination, by reducing the number of susceptible individuals in a population, also reduces the number of potential contacts between infectious and susceptible persons. This reduction is determined by the efficacy of the vaccine in reducing transmission and the amount of vaccination coverage within the population. As a result, the "effective population density" of susceptible people is reduced, and the quantity of pathogen available to infect the nonvaccinated individuals decreases. Each vaccinated individual therefore contributes to herd immunity, and a reduction in the effective reproductive number will occur. If there are insufficient susceptible people in a population, R may drop below 1, transmission will be interrupted, and the disease will eventually be eliminated. The level of herd immunity needed to reduce R to under 1 is called the "herd immunity threshold (HIT)." The HIT is useful in that it provides a target for vaccination coverage. The HIT has been calculated for major infectious diseases. It ranges from 90% to 95% for measles and rubella, to about 85% for diphtheria, and from 70% to 80% for smallpox.

Do not confuse R with lethality of an outbreak. Thus the common cold may spread to many people, have a high R, but not kill anyone.

Immunization

There are two ways of increasing a person's immunity to infection, active immunization and passive immunization.

Active immunization

Active immunization involves the administration of a microbial antigen in the form of a vaccine. This is usually done by injection, but some vaccines such as polio may be given orally, and some influenza vaccines may be squirted up the nose. When the immune system recognizes what it believes is an invader, it will mount an adaptive immune response. Subsequent boosting with the same vaccine will generate large numbers of memory cells and maintain protective immunity at a high level (Fig. 2.2).

The most common vaccines are made from killed whole pathogens (killed vaccines) or from live pathogens that have been attenuated or weakened so that they are no longer virulent (modified-live vaccine). Other vaccines may simply contain the purified antigen or antigens of the pathogen (subunit vaccine). Killed vaccines include the injectable polio vaccine (Salk vaccine), the injectable influenza vaccine, and the rabies vaccine; modified live vaccines include the measles, mumps, rubella (MMR) vaccine, the oral polio vaccine (Sabin vaccine), and the intranasal influenza vaccine; subunit vaccines include the whooping cough and the human papillomavirus (HPV) vaccines. The nucleic acid–based vaccines, used against COVID-19, are a type of subunit vaccine. The injected ribonucleic acid (RNA) enters a vaccinated person's cells and stimulates them to make pathogen proteins. These newly formed proteins then act as antigens and so trigger a protective immune response. To help protect against the lethal toxins that are produced by some bacteria, vaccines can be made by purifying and then inactivating the toxin. These inactivated toxins are called toxoids; examples include tetanus toxoid and diphtheria toxoid.

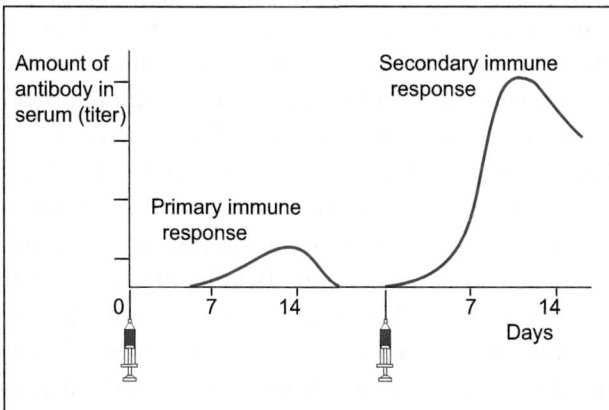

Fig. 2.2 *The characteristic time course of the adaptive immune response to an antigen as measured by serum antibody levels. Note the differences between a primary and secondary immune response. These are the reasons why the adaptive immune responses are so successful.*

Passive immunization

Natural transfer of antibodies occurs when they pass from the mother to her fetus through the placenta before birth. These passively derived antibodies protect the newborn baby while its immune system is developing. Maternal antibodies are very effective in fully or partially protecting newborns against most infectious diseases, but the amount of maternal antibody (the antibody titer) begins to decline over 6 to 12 months. At the same time, these maternal antibodies can interfere with the ability of vaccines to stimulate an antibody response. This is typified by the MMR vaccine. During the first 12 months of life, maternal antibodies protect the infant from these infections but also interfere with production of antibodies by the MMR vaccine. Thus MMR vaccination is not recommended until the child is between 12 and 15 months of age.

Passive immunization was first developed to protect against bacterial diseases caused by bacterial exotoxins (toxins like diphtheria and tetanus toxins secreted by bacteria) [8]. The two most important of these diseases are tetanus and diphtheria. Diphtheria is caused by a gram-positive, nonmotile bacillus called *Corynebacterium diphtheriae*. It is spread by direct contact, coughing, and sneezing. The disease begins with a fever and a sore throat. The organism invades the throat and pharynx of affected children where its toxin kills nearby cells. The dead tissues form a thick gray or white "pseudomembrane" that blocks the throat. This causes a severe cough and makes it very difficult for the patient to breathe or swallow. The bacterial toxin also circulates throughout the body, damaging the kidneys, liver, and especially the heart. Abnormal heart rhythms can lead to heart failure. Diphtheria toxin was discovered in 1888 by Emile Roux and Alexandre Yersin, working in Paris. Two years later, Shibasaburo Kitasato and Emil Behring immunized goats and horses with heat-inactivated diphtheria toxoid (a toxoid vaccine). They then showed that the blood serum (the liquid left after cells and clotting proteins are removed from blood) from these immune animals could, when injected into unimmunized animals, protect them against the infection. They called this an antitoxin. Behring then went on to treat human patients with the horse-derived antitoxin and eventually got it to work. Diphtheria antitoxin production in the United States began in 1895. Kitasato and Behring got the first Nobel Prize in 1901 because this was the first procedure ever shown to cure an infectious disease (Fig. 2.3).

Passive immunization with antitoxin has one significant advantage over active immunization—the injected antibodies confer immediate protection.

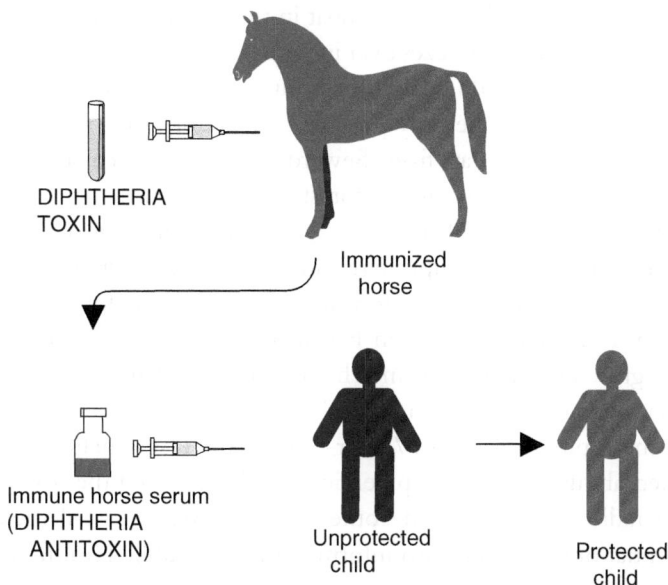

Fig. 2.3 *The principle of passive immunization. For example, antibodies can be induced in horses by vaccination with diphtheria toxin. When the horse is bled, these antibodies may be injected into diphtheria-infected humans to neutralize the bacterial toxin. This product is called diphtheria antitoxin or diphtheria immune globulin.*

However, it also has some disadvantages, the most notable of which is the fact that protection is short lived. For example, if someone receives a deep, potentially infected wound, they are at risk of developing tetanus. In order to prevent this, they may be injected with tetanus immune globulin (tetanus antitoxin). This consists of serum from a horse that has been injected multiple times with tetanus toxoid. The horse serum contains very high levels of antibodies against tetanus toxin. When injected into a patient, these antibodies will neutralize any tetanus toxin produced in the wound. However, these horse antibodies are gradually metabolized by the patient, so the protection conferred only lasts about 2 weeks.

The threat of diphtheria has been largely removed by routine childhood immunization, and diphtheria is readily treated with antibiotics. That was not always the case. One of the most significant outbreaks of diphtheria in the United States occurred over the winter of 1924–1925 in Nome, Alaska.

The serum runs to Nome

Nome is a small city on the west coast of Alaska, not far south of the Arctic Circle. Even today there is no road to Nome from central Alaska. As a result, all supplies are usually delivered by boat in the summer or nowadays by air. Because the Behring Sea freezes over in winter, Nome was effectively cut off from the rest of the world during the winter months prior to the development of air transport. The only possible connection was a rough sled trail, the Iditarod Trail, that ran from Seward on the south coast of Alaska to Nome across 938 miles of tundra, forest, and mountains.

In the summer of 1924, the only doctor in Nome, Dr. Curtis Welch, noted that his stores of diphtheria antitoxin had expired. He ordered more, but it had not arrived before ice closed the port for the winter. In December 1924, several children began to sicken with sore throats. The outbreak grew so severe that some children died. By January, Welch recognized that he was dealing with an outbreak of diphtheria. He administered some of the expired serum to a 7-year-old girl, but it was ineffective. Deeply concerned about a potential epidemic, Welch informed the town council and sent radio telegrams to the other cities in Alaska to tell them about the potential outbreak. He also informed the US Public Health Service.

An epidemic of diphtheria is almost inevitable here STOP I am in urgent need of one million units of diphtheria antitoxin STOP Mail is only form of transportation STOP I have made application to Commissioner of Health of the Territories for antitoxin already STOP.

By the end of January 1925, there were over 20 confirmed cases of diphtheria in the Nome area. The mortality was close to 100%. The city had two choices. One was to fly the serum in by aircraft. Unfortunately, the planes at that time could not reliably operate under Alaskan winter conditions. Likewise, there were no experienced pilots available in Alaska at that time. For that reason, the Nome City Council voted to set up a dogsled relay.

A search for diphtheria antitoxin found about 300,000 units in an Anchorage hospital, enough for 30 patients. It would have to suffice for now! They were first taken by rail to Nenana. The mail route from Nenana to Nome was 674 miles long. It initially followed the Yukon River and then crossed overland to Norton Sound and along the southern shore of the Seward Peninsula to Nome (Fig. 2.4). The weather could not have been worse. Temperatures in the interior were at −50°F. Winds were sweeping across the state and causing huge snowdrifts. Along the shore of Norton Sound, there was essentially no shelter from gales and blizzards.

The best dog mushers in Alaska, who worked for the Northern Commercial Company as mail deliverers, were positioned along the route. Most of them were native Athabaskans. The first dog musher was handed

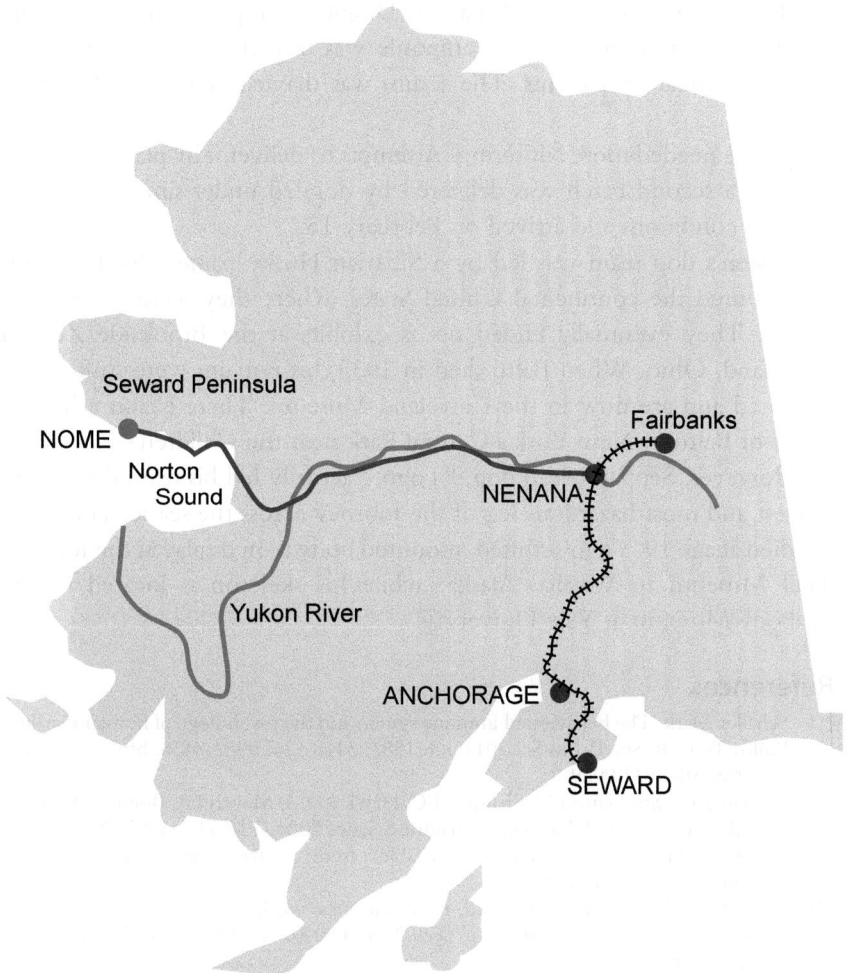

Fig. 2.4 *The route of the serum run to Nome. In 1925 the serum was taken by train to Nenana and then to Nome by dogsled—in the middle of winter. The modern Iditarod race follows a somewhat different route.*

the 20-pound package at the train station in Nenana on January 27 at 9 p.m. It was −50°F. He traveled over 50 miles, lost three of his dogs as a result of frozen lungs, and suffered severe frostbite himself. And so it went, 17 different stages of about 20 to 30 miles each until the serum was handed over to a musher from Nome at Shaktookik on January 31. That musher, Leonard Seppala, and his team undertook a 91-mile leg, crossing the sea ice in a blizzard at night with a wind chill estimated at −85°F and 70 mph winds. Another team, led by Gunnar Kassen, finally delivered the serum to Nome at 5:30 a.m. on February 1. Despite multiple crashes during the 674-mile journey, not a single ampoule was broken, but there was only enough to treat 30 patients. The serum was thawed and administered by noon.

Nome needed more antiserum. Attempts to deliver it by plane failed. As a result, a second batch was delivered by dogsled under similar atrocious weather conditions and arrived on February 15.

Kassen's dog team was led by a Siberian Husky named "Balto." This team toured the continental United States, where they were regarded as heroes. They eventually ended up as exhibits at the Brookside Zoo in Cleveland, Ohio. When Balto died in 1933, his remains were stuffed and mounted and are now in the Cleveland Museum. There is also a bronze statue of Balto in New York's Central Park near the Children's Zoo.

However, Seppala's lead dog, "Togo," actually led his team during the longest and most hazardous leg of the journey across the sea ice and back. He died at age 19. Togo's stuffed, mounted body is on display at the Iditarod Trail Museum in Wasilla, Alaska, while his skeleton is located in the Peabody Museum at Yale University!

References

[1]   Schaller Mark. The behavioural immune system and the psychology of human sociality. Philos Trans R Soc B Bio Sci 2011;366(1583):3418−26. 0962-8436. https://doi.org/10.1098/rstb.2011.0029.

[2]   Stockmaier S, Stroeymeyt N, Shattuck EC, Hawley DM, Meyers LA, Bolnick DI. Infectious diseases and social distancing in nature. Science 2021;371(6533). 10959203. https://doi.org/10.1126/science.abc8881. Available from: https://science.sciencemag.org/content/371/6533/eabc8881.

[3]   Iwasaki A, Omer SB. Why and how vaccines work. Cell 2020;183(2):290−5. 10974172. https://doi.org/10.1016/j.cell.2020.09.040. Available from: https://www.sciencedirect.com/journal/cell.

[4]   Annunziato F, Romagnani C, Romagnani S. The 3 major types of innate and adaptive cell-mediated effector immunity. J Allergy Clin Immunol 2015;135(3):626−35. 10976825. https://doi.org/10.1016/j.jaci.2014.11.001. Available from: http://www.elsevier.com/inca/publications/store/6/2/3/3/6/8/index.htt.

[5]   Quintana-Murci Lluis. Human immunology through the lens of evolutionary genetics. Cell 2019;177(1):184−99. 00928674. https://doi.org/10.1016/j.cell.2019.02.033.

[6]   John TJ, Samuel R. Herd immunity and herd effect: new insights and definitions. Eur J Epidemiol 2000;16(7):601−6. 03932990. https://doi.org/10.1023/A:1007626510002.

[7]   Hedrick SM. Understanding immunity through the lens of disease ecology. Trends Immunol 2017;38(12):888−903. 14714981. https://doi.org/10.1016/j.it.2017.08.001. Available from: www.elsevier.com/locate/it.

[8]   Graham BS, Ambrosino DM. History of passive antibody administration for prevention and treatment of infectious diseases. Curr Opin HIV AIDS 2015;10(3):129−34. 17466318. https://doi.org/10.1097/COH.0000000000000154. Available from: http://journals.lww.com/co-hivandaids/pages/default.aspx.

Modern American vaccines

Contents

Deaths from infectious diseases declined significantly in the United States after 1900. As a result, there was an increase in life expectancy of almost 30 years in White men, from 47 years in 1900 to 75 years in 2000 (Fig. 3.1). The increase in expectancy for Black men has gone from 33 years to 68 years, for White women from 49 years to 80 years, and for Black women from 34 years to 75 years. In 1900 a third of all deaths occurred in children under 5, and the three leading causes of death were all infections—pneumonia, tuberculosis, and enteric (intestinal) diseases. Together, these caused almost a third of all deaths. By the end of the 20th century, infectious diseases accounted for about 5% of deaths. Heart disease and cancer accounted for more than half the deaths. The leading lethal infections in 2000 were pneumonia, influenza, and AIDS.

These incredible improvements were largely a result of public health measures such as the availability of clean drinking water and effective sewage disposal systems, the development of antibiotics and, most importantly, the extensive use of vaccines, especially in children.

Great American Diseases
ISBN: 978-0-443-31404-9
https://doi.org/10.1016/B978-0-443-31404-9.00003-3

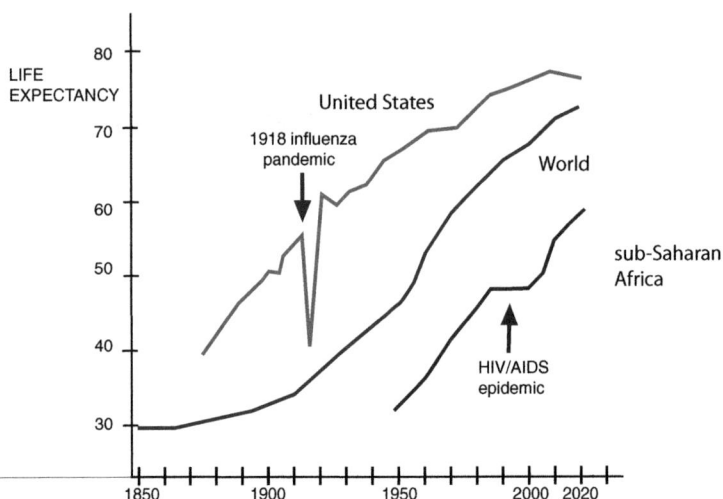

Fig. 3.1 *The average estimated life expectancy in the United States, the world as a whole, and in sub-Saharan Africa from 1850 to the present. Two points are evident from this figure. First, life expectancy varies greatly among nations due to economics, political stability, and public health resources. Second, pandemics adversely affect life expectancy, though only temporarily. It has been projected that life expectancy in 2020 due to COVID-19 will drop by about 1 to 2 years in the United States. This figure is a compilation of data from the CDC, World Bank, and Our World in Data. https:// ourworldindata.org/grapher/life-expectancy.*

In 1798, Edward Jenner, an English physician, demonstrated that material from cowpox (vaccinia) lesions could be substituted for smallpox in variolation (Chapter 9). Because cowpox does not cause severe disease in humans, its use reduced the risks incurred by variolation to insignificant levels. The effectiveness of this procedure, called vaccination (*vacca* is Latin for "cow"), was such that it was eventually used in the 1970s to eradicate smallpox from the globe.

However, the broader implications of Jenner's observations on vaccination and the importance of reducing the ability of an immunizing organism to cause disease were not realized until 1879.

Vaccination diversifies

During the latter half of the 19th century, the new science of microbiology led to the recognition that infectious diseases were caused by microorganisms, especially bacteria. Thus Robert Koch in Germany reported in 1877 that he could transmit the disease to mice using blood from cows with anthrax. He could even see the anthrax bacilli in the bovine blood under his microscope. He could grow the bacterium and showed that pure cultures of the anthrax bacillus could cause typical anthrax in animals.

The French chemist and microbiologist Louis Pasteur followed up Koch's discoveries by collecting, growing, and investigating any bacteria that appeared to be able to cause disease. In 1879 he received a sample from a professor at the Toulouse Veterinary College. This bacterium, now called *Pasteurella multocida,* is the cause of fowl cholera. Inoculation of this bacterium into chickens causes lethal disease. While investigating fowl cholera, Pasteur's assistant, Émile Roux, inadvertently allowed a culture of *P. multocida* to age in the laboratory cupboard for several weeks. When this aged culture was eventually injected into chickens it failed to sicken or kill them. Being economical, Pasteur retained the chickens and, sometime later, injected them with a fresh culture of the bacterium. To his surprise, they remained healthy. Pasteur recognized that what had happened was in many respects similar to vaccination! The organisms had lost their virulence. In honor of Jenner, Pasteur confusingly called the procedure *vaccination* as well. In 1880, Pasteur presented his first vaccination results to the members of the Academy of Medicine and Sciences in Paris in a treatise, "Of Infectious Diseases, Especially the Disease of Chicken Cholera."

Anthrax and rabies

Fowl cholera was not the most significant disease problem affecting French agriculture at that time. Anthrax, in contrast, was a major killer of sheep and cattle (and some farmers) so Pasteur began to develop a vaccine against the anthrax bacterium (*Bacillus anthracis*). Pasteur's working hypothesis regarding the mechanism of immunity was that a vaccine somehow depleted the vaccinated animal of essential nutrients (the same theory believed by Auzias-Turenne when he used multiple doses of "vaccine" to treat or prevent syphilis; Chapter 6). As a result, the pathogenic bacterium could not survive in a vaccinated host. It followed from the depletion theory that protection could only be conferred using vaccines containing live organisms. Pasteur thus concentrated his efforts on attenuating *B. anthracis* so that it could no longer cause disease. For example, he grew the organism at a high temperature (42−43°C) for multiple generations.

In March 1881, Pasteur reported to the Academy that his anthrax vaccine was working [1]. However, he was promptly interrupted by another Academy member, Jules Guérin. Guérin, a surgeon, was a staunch opponent of the germ theory. He repeatedly asked Pasteur to explain how his vaccine worked and then claimed not to understand his explanations. The meeting ended in chaos. These events did, however, provoke a challenge to a public trial of his anthrax vaccine by a group of local veterinarians—a challenge that Pasteur accepted.

In May 1881, Pasteur put on a public demonstration of his anthrax vaccine at Pouilly-le-Fort outside Paris. He used two groups of animals. The vaccinated group, consisting of 24 sheep, 1 goat, and 6 cattle, received two doses of the vaccine 15 days apart. The unvaccinated group of similar composition received nothing. On May 31, 30 days after the first dose, all the animals were challenged with a live anthrax culture. On June 2 more than 200 interested observers turned up to see the results. It was spectacular!!! All the unvaccinated animals were dead or dying while all the vaccinated animals survived. It was a major media event and a triumph of immunology. The media coverage of this experiment made Pasteur famous and introduced the public to the great potential of vaccines in combating infectious diseases (while Pasteur claimed publicly that he had used a live attenuated vaccine). Only he and his colleagues knew that he had actually killed the anthrax bacteria in the vaccine with potassium bichromate [2].

Pasteur and Roux next went on to develop a vaccine against the viral disease, rabies. They infected rabbits with rabid brain tissue (they didn't know it was caused by a virus!). Once the disease developed, they removed the rabbit's spinal cord and dried it for 5 to 10 days—thus attenuating the virus—and then prepared an emulsified injectable. By drying the spinal cords for different times, Pasteur and Roux were able to produce rabies viruses with different degrees of attenuation. They then inoculated dogs with a series of emulsified cords. The first injections contained highly attenuated virus. These were followed by cords containing viruses of increasing virulence. The dogs developed strong immunity to rabies.

Pasteur had been working on this project for some time when in July 1885 9-year-old Joseph Meister was brought to them. Meister had been badly savaged by a rabid dog 2 days previously. Under considerable pressure because the boy was sure to die, a very anxious Pasteur vaccinated him. Meister survived to become the gate keeper of Pasteur's grave and committed suicide rather than permit German troops to open the crypt in 1940. In July 1886, a 7-year-old boy named Harold Newton Newell became the first person to receive Pasteur's rabies vaccine in the United States. He only received the first four inoculations, and his fate is unknown [3]!

Dead vaccines

In 1886, Daniel Salmon and Theobald Smith, working in the United States, were able to protect pigeons against disease by inoculating them with a heat-killed culture of "hog cholera virus," now known as the bacterium *Salmonella choleraesuis*. Thus killed vaccines clearly worked. This eventually led to the rapid development of many killed bacterial vaccines such as typhoid, pertussis (whooping cough), and cholera [4,5].

Antiviral vaccines

Pasteur and his successors succeeded in developing numerous new antibacterial vaccines. However, it was not until 1898 that Loeffler and Frosch showed that the agent of foot-and-mouth disease could pass through a bacteria-proof filter. This was not a bacterium but a much smaller virus. There was great reluctance by bacteriologists to abandon the methodology that had yielded such great results. For example, for many years scientists clung to the notion that influenza was due to a bacterium. Indeed, many severe cases of influenza are characterized by secondary bacterial invasion. As a result, it is by no means difficult to isolate multiple bacteria from patients suffering from the flu. Bacteriologists had no shortage of alternative candidate agents to investigate. During the great influenza pandemic of 1918, many people were immunized with vaccines directed against these bacteria. They had no effect on the course of influenza.

By the beginning of the World War II, vaccine technology was firmly established. Many important infectious diseases had been brought under control. The subsequent development of vaccines centered on diseases such as polio and on increasing vaccine safety. The development of Salk's inactivated polio vaccine after it had been grown in cultured monkey kidney cells was the first step in the development of the modern cell-culture-based vaccine industry (Fig. 13.3). Because of the risks associated with modified live organisms investigators also began to create vaccines based on individual viral components or subunits. Polysaccharide and virus-like particle vaccines began to be produced in the 1980s. DNA and RNA-based vaccines have been produced much more recently. As a result of these impressive advances, most major infectious diseases remain relatively well controlled—thanks to vaccines. There is, however, room for improvement and a continuous push for greater safety.

Smallpox vaccination

Natural cowpox did not occur in American cattle. As a result, the United States was entirely dependent upon the arrival of erratic supplies of the vaccine from Jenner and his colleagues. Physicians were also obliged to rely on arm-to-arm transfer to a great extent. One result from this was that vaccination became expensive. Thus vaccination was largely restricted to the most prosperous members of society.

In the early 19th century, vaccination still carried significant health risks. Once applied, the inoculated virus grew in the nearby skin cells. After 3 to 5 days a papule formed at the inoculation site. It then became vesicular (that is, fluid accumulates inside it like a blister) between days 5 to 8. The fluid filled with cells to form pus (pustular), reaching its greatest size at 8 to 10 days as the immune system (T cells) attacked the virus-infected cells. The pustule then started to heal by drying from the center out and forming a scab. The scab fell off 14 to 21 days after vaccination, leaving an obvious pitted scar. Small satellite lesions sometimes developed close to the central lesion and grew and healed in the same way. Vaccination also induced symptoms such as headache, fatigue, muscle aches, chill, malaise, nausea, and a mild fever. These could be so severe that an individual might have had to take time off work and hence lose earnings.

Eventually, vaccinia was produced in calves in specialized vaccine farms. As the science of microbiology developed, these vaccine farms eventually expanded their activities into making other biologicals such as tetanus and diphtheria antitoxins as well as vaccines. This production was initially unregulated, but a series of unfortunate accidents eventually forced Congress to pass The Biologics Control Act in 1902 to regulate vaccine production. One such accident occurred in Camden, New Jersey.

The Camden Incident

One of the prime reasons for vaccine refusal at the beginning of the 20th century was the fear of vaccine-induced infections. It was obvious that many smallpox vaccines were contaminated with infected pus. For example, in late 1901, nine schoolchildren in Camden, New Jersey, died as a result of tetanus acquired from a contaminated smallpox vaccine. The first tetanus case occurred in November 1901 and another 10 quickly followed. Eventually there were 40 tetanus cases. A formal investigation identified a batch of contaminated smallpox vaccine as the source [6]. A similar outbreak of tetanus occurred in St. Louis in 1901.

Vaccination not only caused sore arms but often much more serious complications. For example, it was also linked to cases of septicemia. Many physicians and pharmacies responded by advertising "fresh, pure, and clean vaccines." These advertisements were widely used. They could even develop into conspiracy theories when, for example, a Memphis paper suggested in 1917 that there was a nationwide German plot to kill American soldiers by poisoning their "vaccine serum" with tetanus germs. The plot was reportedly uncovered after five deaths had occurred [7].

The Biologics Control Act of 1902

On July 1, 1902, President Theodore Roosevelt signed the Biologics Control Act into law. The Act was largely triggered by the two vaccine-associated tetanus outbreaks. The act mandated annual licensing of establishments that manufactured or sold vaccines, serums, antitoxins, and similar products such as blood components. It required accurate labeling of the product and its source. It established a board to oversee its implementation and regulated interstate traffic in these products. The Hygienic Laboratory of the US Public Health and Marine Hospital Service was charged with its enforcement. It established antitoxin standards. The law was followed in 1906 by the Pure Food and Drug Act.

Diphtheria

Diphtheria is a bacterial infection that causes a severe throat infection and then releases a toxin that damages the heart. As a result, infected young children choked to death. Children who survived the initial attack often died young as a result of the damage caused by the toxin.

In 1889, Emil Behring in Berlin and Emile Roux in Paris showed that the serum from immune animals could neutralize the toxicity of diphtheria toxin. In late 1891, two seriously ill diphtheritic children were saved by the administration of serum from an immune sheep produced by Emil Behring. Sheep are small and do not produce a lot of serum. Roux in Paris therefore used horses that produced a lot more serum than the sheep and at a higher titer (antibody level). By 1894, both groups of investigators had treated hundreds of children and shown that this antitoxin treatment had more than halved diphtheria mortality rates [8]. This was the first specific cure for an infectious disease, ever!

In April 1893, Dr. Herman Biggs, the Director of the New York City Health Department, appropriated funds for a bacteriology laboratory [9]. Dr. William H. Park (1863—1939) was appointed to be the bacteriologic diagnostician. At first he simply isolated the organism from suspected diphtheria cases. In October 1894, Biggs traveled to Europe and heard talks by both Behring and Roux regarding the use of antitoxin to treat diphtheria. He was so impressed that he cabled Park from Europe instructing him to begin antitoxin production immediately.

Over the next few years, the New York City Health Department got involved in large-scale antitoxin production. They established stables and filled them with 60 horses repeatedly immunized with the diphtheria toxin so that the horses responded by making large amounts of antibodies (antitoxin, nowadays called diphtheria immune globulin). The horses were bled at regular intervals, and their serum was harvested. The first doses of this serum became available in January 1895 [10]. That fall, an outbreak of diphtheria occurred in the Mt. Vernon branch of the New York Infant Asylum. Park was given permission to administer the antitoxin to every child in the institution. "The epidemic stopped immediately" [11].

The New York City stables were the major source of diphtheria antitoxin for much of the United States during the 1890s. During their first year they produced over 25,000 protective doses. In 1894, there were 2870 deaths from diphtheria in New York City. By 1900, this had dropped to 1400 and it continued to decline for many years thereafter. The NYC Health Department continued to mass produce diphtheria antitoxin until well after the World War II, when commercial companies progressively took over the market.

Measles

Measles is a viral disease characterized by the development of a fever and a characteristic skin rash. While usually mild, some patients may develop lethal complications such as encephalitis. Prior to the 1960s in the United States, almost everyone contracted measles, usually in childhood. Up to 1962, there were approximately 500,000 cases reported annually in the United States, with 48,000 people hospitalized as a result. Of these, 4000 developed encephalitis, a life-threatening inflammation of the brain. Measles-related complications such as pneumonia and encephalitis resulted in about 500 deaths annually. Prior to the introduction of vaccines, measles occurred almost universally among American children, and as a result, more than 90% were immune by the time they reached age 15.

Dr. John Enders won the Nobel Prize for Medicine in 1954 for discovering how to grow the polio virus in monkey kidney cells. In that same year, at Boston Children's Hospital, Enders and his colleague Thomas Peebles switched to studying measles. By collecting blood and throat-washing samples from children with measles and growing the samples on cultured cells, Peebles was able to isolate the virus. One such sample was taken from a 13-year-old student, David Edmonston. This cultured measles virus was named the Edmonston strain and was subsequently used for the development of the measles vaccine.

During the late 1950s and early 1960s, continuing research in the United States and a large trial in West Africa culminated in the development and licensure in 1963 of two measles vaccines: a killed vaccine and a live attenuated vaccine. The killed virus vaccine was produced and administered between 1963 and 1967. However, its use was discontinued because it produced a relatively short-lasting immunity. The attenuated virus strain was produced by repeatedly growing and transferring the Edmonston virus through different cell cultures in the laboratory. This prolonged serial passage in cell culture eventually resulted in changes in the genetic makeup of the virus. It was able to infect cells but could no longer cause disease in humans—it was effectively attenuated. This improved vaccine was released in 1968. One dose is 93% effective in preventing measles while two doses are 97% effective.

Once measles vaccines became available, a nationwide immunization program was initiated. This program was mainly based upon the use of the attenuated virus vaccine for routine vaccination of infants and vaccination of children upon entry to public school. By 1968, only 22,200 cases of measles were reported, a drop of nearly 95% compared to prevaccination levels.

Beginning in 1971, measles vaccine was also combined with two other childhood vaccines directed against mumps and rubella viruses. Before the measles, mumps, rubella (MMR) vaccine became widely available in the 1960s, measles affected about 3000 persons per million. As a result of widespread vaccination, cases fell from about 500,000 cases annually to several thousand in the 1980s and to just a few hundred annually since the mid-1990s. In 2005, a combined measles, mumps, rubella, and varicella vaccine was also licensed. However, vaccination coverage is not universal and was well below the calculated 90% to 95% level required to protect and ultimately eradicate measles from the US population—the herd immunity population, disease prevalence, and available resources. 210,000 children received the placebo and 440,000 received the vaccine. An additional 1,180,000 children served as unvaccinated controls. Children received two doses, 2 to 4 weeks apart, followed by a third 7 months later.

The results of the trials were announced to the press on April 12, 1955, at the University of Michigan. The trials had shown that the vaccine was about 70% effective against type 1 poliovirus and 90% effective against types 2 and 3. It was 94% effective against bulbar poliomyelitis, the most severe form of the disease. The vaccine was licensed by the FDA within hours. The news traveled across the country. Overnight Salk became famous and an instant hero—a scientific superstar. His name was posted across the front pages of newspapers in banner headlines. Church bells pealed across the country. People huddled around their radios, and the news was announced in offices, stores, and factories. In the face of huge demand there was a rush to vaccinate.

While the vaccine had been rapidly licensed, the government had made no arrangements whatsoever to handle the enormous demand. Large-scale vaccination only began in the United States in 1957 and, as a result, the prevalence of polio dropped precipitously from nearly 58,000 cases in 1952 to 3200 cases in 1960.

Meanwhile, a second polio vaccine was also being developed by Albert Sabin. Unlike Jonas Salk, Albert Sabin believed that viruses that grew in the intestine would be much more likely to trigger prolonged strong immunity than injected, killed viruses. As a result, beginning in 1951, he developed an oral polio vaccine (OPV) using a virus that had lost its ability to cause disease. Sabin grew this attenuated poliovirus in monkey cell tissue cultures until they had lost the ability to cause disease in humans. He first tested his vaccine on himself and his family members, members of his research team, and 30 volunteers from a nearby penitentiary. They all developed antibodies, and nobody got sick. The enthusiasm for the newly available Salk vaccine made it very difficult for Sabin to get his vaccine tested in the United States. Sabin therefore had to look elsewhere for a large population of children that had not received the Salk vaccine. Sabin convinced the Health Ministry of the Soviet Union to conduct these studies. By 1957, clinical trials began involving 2 million Russian children. The OPV doses were either mixed with candy or simply dropped directly into the children's mouths. At the end of the year, the results were so good that the Russians decided to vaccinate 77 million individuals under the age of 20. Other trials took place around the same time in Eastern Europe, Singapore, the Netherlands, and Mexico.

Jonas Salk received many international awards and was globally recognized as a hero. Sabin and Salk disliked each other and never cooperated. The feeling between the two men was so intense that some believe this was why neither received the Nobel Prize. Jonas Salk died in 1995 in La Jolla, California; Albert Sabin died in 1993 in Washington, DC. Neither Salk nor Sabin ever sought to patent their vaccines. When asked about the patent, Salk replied, "There is no patent. Could you patent the sun?"

Covid vaccines

It was obvious, almost from the beginning of the COVID pandemic, that the only long-term solution to this disease was the development of effective vaccines. Funding for vaccine production infrastructure has rarely been available in the absence of an imminent disease threat. As a result, coronavirus vaccine development started belatedly. Despite this, the vaccine industry took up the challenge and developed several different and remarkably effective COVID-19 vaccines in an incredibly short time. The first human trial of a vaccine began on March 17, 2020, with the Moderna vaccine being tested at Kaiser Permanente in Seattle, Washington.

Vaccines are tested through a multistage process. For example, a trial batch of vaccine must first be tested in a small group of individuals (Phase 1 trials). If safe and successful, it can then be tested in a small number of human volunteers to assure their safety. This group should have the same characteristics, such as age and health, as the proposed target population (Phase 2 trials). If all appears well and no significant adverse events have occurred, then the vaccine undergoes large-scale clinical trials involving thousands of people (Phase 3 trials). The recipients are monitored, looking for adverse events and to determine whether their antibody levels confirm that they are protected against the natural disease. These trials normally take many months or even years. Once developed, any vaccine will then have to be manufactured in huge amounts, and this may also be difficult to achieve.

On May 15, 2020, the United States government launched "Operation Warp Speed," a national private/public partnership program to accelerate the development of COVID-19 vaccines. The aim of the program was to deliver 300 million doses of a safe, effective vaccine to the US public by January 2021. This program worked well. As a result, on December 11, 2020, the FDA issued an emergency use authorization for the first COVID that some unknown diseases circulated in these populations. The clinical description of these diseases is unclear and confusing, but they may well have been infections acquired from wildlife. What is clear is that Native Americans did not suffer from the great pandemic diseases—plague, small-pox, influenza, and measles—that afflicted contemporary European populations.

Native American diseases
Tuberculosis

As described in Chapter 6, tuberculosis caused by *Mycobacterium tuberculosis* was clearly present in the Americas prior to the arrival of Europeans [1]. Tuberculosis is an heirloom disease that would have affected our earliest human ancestors. It was likely brought by the first migrants from Siberia, and it may have preexisted in wild mammals. Across the Americas, mummies and skeletons have been found with clear evidence of tubercular lesions dating from well before the arrival of Europeans. It was, however, a sporadic disease and unlikely to have caused significant epidemics in such a thinly populated continent.

Viral hemorrhagic fevers

Hemorrhagic fevers are caused by many different RNA viruses, especially arenaviruses (arenaviruses are a family of rodent viruses readily transmitted to humans). What they have in common is that they affect many different organ systems, especially the blood vascular system. As a result, the victim's blood fails to clot and they suffer from severe, possibly lethal, bleeding. All the arenaviruses have natural reservoir hosts such as arthropods or wild animals [2]. Rodents are their main animal reservoirs. These include rats and mice as well as other field rodents. Their arthropod reservoirs are predominantly ticks and mosquitos. Humans become infected when they come into contact with these creatures. Hemorrhagic fever viruses are naturally present in rodent populations across much of Central and South America [3].

Cocoliztli

The 16th-century depopulation of the native peoples of Mexico was one of the most catastrophic events in human history. It is clear that introduced European diseases, such as smallpox and measles, played a major role in this decline. However, by far the most significant losses were attributed to a disease called cocoliztli. Multiple outbreaks of cocoliztli were recorded in the highlands of Mexico in the 16th century soon after the Spanish arrived. The most important of these outbreaks occurred in 1545 and 1576 (Fig. 4.1). For example, the 1545 pandemic was responsible for 800,000 deaths in the Valley of Mexico alone. This may have accounted for as much as 80% of the local indigenous population. The 1576 pandemic killed 2 million out of a population of 4.4 million—almost half the population [4]. The disease showed a clear ethnic preference; the Spaniards were scarcely affected whereas the native people suffered massive mortality. The 1813 pandemic spread across the whole of Mexico but only killed about 3% of the population. Cocoliztli may have caused millions of deaths, yet its cause is unknown.

The reported signs and symptoms of cocoliztli included fever, headaches, vertigo, black tongue, dark urine, severe diarrhea, severe pains in the chest and abdomen, the development of swellings, and jaundice as well as profuse

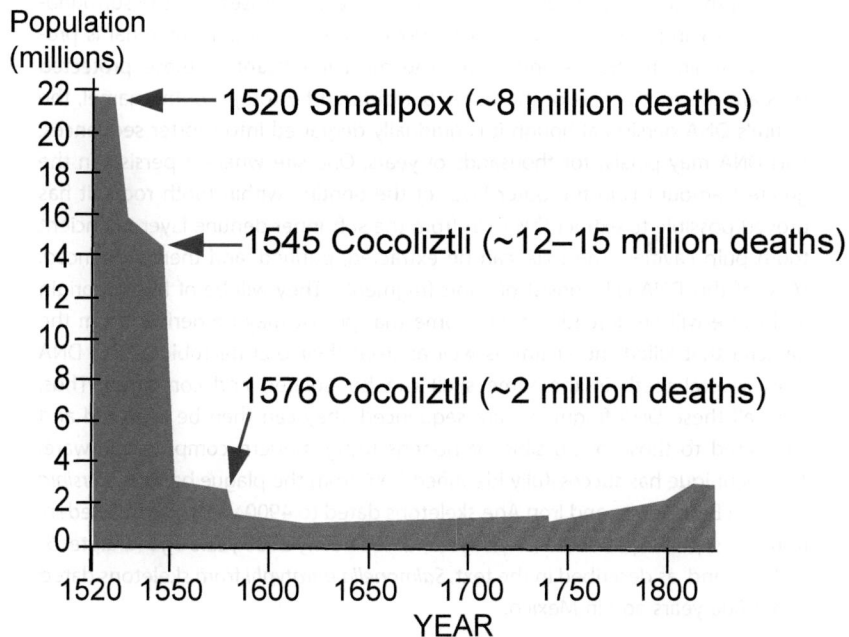

Fig. 4.1 *The 16th-century population collapse in Mexico. The 1545 and 1576 cocoliztli epidemics appear to have been hemorrhagic fevers caused by an indigenous viral agent and aggravated by unusual climatic conditions. The Mexican population did not recover to pre-Hispanic levels until the 20th century. From Acuna-Soto R, Stahle DW, Cleaveland MK, Therrell MD. Megadrought and megadeath in 16th century Mexico. Emerg Infect Dis 2002;8(4):360–62.*

bleeding from the nose, eyes, and mouth. Death occurred within 3 to 4 days! These symptoms are strongly suggestive of a viral hemorrhagic fever. However, like many of these early epidemics, it may also have resulted from a mixture of several different infectious diseases. The word cocoliztli is the Nahuatl word for disease, pestilence, or illness. Physicians at the time, both Europeans and Native Americans, recognized that it was different from any other disease they had seen. For example, they knew that it was not yellow fever (Box 4.1).

In 2018, excavations in a Mixtec Cemetery near Oaxaca in southern Mexico revealed multiple skeletons. Analysis of the DNA in the root canals of the teeth taken from 10 individuals indicated that they all contained DNA

BOX 4.1 Finding ancient pathogens.

When buried bodies decompose, most of their organic material breaks up as a result of the activities of soil bacteria and eventually diffuses into the surrounding soil leaving only skeletal remains. However, some organic material is protected within the bones and teeth. The most important of these protected sites are the root canals of the teeth. Surrounded by tough tooth enamel, the victim's DNA persists although it is gradually degraded into shorter sequences. This DNA may persist for thousands of years. One site where it persists in the greatest amount is in the outer layer of the dentine within tooth roots. It has proved possible to extract this DNA from the soft inner dentine layer of ancient tooth pulp cavities. The DNA can be extracted, purified, and then sequenced. Most of this DNA will consist of short fragments. They will be of human origin and some will be unreadable, but some that persist may be derived from the bacteria that killed the victim as well as from their oral microbiota. The DNA from several teeth in the same skull can be isolated and compared. Thus, once all these DNA fragments are sequenced, they can then be analyzed and compared to those of possible pathogens using modern computer software. This technique has successfully identified DNA from the plague bacillus (*Yersinia pestis*) in Bronze Age and Iron Age skeletons dated to 4900 years ago in Sweden; from the syphilis spirochete (*Treponema*) dated nearly 2000 years ago (Chapter 5) in Brazil; and, as described in the text, *Salmonella paratyphi* from skeletons dated about 800 years ago in Mexico.

Margaryan A, Hansen HB, Rasmussen S, et al. Ancient pathogen DNA in human teeth and petrous bones. Ecol Evol 2018; 8:3534—42.
Spyrou MA, Bos KI, Herbig A, Krause J. Ancient pathogen genomics as an emerging tool for infectious disease research. Nat Rev Genet 2019;20:323—40.

from a bacterium, *Salmonella enterica*, serotype Paratyphi C [5,6]. Records show that this cemetery dated from the 1545 Cocoliztli Epidemic, and it is clear that paratyphoid disease must have been present in that population. It is possible that this was one cause of cocoliztli, but the clinical signs of paratyphoid fever are very different from those reported at the time. Intestinal bleeding is an occasional complication, although it is entirely possible that its clinical manifestations might have been different in a highly stressed, overworked, immunologically enfeebled population. Other infections would have been circulating among malnourished natives at the time, and it is unlikely to be the definitive or sole cause of the disease. It is possible that the paratyphoid may have been a secondary invader in this location.

Cocoliztli epidemics appear to have been associated with severe droughts and their accompanying deforestation. Thus the 1576 epidemic occurred after a continent-wide drought and coincided with wet periods within the droughts. It has been suggested that as the rains ended the drought, a population explosion occurred among the rodents carrying the virus. Deforestation may also have contributed to the problem as the Spaniards began to cut the forests to build cities of wood. Several geographically restricted arenaviruses and hantaviruses continue to be isolated in Mexico and Mesoamerica. These are rodent borne and cause hemorrhagic fevers. For example, there was an epidemic of hemorrhagic fever in the southern Mexican state of Chiapas in 1967. The viral reservoir was the deer mouse, *Peromyscus mexicanus*. The virus was classified as a member of the Tacaribe complex, a group of viruses that have been suggested as the cause of cocoliztli. The 1967 disease outbreak was preceded by large-scale deforestation, and an increased number of rodents were encountered around homes in the affected area. Humans usually become infected with arenaviruses by inhaling droplets of secretions or excretions from infected rodents. Alternatively, they may be acquired by eating infected rodents—something that would readily occur in times of famine. Rodents are caught and eaten in some rural areas of Mexico even today—a risky procedure.

Hantaviruses

In April 1993 a young Navajo woman, Florena Woody, living in the Four Corners region of New Mexico where it meets Utah, Colorado, and Arizona, died after a short illness characterized by severe pulmonary edema (fluid in the lungs). This was followed shortly thereafter by the death of her fiancé, Merrill Bahe. Subsequent investigations revealed that five additional young people had died within the preceding 6 months showing similar symptoms. Twelve more individuals from the New Mexico reservation became infected and died shortly thereafter. One of the physicians involved thought that the disease resembled a hantavirus outbreak he had seen previously in Korea. As a result, the US Centers for Disease Control and Prevention (CDC) tested for and confirmed the presence of a previously unknown hantavirus in the victims [7]. It was known that hantaviruses are spread in the droppings of rats and mice. As a result, the investigators trapped almost 1700 rodents and found that 30% of the deer mice (*Peromyscus maniculatus*) in the Four Corners region showed evidence of infection. The virus, now called "Sin Nombre" virus, is shed in the droppings of the deer mice. Transmission does not occur between humans but is acquired by inhalation of aerosolized mouse fluids. The mice themselves suffer minimal adverse effects. Further investigations uncovered previous cases of what came to be called "Hantavirus Pulmonary Syndrome" (HPS) that had been incorrectly diagnosed. Tribal leaders also recalled previous outbreaks in 1918, 1933, and 1934. The Navajo had identified mice as sources of illness since at least the 19th century. They had also noted that the occurrence of the disease was associated with a higher-than-normal winter and spring rainfall. We now know that this increase, due to an El Niño event in the Pacific, provides increased food for the mouse population and consequently an increase in mouse numbers.

[The Sin Nombre virus was reportedly discovered in 1933 in Muerto Canyon on the Navajo Reservation in Arizona. It was therefore named by the CDC after the location of its discovery—Muerto Canyon Hantavirus. The Navajo Nation objected to the name, so it was renamed the Four Corners Hantavirus. They objected to that too. To avoid giving further offense, it was eventually renamed as Sin Nombre—the nameless virus.]

Whereas the Navajo infected with Sin Nombre virus suffered from respiratory disease, other hantaviruses cause hemorrhagic fevers, especially in the Old World. The mechanisms are, however, similar. The virus attacks blood vessels. In the hemorrhagic disease, this results in bleeding. In the pulmonary syndrome, it results in fluid escaping into the lung. Symptoms are nonspecific, with fever, cough, nausea, and vomiting as well as muscle pain and headaches. The lungs eventually fill with fluid (pulmonary edema). Most cases occur in adults. It is a serious disease with a mortality of about 36%. Between 1993 and 2017 there were 728 confirmed cases of hantavirus disease in the United States, and Sin Nombre virus accounted for about 600 of these.

The appearance of this previously unidentified zoonosis at the end of the 20th century points out the likelihood that many other potential pathogenic viruses are circulating among North American rodents. It is also likely that they have been present since pre-Columbian times and have sporadically affected Native Americans for thousands of years.

The Maya collapse

Nomadic hunter-gatherers never stay long in one place. They readily and frequently move away from camp sites, leaving their feces with their accompanying bacteria behind. Additionally, when the population was very low, contacts with other humans would have been infrequent and transient, so opportunities to acquire infections from others would have been limited. As Native Americans adapted to the New World, they settled down in some fertile areas and began to live as farmers. Farming required living in one place to look after and harvest their crops such as corn. As populations grew, these sites progressed from permanent encampments, to villages, to eventually the large urban centers of Central America and the Mississippi Valley. In Mesoamerica, corn-based agriculture eventually developed into several city-based systems, each with large populations. Thus the city of Tikal in modern Guatemala may have had a population of 90,000, whereas Teotihuacan near Mexico City housed about 125,000. Urban crowding increased the risks of disease from contaminated food, soil, and especially water. These cities had water systems to ensure that plenty was available to drink. However, less attention would have been paid to getting rid of human feces. They did have limited sewage systems, but a combination of a high population density with the difficulties of fecal disposal would have permitted enteric bacterial diseases to thrive and have a significant impact. Fatal diarrheal diseases would have arisen in proportion to the density of the urban populations. In the absence of large domestic animals, the fecal burden would have been less than similar situations in Europe and Asia. Nevertheless, whereas sewage is generally avoided even in societies without any concept of germs, urban diarrheal disease outbreaks would have been inevitable. It is no coincidence that, after malnutrition, the most significant risk factors for disease in the world today are contaminated water, poor hygiene, and inadequate sanitation (Box 4.2).

BOX 4.2 The human waste problem.

It has been calculated that the average healthy human poops about 14 ounces of stool daily. This means that they produce about 312 pounds of feces annually, so by the time they reach 21 years of age they have produced over 6000 pounds, and a 65 year old will have produced over 20,000 pounds. When human populations were low and hunter-gatherers wandered the prairies in small groups, the disposition of human waste was not a problem. When they began to settle permanently in large numbers in villages, towns, and cities, it became a major issue. For example, a city of 90,000 like Mayan Tikal in Guatemala would have generated 28 million pounds of fecal material annually! Where did it go? Into nearby streams and lakes? What about their water supply? We know that some rulers had simple toilets that discharged into nearby streams, but what about the great mass of the population? Similar considerations apply to other high-density Native American populations such as Tenochtitlan (Mexico City) and the Cahokia mounds region in southern Ohio.

Fecal contamination would therefore have posed an enormous disease threat to the population of these newly developed cities. Given a lack of awareness of the nature of infectious diseases, it is almost inevitable that enteric disease cases such as typhoid and paratyphoid would have exploded. It is probable that this was a significant contributory factor to the Maya collapse. It is also a problem that persisted and plagued city dwellers throughout history until the development of effective sewage treatment in Western cities in the late 19th century.

It is also of interest to note that the Maya language had terms for diarrhea of varying severity. They also had words for cough and pulmonary tuberculosis. They had terms for the treponemal diseases, pinta and syphilis. They recognized the phenomenon of contagion. They even had a pair of death gods (or demons), *Xic* and *Patan,* who lived in the underworld (Xibalba) and who "caused people to die by coughing up blood while walking on the road"—perhaps a hemorrhagic fever?

The collapse of these cities occurred fairly suddenly between 750 and 950. Perhaps the best investigated is the collapse of the classical Maya civilization in the 9th to 10th centuries. There was a slowing of new construction late in the 8th century. By 830 most of the lowland Maya cities were beginning to fail. By 950 the great Mayan cities were almost totally abandoned. This was associated with a devastating drought that lasted for several hundred years—in other words, climate change. The cities were abandoned, construction ceased, and fine pottery and carved stelae were no longer produced.

Multiple plausible theories have been put forward for this sudden collapse. They include drought with consequent famine, social upheaval as a result of persistent warfare, and epidemic disease. These explanations are by no means mutually exclusive, and the evidence for severe drought is unequivocal. The most obvious explanation would be epidemics of diarrheal disease associated with large cities and contaminated water supplies. As pointed out above, the demographic catastrophe in the 16th century was in large part due to cocoliztli; could the earlier collapse have a similar origin? Thus could a hemorrhagic fever pandemic triggered by drought have contributed to the classic Maya collapse? This pattern of a disease wave following a period of drought is similar to that observed prior to the 1993—1994 hantavirus outbreak in the Four Corners area. As pointed out above, the virus reservoir, the mouse, *Peromyscus maniculatus,* undergoes a population explosion when it rains after a prolonged drought.

References

[1] El-Najjar MY. Human treponematosis and tuberculosis: evidence from the New World. Am J Phys Anthropol 1979;51(4):599—618. https://doi.org/10.1002/ajpa.1330510412.

[2] Acuna-Soto R, Calderon Romero L, Maguire JH. Large epidemics of hemorrhagic fevers in Mexico 1545-1815. American Society of Tropical Medicine and Hygiene. Am J Trop Med Hyg 2000;62(6):733—9. https://doi.org/10.4269/ajtmh.2000.62.733. Available from: http://www.ajtmh.org/.

[3] Cajimat MNB, Milazzo ML, Bradley RD, Fulhorst CF. Ocozocoautla de espinosa virus and hemorrhagic fever, Mexico. Emerg Infect Dis 2012;18(3):401—5. https://doi.org/10.3201/eid1803.111602.

[4] Acuna-Soto R, Stahle DW, Therrell MD, Griffin RD, Cleaveland MK. When half of the population died: the epidemic of hemorrhagic fevers of 1576 in Mexico. FEMS Microbiol Lett 2004;240(1):1—5. https://doi.org/10.1016/j.femsle.2004.09.011.

[5] Andam CP, Worby CJ, Chang Q, Campana MG. Microbial genomics of ancient plagues and outbreaks. Trends Microbiol 2016;24(12):978—90. https://doi.org/10.1016/j.tim.2016.08.004. Available from: www.elsevier.com/locate/tim.

[6] Vågene ÅJ, Herbig A, Campana MG, et al. Salmonella enterica genomes from victims of a major sixteenth-century epidemic in Mexico. Nat Ecol Evol 2018;2(3):520—8. https://doi.org/10.1038/s41559-017-0446-6. Available from: www.nature.com/natecolevol/.

[7] Hart CA, McCaughey C. Hantaviruses. J Med Microbiol 2000;49(7):587—99. https://doi.org/10.1099/0022-1317-49-7-587.

Syphilis and the Columbian exchange

Contents

Among the diseases suffered by the pre-Columbian inhabitants of the Americas were two closely related bacterial infections. One, called yaws, was a chronic wound infection. The other, syphilis, was a sexually transmitted disease. The latter eventually turned out to be of major social and historical significance across the globe.

The organism

The causal agent of syphilis was first observed in 1905 by Fritz Schaudinn and Erich Hoffmann in Germany. It is a slender helical bacterium (a spirochete). They named it *Treponema pallidum*. Humans are its only natural host. Treponema are a family of helical bacteria that belong to the phylum Spirochetes. They are microaerophilic (need low oxygen) organisms that cannot be grown in vitro and do not stain well with Gram stain because they contain a large amount of a lipid called cardiolipin. They are classified into four species that cause human diseases: *T. pallidum pallidum* causes syphilis; *Treponema pallidum pertenue* causes yaws; *T. pallidum endemicum* causes bejel; and *Treponema carateum* causes pinta. Syphilis is the only one of these that is spread sexually.

Great American Diseases
ISBN: 978-0-443-31404-9
https://doi.org/10.1016/B978-0-443-31404-9.00005-7

Yaws is the oldest of the treponemal diseases and probably ancestral to the others. As an heirloom disease, it almost certainly arrived with the first human migrants to the Americas. Central America plays a special role as the region where *T. pertenue* subsequently evolved into the two other species.

The diseases
Yaws

Native Americans still suffer from yaws, a skin, bone, and joint disease caused by *T. pertenue*. It classically occurs in young children who run around and play together unclothed. The spirochete is transmitted between individuals by fluid leaking from a skin lesion and so infecting cuts and scratches in others. As a result, a round, hard, warty swelling develops at the infection site. This is called the "mother" yaw. It may be surrounded by "daughter" yaws. The yaws then rupture to form ulcers. The fluid from these ulcers contains large numbers of bacteria and is highly infectious. The yaws typically heal within a few months. However, new skin lesions may also develop on the face or moist skin areas and cause chronic, nonhealing wounds. Months or years later, the persistent infection may also cause bone and joint destruction, resulting in arthritis.

Pinta

Pinta is a skin disease that is endemic to Mexico and Central America. It was first described in the 16th century in Native Americans by Spanish conquistadors and missionaries. It is caused by *T. carateum*. It is not sexually transmitted but, like yaws, is probably transmitted by skin contact. It is the least severe of the treponemal diseases because it only affects the skin. Painless papules develop on the arms and legs and slowly heal, but after 3 to 9 months, a second crop appears all over the body, especially the hands, feet, and scalp. These "pintids" may cause significant disfiguration. Like other treponemal diseases, it may be treated with appropriate antibiotics. Pinta is now a very rare disease but may still be endemic in remote areas of Latin America.

Syphilis

Syphilis, like the other treponemal diseases, is characterized by three stages. It initially causes a skin infection—on the genitalia. It is the only one of the three that is sexually transmitted. The primary stage occurs when the treponemes grow at their point of entry. For example, primary syphilis first develops as an ulcer called a chancre on the genitals that heals after 4 to 8 weeks. These lesions are painless and develop slowly. The ulcers weep fluid containing treponemes and are the main pathway by which the disease is transmitted between individuals. The chancre heals in a few weeks to months. However, the treponemes are not restricted to the chancre. They spread from the primary lesion throughout the body in the bloodstream and then cause secondary disease. Victims develop skin rashes and a flu-like illness with a sore throat and a mild fever. The victim recovers within a few weeks or months and then appears to be disease free. Some individuals may indeed be cured. In others, however, after a delay of up to 20 to 30 years, the organism begins to destroy target tissues such as bones or blood vessels, and especially the brain, to eventually cause the tertiary disease characterized by bone destruction, sudden blood vessel rupture, neurodegeneration, and insanity (Fig. 5.1). Infants born to syphilitic mothers may be blind or deaf and have teeth of an unusual shape.

History
Syphilis and the Columbian exchange

There are two theories as to the origin of syphilis. The "Pre-Columbian" hypothesis suggests that syphilis had been present in the Old World for many years prior to the discovery of the Americas. This theory suggests that the lack of European accounts of syphilis prior to 1495 is accounted for by conflation with other skin diseases such as leprosy. However, this theory does not explain why it suddenly appeared and grew in virulence at the end of the 15th century. It has been claimed that there is evidence for the occurrence of syphilitic lesions in excavated European skeletons dated prior to 1493, but this is controversial (Box 5.1).

As described in Box 3.2, it has proved possible to extract DNA from the root canals of ancient skeletons. These can be sequenced and compared with the DNA of disease-causing bacteria. In multiple sites across Europe ranging from France to Estonia, it has been shown that treponemal DNA is present in excavated European skeletons dated long before 1493. However, it remains unclear which of the *T. pallidum* subspecies are present in these individuals. Was it yaws, bejel, or syphilis? It has also been claimed that some of these individuals show skeletal lesions like those caused by *T. p. pallidum*. In

other words, they suffered from true syphilis. This is a very controversial subject!

The alternative "Columbian" hypothesis claims that the causal agent of syphilis, *T. p. pallidum*, originated as a mutant of *T. p. pertenue* in the New World [1]. It is claimed that it was then brought back to Europe by Columbus' crew in 1493 after they acquired it from the natives of the West Indies. Upon returning to Spain, some of the crew joined the Spanish army and participated in the siege of Naples in 1495. When the war was over, the soldiers went home and the disease spread throughout Europe.

Molecular data tend to support the Columbian hypothesis. For example, when the genomes of 26 geographically diverse strains of treponema were sequenced and analyzed, it was found that the Old World species that cause yaws are very ancient and likely were the first to infect humans.

Lesions of yaws have been identified in the skeletons of *Homo erectus* in East Africa dated to 1.6 million years ago. Thus it is likely that yaws emerged from Africa with the first hominids. This long preceded the populating of the Americas and suggests that treponemes probably came to the New World with the first human immigrants. They probably also came to Europe as well.

There is no doubt that syphilis was present in the Americas prior to 1492 [2]. For example, there is a probable illustration of a case of syphilis on a Peruvian jug dating from the 6th century. It depicts a mother holding her child. The mother shows the collapsed nose that results from the destruction of the nasal cartilage in tertiary syphilis. Her upper incisor teeth also have characteristic notches. In contrast to Europe, syphilitic lesions have been found in the bones of Native Americans dating from well before Columbus's arrival. Yaws' lesions have been found in skeletons dating from 2300 BCE in Alabama and from 4300 BCE in Illinois. Conversely, the sexually transmitted strains of *T. pallidum* are of much more recent origin. Skeletal lesions that appear to be syphilitic have only been dated to 1600 to 800 years ago.

Molecular genetic studies on treponemal DNA have shown that the closest relatives of *T. pallidum* are certain South American strains of *T. pertenue*. These strains are found only among indigenous peoples living deep in the rainforests of Guyana [3]. This suggests that they may be related to the ancestral strains of *T. pertenue* and uncontaminated by strains from elsewhere.

Recently, studies on tooth canal DNA from human remains from a burial site (Santa Catarina) in coastal Brazil dated to nearly 2000 years ago have provided evidence for the early existence of treponemes in the New

World. The genome obtained most closely resembled the subspecies that causes bejel (*T. p. endemicum*). Analysis of these genomes using molecular clock techniques suggests that the treponemes may have started diversifying as long as 14,000 years ago [14]. It is therefore possible that treponemal diseases originated in Eurasia or Africa and were brought to the Americas by the first human immigrants around 15,000 years ago [14]. They also indicate that the three treponema strains had diverged long before Columbus' voyages.

When the first European sailors arrived with Columbus in San Salvador in October 1492, there was much sexual contact between them and the native Arawak women. It is believed that his crew acquired syphilis and took it back with them to Spain [4]. Columbus also kidnapped nine native American men and displayed six of them naked at court when he returned to Spain in March 1493. Court gossip suggested that they too may have transmitted the disease, although there is no contemporary evidence of this. When Columbus' crew was disbanded, some, it is believed, traveled to Italy and joined the army of Charles VIII in the fighting in Naples.

Two physicians, Fernandez de Oviedo and Ruy Diaz de Isla, in the early 16th century, wrote a treatise on the origin of syphilis. Diaz da Isla declared that this was a new unknown disease and claimed that it had originated on Isla Espanola, the original Spanish name for the island that contains present-day Haiti and the Dominican Republic. He declared that Columbus' pilot, Pinzon de Palos, and other crew members were already suffering from the disease when they returned home to Spain. He claimed to have treated at least one of Columbus' crew for "The West Indian Disease" at Barcelona. de Oviedo subsequently went to the New World and reported to King Ferdinand that the same disease was known to Native Americans and that they were very familiar with it [4].

In the 1490s, the pope had a bitter disagreement with King Ferdinand of Spain over unpaid debts. As a result, the pope offered King Charles VIII of France the crown of the Kingdom of Naples. In August 1494, Charles invaded Italy with a mercenary army that included many Spaniards and was accompanied by a large number of camp followers including prostitutes. On the way to Naples, they stopped off in Rome for a month of "limitless depravity" including both rape and prostitution. Charles entered Naples unopposed in February 1495, and another bout of debauchery followed. The other Italian princes were concerned about Charles' increasing power, and in July 1495, they defeated him at the battle of Fornovo. After this battle, Italian physicians reported the occurrence of a completely new disease [5]. None had seen it previously. It started with genital ulcers but was followed

by a generalized rash and foul-smelling abscesses. It was obviously transmitted through sexual intercourse. The French were blamed for spreading
it across Italy, and it was known initially as the French disease. As the mer-
cenaries returned home, they spread it across Italy and eventually to France
and the rest of Europe. Voltaire, the great French writer, commented: "On
their way through Italy, the French carelessly picked up Genoa, Naples, and
syphilis. Then they were thrown out and deprived of Genoa and Naples.
But they did not lose everything—syphilis went with them." The disease
reached England in 1496, India in 1498, Poland in 1499, Russia in 1500,
and China in 1505. The disease was obviously very contagious and, at this
stage of the outbreak, may have been spread simply by skin contact as
in yaws.

This initial syphilis outbreak was very severe. Pustules covered the whole
body. The lips, nose, and eyes were destroyed. Rotting flesh fell off the face,
and death occurred within a few months. Victims developed abscesses that
leaked pus and could erode tissues down to the bone. They also developed
extremely painful disabling joint pain. This was much more severe than
syphilis is today and is compatible with a completely new infectious disease
spreading through a population that had absolutely no immunity—a "virgin
soil" epidemic. On reflection, these initial lesions were so very disgusting
that they would have greatly reduced the attractiveness of its victims and
worked against venereal spread. Over time, however, the treponema likely
evolved simply by preferentially spreading those strains that caused less
obvious lesions and were thus more likely to encourage venereal spread.
This disgusting new disease was generally attributed to one's enemies.
The English called it the French disease, the Poles called it the Russian disease, and the Turks called it the Christian disease. Others simply called it the
"Great pox."

In 1530 an Italian physician and poet working in Verona, Girolamo
Frascastorio, wrote an epic poem in Latin in three volumes entitled, "*Syphilis
sive morbi gallici*" (Syphilis or the French disease). In his poem, a shepherd,
Syphilis by name, was upset with the sun god Apollo for drying up his
springs and wilting his trees. So he decided to stop worshiping Apollo.
Apollo got upset in turn and cursed Syphilis with a frightful disease. The disease spread to affect the whole population [6]. The name stuck!

It is difficult to determine exactly which major historical figures suffered
from syphilis. One reason is that syphilis and gonorrhea were frequently
confused. Few people bragged about it, and their enemies were very likely
to suggest that they had acquired it, thus confirming their immorality. For

example, Henry VIII of England claimed that his queen Anne Boleyn gave him the pox, but then it was a bitter divorce! Ivan the Terrible, Czar of Russia, was most likely terrible because he had neurosyphilis [7]. Neurosyphilis, also termed "general paralysis of the insane," was generally considered a result of bad heredity, moral turpitude, or weak character [7]. Because neurosyphilis resulted in mental illness, it was not uncommon to accuse someone who behaved erratically as having syphilis. Mary Todd Lincoln, widow of Abraham Lincoln, was so accused (Box 5.2). It was not until 1913 that a Japanese bacteriologist, Hideyo Noguchi, working at Rockefeller University in New York, demonstrated the presence of *T. pallidum* in the brain of a paralyzed patient. Al Capone, the gangster, died of neurosyphilis during his imprisonment in Alcatraz.

BOX 5.2 The case of Mrs. Lincoln.

Claims by others regarding someone being infected with syphilis must always be regarded with skepticism. In effect, it was both a way of accusing someone of promiscuity and a way of explaining an individual's unusual behavior. President Lincoln's widow, Mary, was such an individual (Fig. 5.2). She had plenty of enemies and she behaved very erratically. She had a bad temper with angry outbursts and delusions. She was probably seriously depressed and grief stricken following the loss of her husband and children. Indeed, such was her behavior that she was examined by a team of eminent physicians in January 1882. Her son Robert had told a jury that his mother was mentally incompetent in 1875, and she was consigned to a private mental asylum. She stayed there for 3 months before her friends had her released.

The team of physicians reported that Mrs. Lincoln had "chronic inflammation of the spinal cord, chronic disease of the kidneys, and a commencing cataract in both eyes." The physicians attributed her sight loss to the spinal disease. She had difficulty balancing and couldn't walk without assistance. The prognosis was bad. Although the physicians never said so explicitly, this collection of signs collectively denote *tabes dorsalis*. Tabes dorsalis is a manifestation of tertiary syphilis resulting from demyelination of the large neurons in the spinal cord. There was other evidence supporting this diagnosis, including her complaints of sudden "lightning pains," saying she was "on fire" or being "cut with knives." The physicians never claimed that Mrs. Lincoln had syphilis but rather declared her condition to be due to an accident. There is no evidence to suggest when or where she might have contracted the disease. Likewise, there is no evidence to suggest that President Lincoln had syphilis.

Fig. 5.2 *Mary Todd Lincoln, the wife of President Abraham Lincoln. After his death, she appeared to be behaving strangely. Some believed that this was a result of neurosyphilis. This cannot be proven and is highly unlikely. (Courtesy Library of Congress.)*

American society made strenuous attempts to control sexually transmitted diseases (STD); however, the country was ambivalent regarding public discussion of such things [8]. Some wanted to publicize the issue in an effort to educate people, whereas others believed that such publicity was inappropriate. During the first World War, aggressive attempts were made to reduce STD prevalence rates in troops in order to maximize their effectiveness. After the war was over, campaigns involving talking openly about STDs were not considered appropriate in a moral society. As a result, the educational campaigns stopped and there was a resulting increase in disease prevalence. By 1930, approximately 10% of Americans suffered from syphilis. Each year there were almost 500,000 new syphilis infections; 18% of deaths from heart disease were a result of syphilis; 20% of all mental patients were suffering from neurosyphilis; and 60,000 children were born annually with congenital syphilis. At that time, the disease was treated with arsenical compounds such as salvarsan. (Salvarsan was the first modern antimicrobial drug. It was discovered in 1910 and was used to treat syphilis and the African protozoal disease, trypanosomiasis. It was not very effective and quite toxic.) Treatment was expensive and not always available. Some hospitals refused to admit patients with STDs on the grounds that they were "morally tainted and thus less deserving of care." The New York State Health Commissioner planned to give a talk about public health but was banned from using the words syphilis or gonorrhea.

The Tuskegee Syphilis Study

Syphilis continued to impact society. One highly significant manifestation was the infamous Tuskegee Syphilis Study, where naturally infected Black men were denied treatment based on a study that claimed to follow the natural course of the disease [9]. In 1928, a Norwegian study was published that had followed the course of syphilis in hundreds of White males by retrospectively reviewing their medical records. A group of US Public Health researchers decided to complement this with a prospective study of Black males. The study began in the fall of 1932 when the researchers recruited 623 African Americans from rural Macon County in Alabama. Many were poor sharecroppers; some were construction workers. About 400 of these men had already contracted syphilis, whereas 201 were disease free. They could not afford health care, but a local nurse, Eunice Rivers, would provide the men with incidental medications (tonics (feel-good medicines!) and aspirin). The men were told it would be a 6-month study and they would get free medical treatment, meals, and burial insurance. None of the men were told that they had syphilis. They were told that they had "bad blood," a vague term that encompassed a variety of disease conditions. The original intention of the study was to follow untreated syphilis for 6 to

9 months and then test various treatments. This goal changed to a long-term, no-treatment study.

The men were never offered treatment, even when penicillin became available in 1947 and was known to cure syphilis. Thus the researchers knowingly withheld effective treatment and appropriate information from these patients. The researchers looked for neurosyphilis by performing spinal taps to sample cerebrospinal fluid. They told their victims that they were receiving "special free treatment." They had to undergo an autopsy after they died in order to qualify for funeral insurance. Not only that, but the investigators prevented the trial subjects from accessing local syphilis treatment programs. Some were given placebos. Some men registered for the draft when war broke out in 1941. The Army determined that they had syphilis and they were ordered to obtain treatment (the researchers tried to prevent this), not entirely successfully. The men were not the only victims; while many participants died of syphilis, 40 wives became infected and 19 children were born with congenital syphilis. The last of the study participants died in 2004.

The existence of the study was revealed in 1972 by Peter Buxton, who had worked as a venereal disease investigator for the Public Health Service. Buxton had protested about this unethical study to his superiors, but they had not responded. He therefore gave his information to *The New York Times* and the *Washington Star* as well as the Associated Press. The resulting outcry led to the termination of the study as well as major changes in the laws and rules regarding experiments conducted on human patients. The National Research Act of 1974 required that human studies obtain informed consent, a communication of the diagnosis, and accurate reporting of the results. Institutional Review Boards were established at research centers to assess proposed human studies and ensure that patients are fully informed and that their interests are protected. On May 16, 1997, President Bill Clinton formally apologized to five of the eight surviving participants of the "experiment" on behalf of the United States. "What the United States did was shameful, and I am sorry."

The Tuskegee Syphilis Study was shameful and embarrassing for the American medical profession as well. At the study's inception in 1932, no standard criteria or guidelines for human medical experimentation had been codified. However, by 1947, the "Nuremberg Code" had been developed as a result of the Nuremberg Doctor's Trials. During the trials of doctors who had experimented on patients in the Nazi death camps, experts from the United States were called on to testify about the doctors' experiments and the absolute absence of any ethical standards. As a result, they also helped to develop universal standards for the use of humans in medical experimentation. The American Medical Association (AMA) appointed Dr. Andrew C. Ivy as the AMA's official consultant to the prosecution. Dr. Ivy's report to the Nuremberg Tribunal and the AMA's Judicial Council became the foundation for the Nuremberg Code and the AMA's rules for human experimentation.

The Nuremberg Code asserted that *"before the acceptance of an affirmative decision by the experimental subject, there should be made known to him the nature, duration, and purpose of the experiment; the method and means by which it is to be conducted; all inconveniences and hazards reasonably to be expected; and the effects upon his health or person, which may possibly come from his participation in the experiment."* The AMA's rules for human experimentation, approved in December 1946 by the AMA House of Delegates, stated: *"the voluntary consent of the person on whom the experiment is to be performed must be obtained; the danger of each experiment must be previously investigated by animal experimentation, and the experiment must be performed under proper medical protection and management."*

Despite these statements, the Tuskegee Study continued for almost three decades after these proclamations were issued. Its results were reported in the medical literature and at scientific meetings without protest. In fact, in the late 1960s the US Public Health Service reviewed the study but did not recommend its termination or obtaining subjects' consent. It was only when the press reported on it that the study was terminated.

The Guatemala story

The Tuskegee Syphilis Study was not the only incident involving inappropriate sexual disease experimentation. For example, in 2010, it was revealed that the US Public Health Service, in cooperation with the Guatemalan authorities, had conducted venereal disease studies including studies on syphilis, gonorrhea, and chancroid in Guatemala between 1946 and 1948 [10]. Over 5500 soldiers, convicts, mental patients, children, orphans, and Guatemalan natives, as well as child and adult prostitutes, were involved in the study. Many were deliberately infected with *T. pallidum* or other pathogens without their consent. Only 700 received treatment. These studies were conducted by a Public Health Service physician, Dr. John C. Cutler, who was also one of the principals in the Tuskegee Study. Investigators tested different methods of inducing these diseases by deliberately inoculating agents by various routes as well as "normal exposure," where sex workers suffering from syphilis were used to transmit the disease to prisoners. (This "normal exposure" group was remarkably unsuccessful.) The intent of the study was to investigate possible postexposure treatments such as genital disinfectant washes. At least 83 participants died. No effort was made to obtain informed consent, and sometimes participants were deliberately deceived. None of the participants or their relatives were compensated. These studies were not publicized until after Cutler's death in 2003 and his notes were investigated.

In 2010 the US Government formally apologized and condemned this violation of human rights. President Obama apologized to Guatemalan President Colom, calling these studies "a crime against humanity."

The present status of syphilis

In 1909, the German chemist/pathologist Paul Ehrlich developed an arsenic-based drug, arsphenamine or Salvarsan, that although toxic, was effective in reducing disease severity [7]. In 1943, however, John Mahoney and his colleagues working at the US Marine Hospital at Staten Island treated four patients with primary syphilis with multiple injections of the newly discovered antibiotic, penicillin, for 8 days and cured them! This revolutionized the treatment of the disease (Fig. 5.3). Penicillin had minimal side effects compared to previous treatments with heavy metals such as

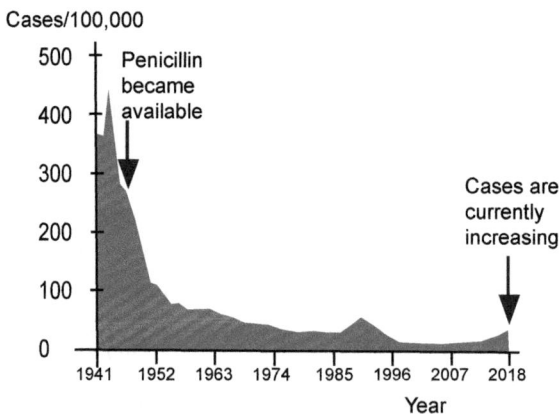

Fig. 5.3 *Syphilis—reported cases—United States. 1941–2018. The decline of syphilis was in response to the discovery of penicillin and its ready availability after 1945. The rise from 1985 to 1995 was associated with the HIV epidemic. Note that the prevalence of syphilis is currently rising. (Data courtesy CDC.)*

mercury or arsenic. The fear of the disease declined as did the publicity given to its severity.

As a result, syphilis persists. Globally, the World Health Organization estimated that there were 8 million new cases in adults in 2022, and more than 90% of these cases were in developing countries. Congenital syphilis is a major cause of stillbirths and neonatal mortality in many of these countries.

Syphilis also persists in the United States despite the availability of antibiotic treatment [11]. As a result of the behavioral changes caused by fear of HIV (Chapter 17), syphilis cases reached a historic low in 1999–2000 when 80% of US counties reported no cases. The Centers for Disease Control and Prevention (CDC) seized the opportunity to begin an eradication campaign designed to drop the numbers still further. Unfortunately, instead, the numbers began to creep up. This may have been due to the advent of successful HIV therapy so that people were no longer afraid of it. While HIV rates dropped, syphilis cases climbed (Fig. 5.3). The opportunity to eradicate it has passed—it has also developed significant resistance to penicillin [12].

In 2019, 129,813 cases of syphilis were reported in the United States. These included 1870 cases of congenital syphilis—a 291% increase from 2015. About half of these occur in Texas and California. Syphilis circulates predominantly in inner-city populations. The disease is prevalent in people living in poverty, the homeless, drug users, and in prisons. It is primarily a male disease, with only 16.7% of cases occurring in women. Nevertheless, according to the CDC, between 2015 and 2019 the number of reported syphilis cases in women rose by 178.6%. Syphilis control depends upon early detection of infected individuals (preferably at the primary disease stage) and their contacts, together with prompt treatment by appropriate antibiotics. Penicillin is still effective, but resistance to other antibiotics has emerged and there is potential for this resistance to increase. There is no vaccine available.

Syphilis infections in the United States are at their highest levels since the 1950s. According to the CDC, they increased by nearly 80% to 207,000 between 2018 and 2022. Over the past decade, cases of congenital syphilis in babies have increased by 973%. They mainly affect racial and ethnic minorities.

It is mandatory to test prenatal women for syphilis and HIV. Treatment of congenital disease requires that mothers receive penicillin at least a month before delivery. However, many impoverished women fail to receive timely prenatal care and syphilis testing.

Testing of a patient's blood for the presence of antibodies came into use in 1906. The presence of antibodies against an antigen called cardiolipin detected by the Wassermann blood test is commonly used to diagnose syphilis. This testing can not only detect infected individuals but also determine its prevalence in a population. It has since been superseded by more modern molecular-based rapid techniques.

Despite the fact that testing is cheap and penicillin is still largely effective, *T. pallidum* will continue to cause human disease, suffering, and death for the foreseeable future.

References

[1]   Majander K, Pia-Diaz M, du Plessis L, et al. Redefining treponemal history through pre-Columbian genomes from Brazil. Nature 2024. https://doi.org/10.1038/s41586-023-06965-x.

[2]   El-Najjar MY. Human treponematosis and tuberculosis: evidence from the New World. Am J Phys Anthropol 1979;51(4):599—618. https://doi.org/10.1002/ajpa.1330510412.

[3]   Mann CC. 1493: Uncovering the new world Columbus created. Vintage Books. New York, NY: Random House Inc; 2011.

[4]   Harper KN, Ocampo PS, Steiner BM, et al. On the origin of the treponematoses: a phylogenetic approach. PLoS Negl Trop Dis 2008;2(1). https://doi.org/10.1371/journal.pntd.0000148.

[5]   Nunn Nathan, Qian Nancy. The Columbian exchange: a history of disease, food, and ideas. J Econ Perspect 2010;24(2):163—88. https://doi.org/10.1257/jep.24.2.163.

[6]   Tampa M, Sarbu I, Matei C, Benea V, Georgescu SR. Brief history of syphilis. J Med Life 2014;7(1):4—10.

[7]   Tramont EC. The impact of syphilis on humankind. Infect Dis Clin North Am 2004;18(1):101—10. https://doi.org/10.1016/S0891-5520(03)00092-8.

[8]   Frith J. Syphilis—its early history and treatment until penicillin and the debate on its origins. J Mil Vet Health 2012;20(4):28—49.

[9]   Brandt AM. No magic bullet. A social history of venereal disease in the United States since 1880. Oxford, UK: Oxford University Press; 1985.

[10] Haller John S, Jones James H. Bad blood: the Tuskegee syphilis experiment. J American History 1981;68(3):739. https://doi.org/10.2307/1902047.

[11] Rodriguez MA, García R. First, do no harm: the US sexually transmitted disease experiments in Guatemala. Am J Public Health 2013;103(12):2122—6. https://doi.org/10.2105/AJPH.2013.301520.

[12] Ghanem KG, Ram S, Rice PA. The modern epidemic of syphilis. N Engl J Med 2020;382(9):845—54. https://doi.org/10.1056/NEJMra1901593.

[13] Stamm LV. Global challenge of antibiotic-resistant treponema pallidum. Antimicrob Agents Chemother 2010;54(2):583—9. https://doi.org/10.1128/AAC.01095-09.

[14] Majander K, Pia-Diaz M, du Plessis L, Arora N, et al. Redefining the treponemal history through pre-Columbian genomes from Brazil. Nature 2024;627:182—8.

CHAPTER SIX

Tuberculosis—another ancient disease

Contents

Of all the plagues described in this book, tuberculosis (TB) is by far the most lethal. Tuberculosis has killed more humans than any other infectious disease in history. Over the last 2000 years, it has killed more than 1 billion people. In 2017 alone, it killed 1.6 million people worldwide. In that same year, it is also estimated that 10 million people became infected with TB. It is the leading cause of death caused by any single infectious agent including AIDS. The continent most severely affected is Africa, and only 3% of TB cases currently occur in the Americas [1].

Tuberculosis is a bacterial disease caused by a group of closely related organisms belonging to the *Mycobacterium tuberculosis* complex. Multiple other mycobacteria belong to the complex and share 97% to 99% identity in their DNA sequences. Within the complex, the bacteria are classified according to their preferred animal hosts. These other organisms include, among others, *Mycobacterium bovis* that prefers cattle, *Mycobacterium caprae* that prefers goats, and *Mycobacterium microti* that prefers mice.

The human pathogen *M. tuberculosis* is by far the most dangerous and the most important of these organisms. It is commonly called the tubercle

Great American Diseases
ISBN: 978-0-443-31404-9
https://doi.org/10.1016/B978-0-443-31404-9.00006-9

bacillus. It is estimated that *M. tuberculosis* infects about 23% of the world's population—1.7 billion people. However, only a very small proportion of these individuals actually develop clinical disease. The reasons for this remain unclear. The bacterium can invade any organ, but the most significant infections involve the lung. Tuberculosis is an heirloom disease with lesions detected in human skeletons from the Middle East dated to about 10,000 years ago. Evidence of spinal and pulmonary tuberculosis has been detected in Egyptian mummies from 2400 BCE. Tuberculosis was mentioned in ancient Indian manuscripts such as the Rigveda of 1500—1000 BCE.

The organism

For a long time, many physicians were reluctant to accept that tuberculosis could be of microbial origin. It was argued that it resulted from poverty, malnutrition, and lack of sanitation. The tubercle bacillus was first identified by the German microbiologist Robert Koch. He couldn't see it when he first examined infected tissues but suspected that TB was caused by a bacterium resistant to conventional tissue stains. Koch therefore experimented with various staining procedures and eventually found one that revealed the presence of huge numbers of rod-shaped bacteria in the disease lesions (tubercles). Their failure to stain was because the waxy coat of *M. tuberculosis* prevented the dyes from binding to the bacteria. Koch also noted that the appearance of these bacteria in tissues preceded the development of tubercles. Koch initially had difficulty in growing the organism but eventually succeeded in culturing it on coagulated blood serum. *M. tuberculosis* grew much more slowly than other bacteria, but Koch was sufficiently patient to let the bacterial colonies grow. He went on to use this cultured organism to produce typical tuberculosis in animals. Tuberculosis causes many different clinical forms of disease, but Koch demonstrated that the same organism was present in all of them. Koch called the bacterium *M. tuberculosis* and reported his findings in March 1882 at a conference in Berlin. *M. tuberculosis* is a thin rod-shaped bacterium with a characteristic cell wall that contains waxes called mycolic acids as well as a complex glycolipid called lipoarabinomannan. These waxes confer the special staining properties on the organism. As described in Chapter 1, in 1882 Paul Ehrlich, a German pathologist, developed a staining procedure using the dye carbol fuchsin and heat that stained the Mycobacterium red despite washing with acid-alcohol. (This makes them "acid-fast," whereas bacteria without mycolic acid do not retain the red stain.) The waxy coat also makes the organism resistant to killing by lethal enzymes produced by white blood cells. As a result, a cell-mediated immune response is required to keep it under control. Antibodies alone are ineffective. Humans whose immune system is suppressed by stress, drugs, or other infectious agents are at increased risk of developing TB. Another important feature of this bacterium is its need for oxygen. As a result, it grows best in well-oxygenated parts of the body such as the lungs.

Robert Koch made an extract of the tubercle bacillus called tuberculin. He tried to use it as a vaccine, but it didn't work. However, he found that when he injected tuberculin into the skin of infected patients, they developed a hard inflammatory nodule at the injection site. (The immune system was attacking the foreign antigen.) This reaction did not occur in uninfected individuals. This procedure is called the tuberculin skin test and is still commonly used to identify those infected with tuberculosis [Box 6.1].

The disease

About 85% of tuberculosis patients have lesions within their lungs. As few as 10 inhaled bacteria can initiate the infection. The organisms can also be found in fluid droplets generated by sneezing, spitting, or coughing so it is

BOX 6.1 The Iowa Cow War.

In 1931, it was the practice of the State of Iowa to test cattle using the tuberculin skin test. Iowa was attempting to get rid of tuberculous dairy cattle and thus eliminate the risk of contaminated milk affecting consumers. All dairy and breeding cattle had to be tested. Those cattle that tested positive (reacted to the tuberculin) were slaughtered, and the farmer received compensation. Opposition was immediate. Farmers were impoverished already by the low price of milk. The Great Depression had started. The testing program was viewed as a threat to their livelihood. In 1931, about 1000 farmers took a train to Des Moines to meet with legislators and force them to make the testing optional. The legislators declined. In response, groups of farmers began to block the testing by veterinarians. The vets were met with thrown eggs, stones, and even blows and kicks. The state responded by calling in the police and sheriffs to provide protection. The fighting came to a head in September when two vets escorted by 65 law officers were confronted by a crowd of about 400 dairy farmers. They destroyed the cars of the veterinarians. The next day, the governor declared martial law and sent in the Iowa National Guard. The violence subsided immediately! Two of the farmer's leaders served short prison sentences.

readily communicated between individuals living in close proximity. When these bacteria are inhaled into the lung, cells called alveolar macrophages readily capture them, but mycobacteria are difficult to kill. The alveolar macrophages can slow bacterial growth and trigger a defensive inflammatory response in the tissues that "walls off" the organisms. If the bacteria are effectively trapped by a layer of macrophages and fibrous scar tissue, then although they survive, they may cause no more problems. This is called latent tuberculosis. If, however, these infected individuals are immunosuppressed later in life, then the bacteria may escape from these constraints, spread through the body, and subsequently cause clinical disease. (That is why AIDS patients may develop lethal tuberculosis.)

Unfortunately, in some patients, their initial cell-mediated immune response may be insufficient to control the organisms. As a result, the mycobacteria will continue to grow within the macrophages. Even when multiple macrophages join together to form "giant cells," they still cannot complete the task. As a result, each focus of infection becomes surrounded by a mass of dead and dying cells as well as lymphocytes and macrophages. These lesions are called tubercles. As the disease proceeds, the tubercles enlarge, fuse together, and replace normal lung tissue. Patients develop a hacking cough and may cough up lumps of dead tissue. The tubercles may erode blood vessel walls and rupture them so that the patient coughs up fresh blood. Even if the body succeeds in walling off the organisms, the tubercles and the fibrous scar tissue that forms around them replace the normal lung tissue. As a result, the patient is deprived of oxygen and ultimately dies from heart and respiratory failure.

In about 2% of patients, the mycobacteria leave the lungs and settle in the joints between the spinal vertebrae. Here, the tubercles erode nearby bones so that the spine collapses, possibly damaging nerves and leading to paralysis and death. Alternatively, the patient may develop a characteristic "hump." This form of tuberculosis is called Pott's disease, after Percival Pott (1714–1788), who identified its cause. In other patients, the mycobacteria may leave the lungs and spread through the bloodstream to form tubercles in other organs. These can include the abdominal cavity, bones, brain, kidneys, and heart.

A slightly different form of tuberculosis is called scrofula or The King's Evil (technically it is "tuberculous cervical lymphadenitis"). In this form of the disease, the mycobacteria invade the lymph nodes of the neck, resulting in unsightly swellings. These swellings may rupture and discharge smelly pus. They may be sufficiently large to interfere with breathing. Scrofula results from drinking raw milk from tuberculous cows. *M. bovis*, the organism that affects cattle, is also a member of the *M. tuberculosis* complex and will therefore infect humans. Heat treatment of milk will kill these bacteria. Pasteurization (short-term heating of milk) was specifically designed to kill *M. bovis* and eliminate scrofula. Scrofula got its other name, the "King's Evil," because of the belief that it could be cured by the touch of a king. A king merely had to touch the swelling for it to go away. The first recorded touching procedure began with Clovis, king of the Franks (466−511), and persisted up until the 19th century. It became especially popular in England in the 17th century. King Charles II of England (1630−1685) touched 92,102 victims during his reign. King William III (1689−1702) refused to participate in such a superstitious activity. When asked to touch he replied, "God grant you better health and better sense." Queen Anne (1702−1714), his successor, was an enthusiastic toucher but was the last English monarch to do so. George I, her successor, abandoned the practice. French kings only abandoned the practice after 1825 when Charles X touched 121 scrofulous patients.

History
Pre-Columbian America

Tuberculosis was present in North America before the advent of humans. Lesions of tuberculosis have been found in the skeletal remains of ice-age bison (*Bison antiquus*), musk ox, bighorn sheep, and mastodons. The oldest proven case of tuberculosis is in a bison found buried in sediments from Natural Trap cave in Wyoming, dated to $17,870 \pm 230$ years ago [2]. This was confirmed by sequencing its DNA as well as the detection of its characteristic complex lipids.

On the other hand, the oldest human skeletons with evidence of tuberculosis have been found in Eurasia and North Africa and all date from less than 9000 years ago [3]. The oldest such human skeletons in North America date from 2100 to 1900 years ago. This has raised the possibility that tuberculosis was initially an animal disease and it subsequently spread to humans. In other words, it was a zoonotic disease. DNA sequencing has, however, shown that *M. bovis* has a much smaller genome than *M. tuberculosis,* suggesting that it most likely lost some genes when it evolved from an *M. tuberculosis*-like precursor.

Because tuberculosis is so common and because it often either does not cause severe disease or kills its victims very slowly, it is likely that it was introduced into the Americas with the arrival of the first human migrants from Siberia. Unlike other acute infectious diseases that kill victims fast, tuberculosis can remain asymptomatic at low levels in human populations. These infected carriers would not have slowed immigration significantly. The *M. tuberculosis* complex has been detected in pre-Columbian mummies in Peru and northern Chile. An autopsy of a mummified child dated from 700 has revealed characteristic lesions of tuberculosis. The child had lesions in their pleura and lungs that appeared to be small tubercles. Likewise, tubercles were present in its heart, pericardium, and lumbar vertebrae. Others have reported evidence for early tuberculosis in human remains dated around 260. To the surprise of the investigators, DNA sequencing of Peruvian mummy samples indicated that the bacteria shared a recent common ancestor with *Mycobacterium pinnipedii* from seals and sea lions rather than from humans [4]! The organism was estimated to have originated 4000 to 5000 years ago, long after the first humans had crossed the Bering land bridge from Asia to North America. Perhaps the disease was introduced by marine mammals on the coast of Peru.

Tuberculosis was not confined to the Andes of Peru. Mycobacterial DNA has been isolated from human remains found at archeological sites in eastern North America and South America. Additionally, clay figurines have been found in Yucatan showing the characteristic hump on the spine associated with Pott's disease. Similar figurines have been found in the Mississippi region in Arkansas, Mississippi, Missouri, and Tennessee. Skeletons from Tennessee and New York also have spinal lesions that appear to be tuberculous. By around 1000, many of the tribes in the Mississippi region had begun to farm corn and develop settled communities. As a result of these larger, denser populations, opportunities for disease spread increased as did the archeological evidence found in skeletal remains [5]. Remains dating from 1100 to 1400 in Ohio also appear to have lesions characteristic of tuberculosis of the spine. The arrival of Europeans carrying tuberculosis and the resulting decline in the living conditions of Native Americans worsened the tuberculosis situation, and, as in Europeans, it became a very serious problem in Native Americans by the beginning of the 19th century.

The ancient world

Tuberculosis was well recognized in ancient Europe and Asia. Pathological evidence dates back to 8000 BCE. It is mentioned in biblical texts from Deuteronomy and Leviticus. Homer mentioned it in the *Odyssey*, dating from around 700 BCE. The Greek physician Hippocrates recognized TB and called the disease "phthisis" from the Greek *phthein*, meaning to dwindle or waste away. Hippocrates claimed that its preferred victims were usually aged between 18 and 35. Because it frequently affected several members of the same family, he considered it to be a hereditary disease [6]. Later physicians preferred the term "consumption." This described the progressive wasting away of patients as the bacteria gradually spread through their lungs and body, destroying normal tissues in their way. When tuberculosis became an epidemic in Western societies in the 18th and 19th centuries, it received other names, such as the "White Death" or the "White Plague," reflecting the anemia associated with these chronic infections. In 1679, a Dutch physician, Sylvius de la Boë, used the term *tubercula glandulosa* to describe the lung lesions in consumption. He also demonstrated that the enlarged lymph nodes in patients with scrofula were also forms of tuberculosis. Nevertheless, the term tuberculosis was not widely used to describe the disease until the 19th century.

The tuberculosis epidemic

Environmental issues have a significant impact on disease susceptibility. Malnutrition, crowding, and other stresses suppress immunity and permit tuberculosis to spread and prosper. As populations grew and people moved from rural areas to densely populated cities the prevalence of tuberculosis grew dramatically. Tuberculosis thus increased in significance in both Europe and North America beginning in the 17th century. By the late 18th to early 19th century, conditions came together to permit a massive epidemic to develop. Tuberculosis mortality peaked in London about 1740, when it caused about 900 deaths per 100,000 people annually.

Cities expanded rapidly. The influx of people from the countryside resulted in massive overcrowding. As a result, the R_0 of *M. tuberculosis* greatly exceeded 1 and possibly reached as high as 10. Lack of clean, fresh air, especially in winter, resulted in increased respiratory transmission. Large families lived in single rooms. Physicians advised against opening windows to prevent miasmas from entering. Cool, damp conditions permitted the bacterium to survive and persist in the environment. Poor nutrition resulted in immune suppression. There was an absence of effective medical treatments and all these factors contributed to this explosion. The disease ravaged Western industrialized societies. It has been estimated that up to one in four deaths in England around 1800 were due to tuberculosis.

The relationships between the poor and factory owners changed in the early 1800s, especially in the textile industries. As factories were built and urban poverty increased, so tuberculosis increased among the poorest. Working conditions deteriorated. Child labor was prevalent. Cold and hungry individuals worked in dark, crowded spaces, breathing smoke and coal dust. Death rates exceeded 1% per year.

Tuberculosis was the most common cause of death in American adults from the colonial period to the Civil War [7]. Between 1810 and 1815 it accounted for more than a quarter of deaths in New York City and Boston, killing between 500 and 900 per 100,000, depending upon race. By the early 1900s, it still remained one of the most significant diseases in the great cities of America, especially along the East Coast. The growing impoverished immigrant population was especially hard hit. African Americans in the large cities had a much higher tuberculosis mortality than Whites. For example, in 1900 the death rate in White Americans was 194 per 100,000, making it the third most common cause of death after heart disease and pneumonia. The rate among Black Americans was double this, 400 per 100,000. Between 1911 and 1920, 25% to 35% of deaths among Native Americans were due to TB [8]. In US penitentiaries, about 12% of the White inmates died of TB between 1829 and 1845. The proportion would have been much higher among African American inmates.

Death rates eventually started to fall for multiple reasons. The main factor was the gradual improvement of socioeconomic conditions in urban areas, especially nutrition and living standards. Governments had to do something, especially because the diseases of the poor urban workers were likely to spill over to the richer and more affluent. Responses were not driven by concern for the "downtrodden." By 1889, following Koch's discoveries, the public health authorities recognized that TB was contagious and thus preventable. As a result, in 1893 the NY Public Health Department added it to its list of communicable diseases. They required physicians to report such patients to the Health Department within 12 days. Registered patients were visited by inspectors, and their homes were disinfected. This voluntary program did not work, so in 1897 reporting was made mandatory. Despite resistance from physicians who considered it an invasion of privacy, the regulations were enforced. Studies showed that tuberculosis was associated with overcrowding, poor nutrition, and alcoholism. Thus improvements in housing effectively reduced overcrowding. Public education also played a role, such as the recognition that spitting could spread the infection. In fact, at the turn of the 20th century, New York City instituted antispitting legislation, beginning a decades-long drive to change public perception and habits.

These findings also resulted in the quarantine of infected individuals and the increased availability of hospitals and then isolation facilities (sanitoriums). Although the mortality due to tuberculosis began to decline around 1800, it still remained a significant cause of premature death. After 1900, improvements continued and mortality eventually dropped significantly, reaching 46 deaths per 100,000 by 1945. This improvement was due in large part to isolation of patients in sanitoria, vastly improved hygiene, decreased urban crowding, and exposing patients to sunlight. (This elevated their vitamin D levels which enhanced immunity.)

Prominent Americans who suffered from tuberculosis included Nathaniel Hawthorne, Henry David Thoreau, and almost all of the family of Ralph Waldo Emerson. It killed Emerson's first wife, her father and mother, as well as her brother and sister. Thoreau decided to move to Walden Pond, where he could rest and write. It killed Andrew Jackson's adopted son Lyncoya. The film star Vivien Leigh died in a sanitorium aged 54. Eleanor Roosevelt, the wife of Franklin D., developed tuberculosis at the age of 12 but the disease went into remission. In her old age the disease resurfaced. She developed miliary TB, where multiple very small tubercles developed in most organs, and she died of disseminated TB at 75. (Miliary is derived from the observation that these tubercles were the size of millet seed.) With so many writers and artists of the 19th century making the disease—and their personal struggles to overcome it—the subject of their work, tuberculosis became a ubiquitous part of Western culture.

For example, Edgar Allen Poe described his wife Virginia's suffering from tuberculosis. In 1842 during dinner, she began to cough and began to bring up fresh blood. What had happened was that an expanding tubercle in her lungs had eroded an artery that had suddenly ruptured. Poe's mother Eliza, his brother William, and his foster mother Frances had also died of tuberculosis. It is likely that the remembrance of this event was the inspiration behind his horror fantasy "The Masque of the Red Death," published that same year. In this story, a prince and his courtiers lock themselves away in a castle to avoid a plague—the Red Death. However, while they were partying, death arrived!

And was now acknowledged the presence of the Red Death. He had come like a thief in the night. And one by one dropped the revelers in the blood-bedewed halls of their revel, and died each in the despairing posture of his fall… And Darkness and Decay and the Red Death held inimitable dominion over all.

Tuberculosis treatments

Prior to the discovery of its cause, any attempted treatments of tuberculosis were essentially ineffective. Since then, slow progress has been made in finding a solution. An important milestone was the discovery of X-rays in 1895 by Wilhelm Röntgen. By the early 20th century, doctors were able to diagnose the disease with confidence by the use of the tuberculin test, chest X-rays, and the detection of the tubercle bacilli in the patient's sputum.

BCG vaccine

Albert Calmette and Camille Guérin produced a vaccine against tuberculosis by passing a strain of *M. bovis* every 21 days (230 times) between 1908 and 1921 in an alkaline medium containing sterile bovine bile. This eventually led to mutations in the organism that caused its loss of virulence for humans. This is now called the Bacille Calmette-Guérin (BCG) strain of *M. bovis*. BCG is used as a vaccine against tuberculosis, but its effectiveness varies between countries, possibly due to genetic differences between strains. At one time it was the world's most commonly employed vaccine. Overall, it is about 50% efficacious. It is used in Europe but not in North America. It is mainly used in countries where tuberculosis is endemic as well as in developing countries. BCG appears to work best in infants and children under 15 years of age and protects them against the most severe forms of the disease.

The sanitorium movement

Although there was no known cure for tuberculosis in the 18th century, thoughtful physicians perceived its association with cold climates and indoor activities. The early Greek hospitals espoused rest and fresh air. The Romans recommended a sea voyage on the basis that the fresh air would help victims. In 1781 an English physician, John Lettsom, founded the "Royal Sea Bathing Infirmary for Scrofula" in the seaside town of Margate. Patients were encouraged to sleep in the open, take gentle exercise, and get sleep. In 1840 another English physician, George Bodington, advocated the "open-air cure." He opened a "sanitorium" where patients were not only exposed to fresh air but encouraged to exercise and were given a healthy diet. Other physicians at the time were very skeptical of this novel approach. Referrals dropped, as did the number of his patients, and Bodington got discouraged. His sanitorium soon closed and was converted into an asylum.

Meanwhile in Germany, Dr. Hermann Brehmer opened a small sanitorium in the Bavarian Alps in 1854. Again, patients were exposed to lots of fresh air, a healthy diet with prescribed exercise, and rest. Patients had to go outside irrespective of the weather. These sanitoria were beneficial in that the rest reduced stress and so may have benefited their patient's immune systems. They also effectively quarantined the patients and so minimized transmission. Patients would stay in a sanitorium for many months or years depending upon their condition. Patients spent the day wrapped in blankets outdoors getting lots of sunshine, rest, good nutrition, and fresh air at all seasons [7].

The first and most famous American sanitorium opened in 1884 at Saranac Lake in the Adirondack Mountains of New York (Fig. 6.1). It was founded by Dr. Edward Trudeau (Fig. 6.2). Trudeau had acquired tuberculosis as a young man while looking after his sick brother. He believed that his symptoms improved after staying in the Adirondacks in the 1870s. In 1882, Trudeau read about the Brehmer "rest cure" and, because it corresponded to his own experiences, tried to copy it. Patients staying at Trudeau's sanitorium were prescribed outdoor exercise and healthy meals. They were obliged to undertake horseback rides and long walks. When not exercising, they spent the day reclining on loungers situated on the broad patios that surrounded the "cure cottages." Trudeau raised sufficient funds to build over 20 of these cottages. He also conducted some scientific research on a model of tuberculosis in rabbits. He showed

Fig. 6.1 *Adirondack Cottage Sanatorium, Saranac Lake 1902. Courtesy Library of Congress.*

Fig. 6.2 *Edward L. Trudeau, the founder of the sanitarium movement in the United States. Courtesy US National Library of Medicine.*

that rabbits deprived of fresh air, light, exercise, and poor nutrition died more rapidly from tuberculosis than rabbits provided with all these things. (But he used a very small number of rabbits in each group.) Trudeau tried to determine if the "open-air" cure worked in human patients but got equivocal results. He was, however, persistent and continued this treatment for many years. He believed himself cured but ultimately died of tuberculosis in 1916. Trudeau's Adirondack Cottage Sanitorium became the largest tuberculosis treatment center in the United States. (It was renamed the Trudeau Sanitorium after his death.)

Although many physicians truly believed that the sanitorium approach was beneficial, public health officials saw sanitoria as useful quarantine facilities. Current law enacted in 1942 states:

Such regulations may provide that if upon examination any such individual is found to be infected, he may be detained for such time and in such a manner as may be reasonably necessary. Section 361. US Public Health Service Act (Title 42 USC 264)

There was always a strong element of coercion associated with the sanitorium movement. Every aspect of a patient's daily life was regulated. Patients were sometimes forced to enter these facilities against their will. Once inside, they were obliged to eat large meals, drink milk, dress appropriately, and refrain from swearing, drinking, and smoking. It is clear that many tuberculosis patients avoided the authorities or lied about the disease to avoid being separated from their families. Because of their quarantine effect, public health officials enthusiastically adopted the movement and set up their own sanitoria. New York set up one in the Catskills and one on Hart Island. Others were established in Massachusetts, Arkansas, Texas, and Michigan (Fig. 6.3).

Private sanitoriums also flourished. By 1900 there were 34 sanitoriums housing 445 patients. By 1925 there were 535 sanitoriums housing 73,338 patients. In those years, more than half the hospital beds and medical facilities in the United States were devoted to tuberculosis patients (Box 6.2).

Although results were unclear, the sanitorium approach was the only method that seemed to impact the disease, and the quarantine unquestionably helped reduce the spread of the infection. As a result, large numbers of sanitoria were established across North America and Europe, where patients were isolated, rested, fed, and hoped to be cured. Because of the sheer number of tuberculosis patients, sanitoria could only take a small fraction of infected individuals. As a result, they often refused to accept patients with advanced disease or who couldn't afford to pay. Poor patients simply died at home. Recent analysis of survival/cure rates in sanitoria versus home confinement suggests that the death rate was similar. About half the patients died in each—but the quarantine effects were probably not insignificant.

Several antituberculosis campaigns were mounted in the 1900s [9]. The most famous was the Christmas Seals of the National Tuberculosis Foundation. These were first issued in 1907. This later became the American Lung Association. While promoting TB eradication, the campaign never achieved the successes of the March of Dimes in eradicating polio (Chapter 13).

Once streptomycin and para-aminosalicylic acid (PAS) were discovered in 1944 and shown to cure the disease, the sanitoria rapidly closed. The Trudeau Sanitorium closed in December 1954. In 1956 a long-term study was

State Tuberculosis Sanitorium, near San Angelo, Texas

(A)

(B)

Fig. 6.3 *(A) The first Texas State tuberculosis sanatorium, Anti-Tuberculosis Colony No. 1, was established in Carlsbad, Texas, 17 miles northwest of San Antonio, Texas. During its 60 years of activity, 1911–1971, over 55,000 people were treated for tuberculosis. From 1919 until October 7, 1965, the sanatorium had its own post office (B), along with a library, public school, and other infrastructure for self-sufficiency. (A) Courtesy US National Library of Medicine. (B) Courtesy Ms. Shannon Green.*

conducted in Chennai, India, to compare the effects of treatment at home as compared to treatment in sanitoria. It clearly showed no advantage to staying in a sanitorium. Rest, fresh air, and good nutrition were not necessary for recovery. Only two specialized sanitoria persist in the United States today,

---

BOX 6.2 Coercion and tuberculosis.

The islands off New York City have often served as convenient isolation facilities for the unwanted and diseased of the city. For example, a small island off the Bronx has served such a purpose for hundreds of years. Hart Island has served as a barracks for Black Union soldiers during the Civil War, a prison for delinquent boys, and an insane asylum. It served as a quarantine station during the 1870 yellow fever epidemic. The largest public cemetery in the world is located on the island. It houses the remains of those unidentified individuals who died, colloquially called the Potter's field. Visits are actively discouraged.

Of special interest to readers of this book is the fact that Hart Island once housed a "Tubercularium"—a fancy name for a tuberculosis sanitorium. However, unlike most such facilities, the Hart island tubercularium appears to have made no pretense of being a voluntary institution. Thus *The New York Times* on November 10, 1917 reported the results of a Grand Jury deliberation that found that the tubercularium was unsuitable. They concluded "... after an investigation into the escape of four prisoners from the tuberculosis ward of the Hart's Island Hospital about a month ago, that the provision of care and custody of decrepit elderly prisoners was unsuitable in every respect."

---

one each in Florida and Los Angeles, to house and quarantine small numbers of patients with drug-resistant tuberculosis, thus limiting the spread of these lethal organisms.

The move to the west

With the opening of the transcontinental railroad in 1869, major efforts were made to attract new settlers. One of the prime promotional strategies was to advertise the healthy climate of California. This was obviously in marked contrast to the squalid industrial cities on the East Coast. The warm, dry, sunny climate was often contrasted with the dark, wet, and cold northeastern cities. The California boosters suggested, why confine yourself to a sanitorium when you can simply move to California? People doomed to die on the East Coast could be assured of a long life in the warm, healthy climate on the West Coast. An outdoor life, farming in California, was more preferable to an early death from working in a cold, dark factory. This was a highly effective campaign, and it was estimated in 1900 that a quarter of California immigrants did so for their health and 10% of immigrants to Colorado did so for the same reason. Much of the enthusiasm stemmed from resistance to incarceration in sanitoria. California was an attractive alternative. Other sunny, dry states such as Arizona and New Mexico also capitalized on their climate to attract tuberculosis patients. It should be noted that Florida also took the opportunity to present its healthy climate as a reason for consumptives to move south. Unfortunately, malaria and yellow fever hurt its image, so the marketing and subsequent growth of health tourism in Florida had to be postponed until these tropical diseases were brought under control.

Texas was also a popular destination for "lungers." San Antonio was said to be "The Sanitorium of the West." Many settled in south-central Texas towns such as Fredericksburg and Boerne. The railroad magnate Jay Gould moved to El Paso in 1892 for "health" reasons but it was too late. He died of tuberculosis later that year. (He also moved out of New York to avoid a rumored band of anarchists who had pledged to assassinate the capitalist class!)

Perhaps the most famous of these health immigrants was John "Doc" Holliday. He was a dentist in Georgia when diagnosed with tuberculosis, so he moved his practice to Dallas, Texas, for his health. After leaving Dallas, where he was indicted for illegal gambling, he moved further west and became a professional gambler, skilled gunfighter, and a friend of US Marshall Wyatt Earp. In 1881 in Tombstone, Arizona, Doc Holliday was involved in the infamous "shootout at the OK corral." Holliday survived that gunfight only to die in his bed of tuberculosis in 1887 at age 36.

Surgical treatments

The human lung is an elastic organ that is maintained in an expanded state by an effective vacuum within the pleural cavity. If air is let into the pleural cavity, the vacuum is lost and the lung collapses. It was therefore suggested that if the lung collapsed, tubercles would collapse too. This would then permit any scar tissue a chance to contract and so shrink the lesions. As a result, "collapse therapy" was performed on huge numbers of tuberculosis patients prior to the availability of effective drugs [10]. Patients generally spent a year in a sanitorium, and during that time up to 70% of their patients had some form of collapse therapy. Collapse therapy could be performed in a number of ways. One popular method was to simply crush a phrenic nerve. These nerves innervate the diaphragm; if one is damaged, the diaphragm on that side would remain paralyzed for 6 to 8 months and so give the lung a chance to "rest." In the 1930s thousands of sanitorium patients had their phrenic nerves paralyzed. Its effectiveness is difficult to evaluate although it was performed enthusiastically.

A more drastic collapse procedure was the production of a pneumo-thorax [11]. Simply put, a measured amount of air was let into the pleural cavity until the lung was deemed to have collapsed sufficiently. The air had to be refilled at weekly intervals. Once achieved, the pneumothorax was maintained for at least 2 years! When discontinued, the lungs reinflated while, ideally, the healed tubercles remained contracted.

Another surgical procedure used to collapse a lung was extrapleural pneumonolysis or plombage. This became popular in the 1940s. The surgeon opened the thoracic cavity and then filled the pleural space with some foreign material, or plombe, to compress the lung lesions and prevent the lung from re-expanding. For this purpose, they used muscle, fat, paraffin wax, or even mineral oil. Unfortunately, tuberculous lesions sometimes ruptured into the pleural cavity. This became obvious when the patient began to cough up wax or the lungs began to fill with oil! Other objects used for plombage to maintain the collapse included plastic spheres, rubber sheets, and even ping-pong balls. Some surgeons removed the plombe after a few months while others let it remain unless some problem developed. Despite its popularity, mixed results were obtained and complications were common. Removal of a few ribs, thoracoplasty, was also used to collapse lungs, whereas it was possible in some cases to simply remove the tubercular lesions by surgery.

Antitubercular drugs

In January 1944 a soil microbiologist, Dr. Selman Waksman, and his colleagues reported the discovery of streptomycin. Waksman was a Ukrainian immigrant to the United States. As Professor of Microbiology at Rutgers University he discovered a new antibiotic produced by a soil bacterium called *Streptomyces griseus*, which he called streptomycin. Two other scientists, Corwin Hinshaw and Hugh Feldman, were investigating possible treatments for tuberculosis and read about Waksman's discovery. They tested streptomycin on guinea pigs infected with *M. tuberculosis* and found that the treated animals showed a remarkable improvement when compared to control animals. Streptomycin inhibited the synthesis of the waxy cell wall of *M. tuberculosis*.

At the same time as streptomycin was being tested, a Swedish scientist, Jörgen Lehmann, developed a chemical—PAS—that could also kill *M. tuberculosis*. In October 1944 Lehmann treated a young woman dying of tuberculosis with PAS, and she made a "dramatic recovery." In

November 1944 Hinshaw, Feldman, and Karl Pfuetz treated a young woman with streptomycin who also survived.

Although both streptomycin and PAS were effective, they had significant side effects. More importantly, *M. tuberculosis* rapidly developed resistance to both of them. In 1952 a third drug was introduced, called isoniazid. Although also very potent, patients treated with isoniazid alone relapsed after a few months as a result of the development of drug resistance. However, a 1950 trial by Feldman of a treatment involving all three drugs, isoniazid, streptomycin, and PAS, showed greatly superior results. Most notably, the combination greatly slowed the development of drug resistance. Combination therapy relies on the fact that the development of resistance to each drug by the mycobacterium is an independent rare event. If two drugs are used together, the chances of two rare events occurring at the same time are very much reduced, and a combination of three drugs is even better. This became the standard treatment of tuberculosis for many years (Fig. 6.4).

Nevertheless, problems remained—patients stopped taking the drugs when they felt better and so permitted drug-resistant strains to emerge, and there were also some side effects. As a result, studies on antitubercular drugs continued. Thus ethambutol and rifampin were introduced in the 1960s. Since then, numerous other such drugs have been developed.

The discovery of powerful antibiotics such as streptomycin resulted in major research efforts to find more. These efforts were very successful, and as a result, many other bacterial diseases were brought under control as well. Unfortunately, as the prevalence and significance of bacterial diseases declined, so did public interest. People were no longer scared and perhaps did not know anyone who had suffered from such an infection. Tuberculosis was considered conquered. The treatment of tuberculosis was no longer a priority. It was treated in general clinics as just any other infection.

As a result of this decreased political attention and a consequent lack of investment in public health, the prevalence of tuberculosis reached a low in 1979 but then began to increase again, especially in the impoverished inner cities [12]. Between 1985 and 1992 the number of cases of tuberculosis in the United States increased by about 20%. Much of this was related to the development of HIV/AIDS in these areas. To make matters worse, multidrug-resistant tuberculosis (MDR-TB; that is, organisms resistant to both rifampin and isoniazid) became an increasing problem. Some health workers believed that some treatment was better than no treatment! In some areas of New York City, only 60% of patients were completing their treatments. In another study, more than half the patients did not arrive for

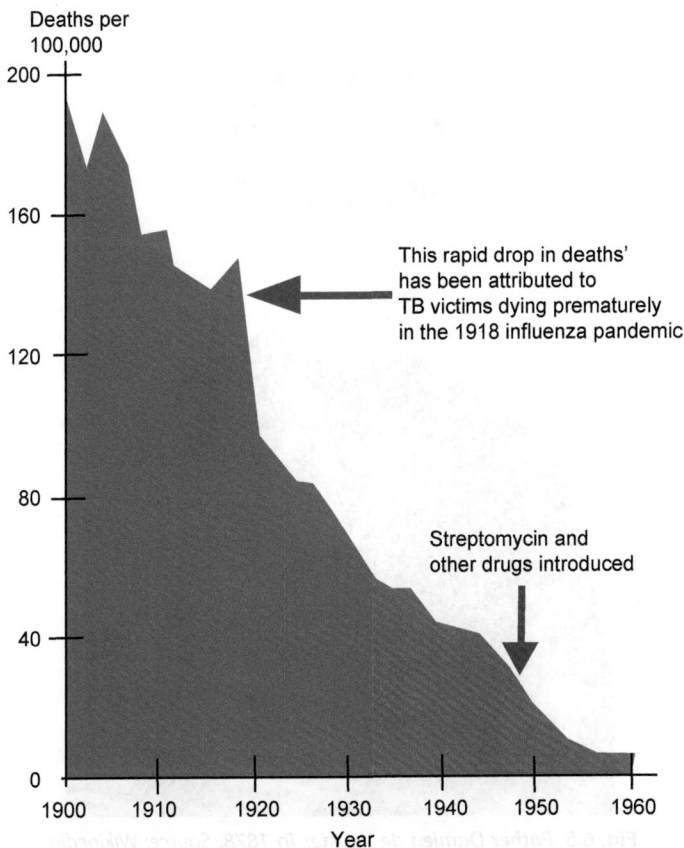

Fig. 6.4 *The progressive decline in tuberculosis cases in the United States during the 20th century. Note that it had declined significantly long before the introduction of effective drugs as a result of improved living and working standards. Data courtesy the CDC.*

their first appointment, and within 12 months 27% were readmitted with tuberculosis. The prevalence of MDR-TB isolates rose from 6% in 1987 to 14% in 1991. The realization that the disease was reemerging has fortunately led to a renewed focus on this infection.

The prime cause of the emergence of MDR-TB was a lack of attention to dosing and compliance. Treatment was either not completed or simply taken at irregular intervals. As a result, the Public Health Authorities resorted to directly observed therapy (DOT). In other words, every dose of antitubercular drugs administered is observed by a health worker and recorded. This is now considered the standard of care in the United States. It ensures

Fig. 6.5 *Father Damien de Veuster in 1878.* Source: Wikipedia.

compliance and hence successful treatment, and it reduces the development of MDR-TB. Experience has shown that DOT has significant cost benefits when compared to unobserved self-treatment.

Thus, beginning in 1992, the DOT program was expanded and aggressively enforced. Detention in hospitals was used to force compliance. As a result, the prevalence of TB declined by 21% between 1992 and 1994. It continues to decline, but the control of TB has required coercion! In 2018 the prevalence of TB in the United States was 2.8 per 100,000, the lowest rate on record, and the number of new cases was 9024. The numbers are good, but progress has slowed.

Another reason for the re-emergence of tuberculosis was the onset of AIDS. As mentioned earlier, immunity to *M. tuberculosis* is cell mediated. T cells are needed to resist these bacteria. Unfortunately, HIV targets and kills these T cells. As a result, immunity to mycobacteria is profoundly suppressed in AIDS patients. The bacteria can grow readily in AIDS patients without the unwanted attention of T cells. This loss of T cells reduces immunity, not only to *M. tuberculosis,* but also to normally nonpathogenic mycobacteria such as *Mycobacterium avium* and *Mycobacterium intracellulare* (Chapter 17).

Today, tuberculosis still kills more people worldwide than any other infectious disease (COVID excepted!). The World Health Organization (WHO) estimates that about 23% of the world's population is infected. Nearly 8 million people worldwide develop disease each year, and of them more than 1 million die annually. In fact, there are more people infected with tuberculosis at the present than ever before. Ninety-five percent of global cases and 98% of the deaths occur in low-income countries. Twenty-seven percent of cases occur in India. In most countries, the death rate from TB is less than 5 per 100,000. In South Asia the rates can reach 25 to 50 per 100,000, while the highest rates are in sub-Saharan Africa where they range from 50 to more than 250 per 100,000. It is important to note also that the COVID-19 pandemic in Africa, Asia, and Latin America has forced the diversion of resources to control that disease, causing tuberculosis control programs to be neglected and permitting the disease prevalence to increase (Box 6.3).

BOX 6.3 Leprosy in America.

Leprosy is a bacterial skin disease caused by two mycobacteria related to *M. tuberculosis.* They are *M. leprae* and *M. lepromatosis.* Like tuberculosis, leprosy is an ancient disease. It is believed, probably correctly, that leprosy was brought to the Americas by Europeans and the slave trade. There is no evidence for the occurrence of leprosy in the Americas prior to 1492. Currently circulating strains appear to have come from either Europe or East Asia.

The leprosy mycobacteria invade the skin and destroy nerves. As a result of local desensitization, patients can no longer feel touch, pain, or pressure. When victims inadvertently injure themselves, the lack of pain sensation can result in secondary traumatic damage. Rotting flesh and even gangrene as well as deformities may develop without the patient even noticing. As a result, leprosy eventually becomes an obviously disfiguring disease. Patients no longer blink, resulting in blindness. The causal agent, *M. leprae,* was first identified by the Norwegian physician Armauer Hansen in 1873. Because of the stigma associated with the term leprosy, its name has been formally changed to Hansen's disease.

(Continued)

BOX 6.3 Leprosy in America. *(cont'd)*

Leprosy was once believed to be highly contagious, but this is incorrect. The infection is not easily spread between individuals. It is believed to be transmitted via the respiratory system to close contacts. It is a slowly developing disease taking up to 20 years to fully develop. It occurs mainly in tropical climates such as Brazil, the South Pacific, India, and Southeast Asia. Small numbers of cases occur in the southern United States. Although effective treatments are available, leprosy persists in poverty-stricken disadvantaged societies.

Leprosy has never caused significant epidemics in the United States with the exception of Hawaii. However, the disease was recorded in Louisiana as early as the 18th century. It is believed to have arrived during the period of Spanish rule as a result of the African slave trade. Consequently, there was an attempt by the Spanish Governor Antonio de Ulloa in 1776 to exile all lepers from the colony. Given the social stigma and disfiguring lesions associated with leprosy as reflected in the Bible, it is unsurprising that the State of Louisiana eventually tried to isolate infected individuals by housing them in the "Louisiana Leper home" located at Carville, Louisiana between 1894 and 1999.

Leprosy was introduced into Hawaii around 1823, probably from Asia. It spread to such an extent that a leprosy hospital had to be established on O'ahu in 1865. The next year, all the patients from the hospital were sent into forced quarantine at Kalawao on the island of Moloka'i. (The parallels between this and the sanitorium movement are obvious!) The Kalawao site totally lacked medical facilities, doctors, or nurses. In 1874 Father Damien de Veuster, a Belgian Catholic priest, arrived on the island. He was horrified at what he found there. Despite the smell of rotting flesh, he began to care for the patients himself and eventually built adequate facilities including hospitals, orphanages, and homes. In 1889 Father Damien died of leprosy (Fig. 6.5). In 2009, Pope Benedict canonized Father Damien as Saint Damien of Moloka'i.

Effective treatments for leprosy were developed in the 1940s and 1950s. It is currently treatable with a multidrug regimen. The Moloka'i facility closed at the same time as the Carville facility.

In the Americas, the leprosy mycobacteria infect not only humans but also nine-banded armadillos (*Dasypus novemcinctus*). These animals develop both skin disease and nerve damage as do humans. Molecular analysis of their bacteria indicates that armadillos probably acquired the infection from humans, but it is quite unclear how, when, or where this might have happened. Although there is anecdotal evidence of people becoming infected from armadillos, this has not been conclusively proven.

(From Pawlikowski J, Banasiuk J, Sak J, et al. Damien de Veuster (1840—1889): a life devoted to lepers. Clin Dermatol 2018;36:680—885.)

WHO has reported that globally an estimated 10.6 million people fell ill with TB in 2022. The total number of TB-related deaths (including those among people with HIV) was 1.3 million in 2022, down from 1.4 million in 2021. However, during the 2020–2022 period, COVID-19 disruptions resulted in nearly half a million more deaths from TB. TB continues to be the leading killer among people with HIV.

References

[1] Daniel TM. The impact of tuberculosis on civilization. Infect Dis Clin North Am 2004;18(1):157–65. https://doi.org/10.1016/S0891-5520(03)00096-5.

[2] Lee OYC, Wu HHT, Donoghue HD, et al. Mycobacterium tuberculosis complex lipid virulence factors preserved in the 17,000-year-old skeleton of an extinct bison, bison antiquus. PLoS One 2012;7(7). https://doi.org/10.1371/journal.pone.0041923.

[3] Mackowiak PA, Blos VT, Aguilar M, Buikstra JE. On the origin of American tuberculosis. Clin Infect Dis 2005;41(4):515–8. https://doi.org/10.1086/432013.

[4] Mark Samuel. Early human migrations (ca. 13,000 years ago) or postcontact Europeans for the earliest spread of *Mycobacterium leprae* and *Mycobacterium lepromatosis* to the Americas. Interdiscip Perspect Infect Dis 2017;2017:1–8. https://doi.org/10.1155/2017/6491606.

[5] Donoghue HD. Paleomicrobiology of human tuberculosis. Microbiol Spectr 2016;4(4). https://doi.org/10.1128/microbiolspec.PoH-0003-2014.

[6] Frith J. History of tuberculosis. Part 1—phthisis, consumption and the white plague. JMVH 2014;22(2):29–35. Available from: http://jmvh.org/issues/.

[7] Frith J. History of tuberculosis. Part 2—The sanatoria and the discoveries of the tubercle bacillus. JMVH 2014;22(2):36–41. Available from: http://jmvh.org/issues/.

[8] Rieder HL. Tuberculosis among American Indians of the contiguous United States. Public Health Rep 1989;104(6):653–7.

[9] Murray JF. A century of tuberculosis. American Lung Association. Am J Respir Crit Care Med 2004;169(11):1181–6. https://doi.org/10.1164/rccm.200402-140oe.

[10] Murray JF, Schraufnagel DE, Hopewell PC. Treatment of tuberculosis: a historical perspective. Ann Am Thorac Soc 2015;12(12):1749–59. https://doi.org/10.1513/AnnalsATS.201509-632PS.

[11] Steele JD. The surgical treatment of pulmonary tuberculosis. Ann Thorac Surg 1968;6(5):484–502. https://doi.org/10.1016/S0003-4975(10)66061-4.

[12] Ryan F. The forgotten plague: how the battle against tuberculosis was won—and lost. Little, Brown and Company. New York, NY; 1993.

Malaria—1492

Contents

For most people living in North America, malaria is a rare and exotic disease that we only think about when traveling to parts of Africa or the Amazon. About 1500 cases of malaria, causing fewer than 5 deaths, occur annually in the United States. These are usually in travelers who acquire the infection overseas.

But malaria remains a heavy burden around the world. The World Health Organization (WHO) estimates that in 2018 there were 228 million cases of malaria worldwide, resulting in 405,000 deaths. Two-thirds of these deaths occurred in children under 5 years of age. That number of cases is equivalent to about 70% of the US population or to all the people in the United Kingdom, France, Italy, and Spain combined. The number of deaths is equivalent to the entire population of Tampa or New Orleans. Since the beginning of the 21st century, about 10 million people are estimated to have died from malaria!

Malaria is caused by mosquito-borne protozoan parasites. The disease is of African origin and almost certainly was not present in the Americas until after the arrival of Europeans. It was introduced several times over the centuries, but the most significant introductions were by way of the slave trade. Once the parasite arrived, mosquitoes spread it throughout the warmer

Great American Diseases
ISBN: 978-0-443-31404-9
https://doi.org/10.1016/B978-0-443-31404-9.00007-0

regions of both North and South America, and it is still present in the Amazon rainforests. Endemic malaria was eradicated from the United States in 1951.

Malaria is an ancient heirloom disease that affects tropical primates. It has likely infected humans since our ancestors descended from the trees. The earliest written records stem from China around 2600 BCE. It is described in early Hindu texts and Egyptian papyri. The ancient Greeks, including Homer and Hippocrates, knew the disease well. They observed that its occurrence was associated with wet areas such as marshes and swamps. It was therefore believed that it was caused by poisonous miasmas arising from the stagnant water. The word malaria comes from Italian for "bad air." Other names for malaria include ague, swamp fever, and malignant fever. That said, it is often difficult to determine from early descriptions just what specific sort of fever was being discussed.

Malarial parasites

There are more than 100 species of protozoa in the genus *Plasmodium*. They cause malaria in humans, other mammals, birds, and reptiles. Human malaria is caused by only five of these species: *Plasmodium falciparum*, *Plasmodium vivax*, *Plasmodium malariae*, *Plasmodium ovale*, and *Plasmodium knowlesi*. In general, malaria is a host-specific disease; the malaria parasites that infect humans only infect humans. Consequently, there is no wildlife reservoir for human malaria. *P. knowlesi*, the most recently described human malaria parasite, is the exception; it infects some species of macaque and is now the most common human malarial parasite in northern Borneo. It is found across Southeast Asia.

The human malaria parasites are transmitted solely by blood-sucking mosquitoes of the genus *Anopheles*. Only the female mosquitoes feed on blood, which they need for their eggs to develop. Male mosquitoes, in contrast, drink nectar. The infective stages of the parasite, called sporozoites, live in the mosquito salivary glands. When an infected female mosquito feeds, these sporozoites are injected into the victim through her saliva (Fig. 7.1). They are carried by the bloodstream to the liver, where they penetrate the liver cells and rapidly multiply. No clinical signs are apparent during this "liver stage." After about 8 days, however, the parasites, now termed merozoites, are so numerous that they rupture the liver cells and escape into the bloodstream, where they enter red blood cells. Once inside the red blood cells, the merozoites continue to multiply until the red blood

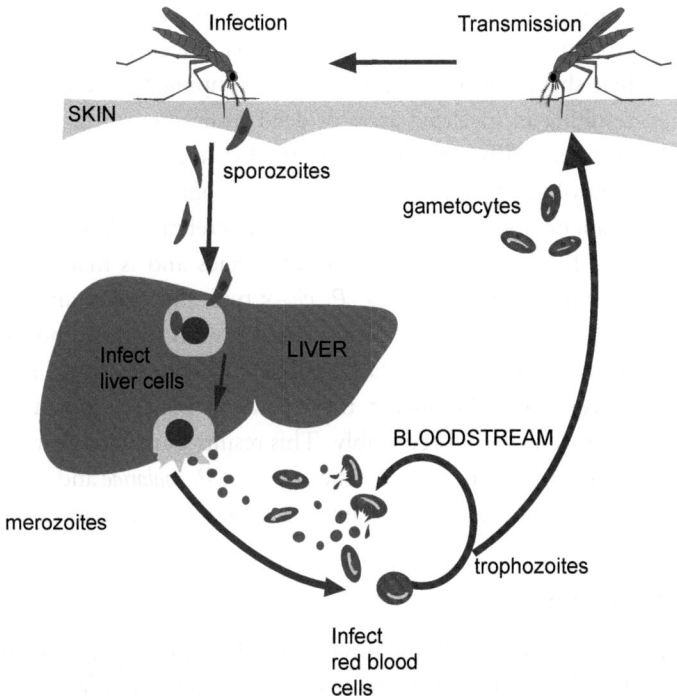

Fig. 7.1 *A simplified view of the stages in the life cycle of the malaria parasite within an infected human. The periodicity of fevers is determined by the timing of the cycle in blood red cells.*

cells eventually rupture, releasing more merozoites that then enter yet more red blood cells. Over a period of days, this cycle continues, increasing the number of parasites in the blood. The rupture of the red blood cells quickly becomes synchronized, occurring every 48 to 72 hours. Symptoms of malaria coincide with the rupture of the red blood cells.

When another mosquito feeds on a person with the malaria parasite in their blood cells, that mosquito will also ingest the parasite. Once ingested, the parasites multiply in the mosquito's midgut and within 7 to 10 days migrate to the mosquito's salivary glands—ready to infect another person.

It is important to note that offspring of a mosquito with malaria do not themselves contain the protozoan; the mosquito is not born with the parasite. To become infected, a female *Anopheles* mosquito must first feed on a person that has the parasite circulating within their bloodstream.

Of the five species of *Plasmodium* that infect humans, *P. falciparum* is both the most common and the most virulent and, as a result, is responsible for most malaria deaths worldwide. It is especially prevalent in sub-Saharan Africa.

It kills up to half of those infected but not treated and is the cause of cerebral malaria, a lethal brain disease. When *P. falciparum* infects red blood cells, the cells alter their shape and become prone to clumping. These clumps block capillaries, resulting in localized areas of dead tissue due to the interruption of the blood supply. When this occurs in the brain, the result can be severe, long-term neurological impairments, such as cognitive defects, epilepsy, behavioral changes, and coma, hence the term cerebral malaria.

P. *vivax* is the second most important species and is mainly found in Southeast Asia and South America. *P. vivax* typically causes an acute self-limiting, debilitating febrile illness, but usually with few complications or deaths. *P. vivax* and *P. ovale* also cause persistent liver infections. Following apparent recovery from the disease, these parasites may stay dormant in the liver and then reactivate unpredictably. This results in malaria that can recur yearly and is difficult to cure. Two other species, *P. malariae* and *P. knowlesi*, are much less common and responsible for only a few human cases.

The disease

The incubation period, the time from when a mosquito infects a person until clinical signs appear, varies from 7 to 30 days. *P. falciparum* has the shortest incubation period, whereas *P. vivax* has the longest. A "normal" malarial attack lasts for 6 to 10 hours. It is triggered by the simultaneous bursting of the infected red blood cells and the release of the parasites and cell products. These attacks consist of three stages, a cold stage, a hot stage, and a sweating stage. During the cold stage, the parasite products fool the brain into thinking that the body is too cold (in effect, the body's thermostat is turned up). As a result, the patient feels chilled, and their body responds by shivering in an effort to generate more heat. The patient gradually develops a high fever. This fever then persists, resulting in the hot stage. This stage is associated with headaches, vomiting, sweating, diarrhea, anemia, and possibly even seizures. Some patients develop a yellow pallor as a result of red cell destruction and resulting high levels of a yellowish hemoglobin breakdown product, bilirubin. Their body temperature may oscillate, causing chills and fevers in succession. Eventually, however, the malarial products are removed, the effects on the temperature centers of the brain fade, the thermostat returns to normal, and the patient sheds heat as rapidly as possible by profuse sweating. All of this uses a lot of energy, and patients are exhausted by the end of the attack. Once an individual is infected, these attacks occur at regular intervals. *P. vivax* and *P. ovale* cause fevers that occur every 48 hours (coinciding with the rupturing of the red blood cells due to the replication of the parasite). This 48-hour cyclicity of fever (e.g., Monday—fever; Tuesday—no fever; Wednesday—fever) gives the name "Tertian fever" to these diseases. (The ancient Greeks gave it this name because they counted both the day of the first attack of chills and the day of the second attack! Attacks thus occur on the first, third, and fifth day, etc.) In the case of *P. malariae*, the life cycle is slower, and attacks occur every 72 hours on the first, fourth, and seventh days (Quartan fever). Conversely, *P. falciparum* can cause severe, often lethal attacks that may occur daily (Quotidian fever), and this continuous disease is said to be malignant. Death is especially common in children and nonimmune individuals during the first attacks. If the victim survives the first attack, they will develop limited immunity and, as a result, the severity of subsequent attacks may gradually decline to chills and a slight fever. Chronic ill health may also result from anemia and its associated debility. The other *Plasmodium* species generally cause less severe, "benign" disease.

History

Classical symptoms of malaria—the chills, fevers, and headaches—are described in the *Nei Ching*, a Chinese medical text dating from 2600 BCE, probably the oldest known medical text. It describes the three demons of great cold, fever, and massive headaches—clinical symptoms that resemble malarial infections. Malaria was also endemic in ancient Egypt. In the Upper Egyptian Dendera Temple complex, there is a warning not to leave home after sunset in the weeks following the annual flood of the Nile. This warning is probably due to increased mosquito populations and increased malaria transmission. Plasmodium DNA has been found in Egyptian mummies dating from 1500 to 500 BCE.

The Greek philosopher and physician Hippocrates and the Roman medical writer Celsus both described the classical intermittent fever paroxysms of malaria, chills, fevers, and sweats. The disease, known as "the fever" in Greece and "intense burning heat," "marsh vapor disease," or "Roman fever" in Rome, was endemic in the Mediterranean region. Hippocrates suggested that malaria might be caused by drinking bad water, as there was less disease in people living in mountainous regions who drank mountain spring water than in lowland people who drank swampy, stagnant water. It was also believed that miasmas, the poisonous vapors or mists that arose from decomposing matter, such as that found in swamps, were a cause of malaria.

In Rome and Greece, the association between swamps, stagnant water, and summer months with the fevers of malaria was well recognized. This association gave rise to what is thought to be the first organized attempt to intervene against this or any disease. Engineers built extensive drainage systems in the Pontine Marshes and Campania regions south of Rome that essentially eliminated malaria as a major problem in the drained areas. Some believe that years later the disrepair of these drainage systems and the rise in malaria was a contributing factor to the fall of the Roman Empire.

Malaria was thus well established in southern Europe for centuries before the discovery of America, especially in the marshes and wetlands around the Mediterranean [1]. For example, *P. vivax* was common in the marshes of southern Andalusia close to the great Spanish ports of Seville, Cadiz, and Palos de la Frontera, Columbus' departure point. As a result, it is probable that some of Columbus' crew were carriers of the malaria parasite. In addition, the practice of filling drinking water barrels from the nearest convenient source of fresh water makes it likely that some mosquito larvae were present in his ship's water supply.

Malaria is not believed to have been present in the Americas prior to the European invasion, but the vector of malaria, the female *Anopheles* mosquito, certainly was.

Columbus's first voyage visited the coasts of Cuba, Puerto Rico, and Hispaniola, where he left some sailors behind. Overall, he and his crew had extensive contact with the islanders and, no doubt, with their mosquitoes. Malaria had arrived in the Americas.

On his fourth voyage in 1502, where Columbus explored the coast of Central America, he complained bitterly about the plague of mosquitoes. This region, encompassing modern Nicaragua and part of Honduras, was subsequently called "The Mosquito Coast." Columbus was later shipwrecked in Jamaica. One of his crew members, Diego Mendez, came down with a "quartan ague" after being rescued. Columbus himself may have also suffered from malaria at the same time.

When the English adventurer Sir Richard Grenville raided Puerto Rico in 1585, he was glad to leave because his men had been badly "stoong by Muskitoes" when ashore.

The English

The two English settlements in North America had distinctly different climates, origins, philosophies, and social consequences. They were also very different with respect to the impact of malaria. The Massachusetts colony, being further north and much cooler, had few issues with malaria. European settlers could thrive in the healthy climate. The Virginia colony, on the other hand, suffered seriously from the disease as a result of its mild, warm, mosquito-friendly climate. The climatic boundary that determined the presence of malaria and the northern boundary of the malaria-endemic zone was, not coincidentally, approximately the Mason-Dixon line. Mosquitoes survive further north, but the malaria parasites are temperature sensitive and will not survive the cold winters. As a result, malaria was primarily a disease of the southeastern states and played a major role in their development and history. It was, in effect, both a cause and consequence of chattel slavery as discussed later.

Malaria was endemic in the low-lying fens of eastern England in the 16th century. As a result, malarial fevers affected many Englishmen. When the first English arrived to establish the Virginia colony in 1607, they brought malaria, probably *P. vivax*, with them and introduced the parasite along the Virginia coast. Here, it made life miserable. It became endemic in the Jamestown settlement. The first reference to intermittent fever or ague, the terms used by the English for diseases like malaria, was reported in 1610. The disease spread to Native Americans and, as a result, spread north to Maryland and south through the Carolinas and Georgia to Florida. Originally it was sustained by a continuous influx of susceptible English settlers to the Virginia colony. However, when it became apparent that the climate of Virginia was "unhealthy," the supply of cheap English labor dried up. The settlers needed more manpower. Malaria-resistant labor had to be acquired from somewhere. The Native Americans in Virginia were also decimated by malaria and had retreated to the west, well away from the European settlements, so they were no longer available to enslave. The obvious source of such labor was West Africa.

In 1619 John Rolfe, the widower of Pocahontas, reported "About the last of August came a Dutch man-of-war that sold us twenty negroes"—the first reference to slavery in the English colonies. Within 40 years, the Virginia slave population outnumbered the Europeans. Coming from West Africa ,where *P. falciparum* was endemic, the slaves were, in general, more resistant to its effects [2]. Early urbanization ensured that mosquitoes did not have to fly far to seek a meal. As a result, *P. falciparum* became common in the Chesapeake Bay area of Maryland and Virginia and the tidewater regions further south. The establishment of rice as a crop along the southeast coast provided an ample supply of stagnant shallow water for breeding.

However, malaria outbreaks also occurred regularly each summer in the cities of Boston and New York and persisted along the Atlantic coast for several hundred years. In effect, the distribution of malaria corresponded to the need for slaves to operate the plantation economy of the South. There was a strong correlation between the presence of slaves and the prevalence of malaria across the United States. Higher prices were paid for slaves when they were believed to be more resistant to malaria.

The slave trade

Genetic analysis of P. falciparum strains from South America has revealed significant information about its origins. First, all the South American strains are closely related to African strains, supporting the belief that malaria originated in Africa and spread to South America. They also revealed two main genetic clusters of the parasite, a northern one in Central America and Colombia and one in the south in Brazil. This is consistent with two separate importations, one through the Spanish slave trade in the north and the other through the Portuguese slave trade to Brazil [3]. This divergence probably occurred in the early years of the slave trade. For 300 years, between 1560 and 1852, it is estimated that almost 5 million individuals were kidnapped from Africa and brought as slaves to Brazil. They were generally disembarked at either Rio de Janeiro or Bahia. This amounted to about 40% of the Atlantic slave trade. The Spaniards, in contrast, disembarked their slaves in Veracruz in Mexico or Cartagena in Colombia.

American independence

George Washington was not a stranger to malaria, first becoming infected with ague in 1749 at the age of 17. He suffered recurring bouts of fever thereafter, which suggests that he was infected with P. vivax, as this species can cause recurring fever for months to years. During his presidency, he was treated with the antimalarial remedy quinine in August and September 1786. His physician gave him eight doses of the "red bark." Quinine, as discussed later, is derived from the red bark of the Cinchona tree. As commander of the Continental Army, Washington urged the Continental Congress to purchase as much bark of the Cinchona tree as possible. The Cinchona tree, which at that time only grew in the Andes mountains of Peru, Ecuador, and Bolivia, was known as a source of a powdered bark that cured intermittent fevers such as malaria. The bark had long been used by the indigenous people of the Andean region to treat fevers and other ailments. By the 1630s, the Spanish colonists in Peru were ingesting the powdered bark as well. The *Cinchona* tree bark was exported to Europe and by 1640 was widely used as a treatment for fevers. It contains the antimalarial compound quinine.

Soldiers on both sides of the American Revolution were susceptible to malaria, but not all suffered equally. Americans from the southern colonies had a greater exposure to malaria and thus had acquired a partial immunity. Unlike smallpox, malaria does not produce lifelong immunity. Adolescents and adults, having survived malaria during childhood and experiencing repeated infections, acquire a partial protective immunity against severe disease; they still can become infected numerous times, but the subsequent clinical symptoms can be less severe. In contrast, a large percentage of the British and Hessian troops, especially those participating in campaigns in the southern colonies, had no previous exposure to malaria. Thus, when infected with malaria, these troops were prone to have more severe disease and were incapacitated for longer [4].

These factors played a major role in the British strategy for the southern colonies during the last years of the war. By the summer of 1780, the military situation in the northern states was at a stalemate. As a result, the British decided to try a "southern strategy." In effect, the British army under General Cornwallis attempted to secure the southern colonies of South Carolina, North Carolina, and Georgia. This region was, however, rife with disease, especially malaria. As a result, British troops suffered increasingly from illnesses, especially the "intermittents" or "agues and fevers," most likely malaria. Cornwallis moved his troops and camps many times during that summer and autumn, in unsuccessful attempts to outmaneuver the Continental forces, and also to find a healthier environment for his troops. What he could not understand was that the environment was only part of the issue. His men were the reservoir for *Plasmodium* and carried the malaria parasite with them wherever they went. All that was needed was for the vector, the *Anopheles* mosquito, which was ubiquitous in the south, to feed upon an infected soldier and spread the parasite from soldier to soldier. Cornwallis himself became infected in October 1780. He became severely ill, unable to write or move, for weeks. Consequently, in April 1781, Cornwallis decided to move his army north into Virginia to link up with the British army there and to protect his troops from the sicknesses.

By June 1781, Cornwallis and his army had established a base at Yorktown, Virginia, and by the beginning of August, they were well entrenched. The siege began on September 28, and Cornwallis surrendered to General Washington on October 19, 1781. This rapid collapse was in large part due to malaria. Malaria spread through the British troops with the result that only 3800 of the 7700 men in the Yorktown garrison were fit and able to fight. Thus malaria, along with the French defeat of the British fleet, left Cornwallis' army isolated. The surrender at Yorktown effectively ended the American Revolution. In explaining his surrender, Cornwallis credited malaria as a major factor as "the troops being much weakened by sickness …. Our force diminished daily by sickness." The rapid surrender was also very fortunate for the Americans, as the troops from New England that had been marched south for the battle were also falling ill.

Malaria continued to be a factor in the development of the new republic. John Wilson, signer of the Declaration of Independence and the Constitution and a Supreme Court Justice appointed by President Washington, died in August 1798, at the age of 55, of malaria while visiting a friend in North Carolina. The lower south could be unhealthy for all. A statement attributed to a British soldier summed it up: "South Carolina is in the spring a paradise, in the summer a hell, and in the autumn a hospital." The future president, James Monroe, contracted malaria while traveling through swamps near the Mississippi River in 1785. He suffered from recurrent fevers and weakness episodes for many years afterward. Although malaria has long been assumed, the fact that these episodes were accompanied by a severe cough points to pulmonary tuberculosis as an alternative explanation.

The Seminole wars

In the 1830s, the US Army entered Florida to protect White settlers against the Seminole Indians whose land they were taking. From 1838 to 1842, the army penetrated deep into the swamps of Florida. As a result, they suffered from a growing number of intermittent and remittent fevers. Northern soldiers with no resistance to the disease were severely affected. The severity of the disease and its effects included not only direct deaths but also mental breakdowns and an unusually high suicide rate among the American troops as a result of the depressive effects of the disease. About midway through the conflict, cinchona bark (quinine) became available, which helped considerably. It also ensured that quinine would be widely employed in future wars. Andrew Jackson led the military campaigns to suppress the Seminoles in Florida and acquired malaria at that time. For years afterward, Jackson complained of recurring chills and fever. He also suffered chest pains and coughed up blood. Tuberculosis and heart failure are other possible causes of his lung problems.

Malaria in the west

In 1803, the United States purchased France's land claims in North America, the Louisiana Purchase. The Corps of Discovery, headed by Meriwether Lewis and William Clark, was tasked by Thomas Jefferson with exploring and mapping this new territory. In assembling the medicine chest for the expedition, the item of largest quantity and most expense was Peruvian bark, the bark of the *Cinchona* tree. Meriwether Lewis was familiar with the tonic of bark that reduced and relieved the "seasonal fevers" of malaria. This single item attests to their concerns with malaria, which was endemic in the southern United States and along the waterways of the Midwest at that time. However, it is interesting to note that in the journals from the Corps of Discovery, neither ague nor seasonal fevers were mentioned as causing problems.

Malaria reached the Pacific Northwest in 1830. In the ensuing epidemic, huge numbers of Native Americans died. It appears that *Anopheles* mosquitoes were abundant in the swampy areas of western Oregon [5]. The natives began to sicken and die. The fever and ague suddenly appeared in native villages near Fort Vancouver. It was possibly introduced by an American ship, the *Owyhee,* that was involved in the salmon trade but had visited some malarial areas in Central America before arriving in Oregon. Mortality was enormous and continued until the first frosts in November 1830. It reappeared each summer for the next 3 years, spreading to involve wider areas each year. Based on population estimates made by Lewis and Clark, it is estimated that more than 90% of the native population in the Portland area and the Willamette Valley died. (The population dropped from 32,000 to about 2100.) Major social breakdown and cultural losses followed. Some historians have questioned whether this disease really was malaria and have suggested alternatives such as influenza or typhus. However, its seasonality and the characteristic clinical signs make malaria the most likely cause.

Malaria became a significant disease in the United States during the 19th century, especially in the southeastern states below the Mason-Dixon line and up the waterways of the Mississippi watershed (Fig. 7.2). The westward progression of explorers, settlers, and settlements, with rivers and water sources being important to both travel and settlements, contributed to the spread of the disease. Malaria, along with measles, smallpox, and other infections, was the major factor in reducing Native American populations in the western United States. During the immigration west on the Oregon Trail (late 1830s to 1869), malaria was most likely one of the most prevalent diseases

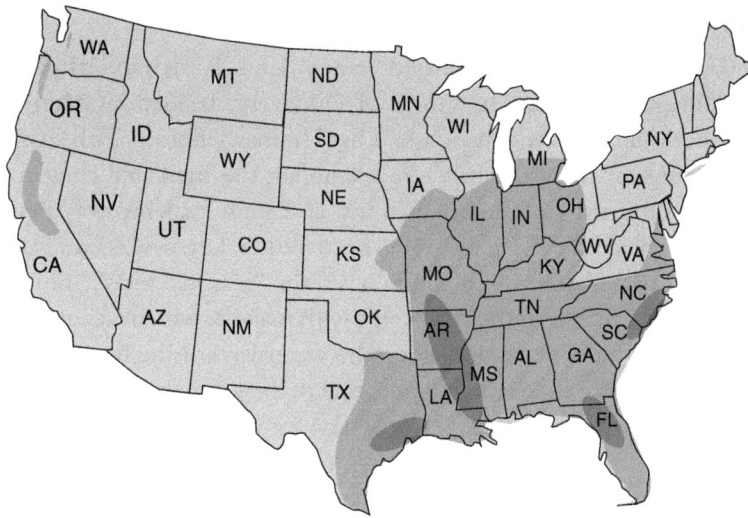

Fig. 7.2 *The distribution of malaria at its greatest extent across the United States. The darker areas* had the greatest prevalence. These were especially significant during the *course of the Civil War.*

among migrant children. What is interesting is that there is relatively little mention of malaria in the diaries of the travelers. The question arises if this was possibly due to the pervasiveness of the disease and fevers so that they were not considered noteworthy.

It is difficult to visualize the environmental conditions and the amount of disease present when the first American settlers arrived in the Great Plains, the Midwest, and the West. A masterful article by Roger Winsor, "Environmental imagery of the wet prairie of East Central Illinois, 1820–1920," describes the formidable quagmire of wetlands, swamps, and the presence of malaria that faced the early pioneers [6]. Until drainage projects changed the landscape, malaria was endemic, so much so that Illinois was considered "a gigantic emporium of malaria" in the mid-1800s.

So widespread was malaria on the Great Plains that Laura Ingalls Wilder's *Little House on the Prairie* devotes a whole chapter to it, Chapter 16 ("Fever 'N' Ague"). It describes how the whole Ingalls family became stricken with the fever "n" ague, malaria. *"All the settlers, up and down the creek, had fever 'n' ague. There were not enough well people to take care of the sick."*

Abraham Lincoln acquired malaria soon after arriving in Illinois in 1830 along with other family members. Five years later he had a second, severe episode while living in New Salem. In the summer of 1835, he developed a tertian fever and had to take high doses of Peruvian bark. He was sent to a neighbor's house to be treated and cared for.

James Garfield acquired malaria in 1848 while working as a young man on the barges on the Ohio Canal. In 1881 Garfield's wife Lucretia was bitten by a mosquito in the White House and contracted the disease. She eventually recovered. He was assassinated. Theodore Roosevelt acquired malaria during an expedition exploring remote regions of the Amazon in 1913. It seriously weakened him to the extent that he became delirious and lost a quarter of his body weight. He suffered chronic ill health for the rest of his life.

In the northeastern portion of the United States, malaria was seasonal, but by 1860, its prevalence was declining. That is, until the Civil War erupted. As described in Chapter 14, malaria was a major problem and cause of both morbidity and mortality during and after the Civil War.

As a direct consequence of the high prevalence of malaria in the southern states, Union soldiers returning to the northeast acted as a source of local epidemics for many years after the war. By the beginning of the 20th century, however, malaria outbreaks diminished across most of the north, most likely due to a short transmission season, increased agricultural drainage, which reduced the available breeding areas for the *Anopheles* mosquitoes, and improved economic status of the general population and the use of mosquito screens. Malaria remained concentrated in the southern states and states along the Mississippi River, such as western Kentucky, eastern Missouri, and southern Illinois. The continued malaria problem in the south was partially due to reduced agricultural manpower contributing to a reduction in land cultivation. Land cultivation and drainage of swampy soil promote an environment that is less conducive to mosquito breeding, thus reducing their numbers.

New discoveries

A French army physician, Charles Alphonse Laveran, was the first to observe plasmodia in the red cells of a malarial soldier in Algeria in 1880. Meticulously observing hundreds of blood smears, Laveran also worked out much of the life history of the organism. This was a remarkable achievement given the poor-quality microscopes provided to him by the French Army. Like others before him, many of his colleagues were unimpressed and considered that he had only observed dead and dying red blood cells. However, he persisted and, by 1884, had persuaded a group of Italian scientists to his way of thinking. Not only that, but he also persuaded the leading microbiologists in France, Louis Pasteur, Emile Roux, and Edouard Chamberland.

A year prior to Laveran's discovery, two other scientists, Edwin Klebs and Ettore Marchiaflava, had claimed that the cause of malaria was a bacterium that they called *Bacillus malariae*. The US Army physician and bacteriologist George Sternberg was given the task of reviewing the conflicting claims. Sternberg therefore visited Europe and checked out the competing laboratories. Sternberg returned to the United States, persuaded that Laveran was correct [7]. As a result of Sternberg's public endorsement of Laveran's claims, they were rapidly accepted by the American medical community. It is good to record that Laveran received the Nobel Prize in 1907 for his discoveries.

In the meantime, further advances were occurring in Russia. Thus a biologist, Vassily Danilewsky, found similar parasites in the red blood cells of birds. However, he published exclusively in Russian journals, and it was not until 1889 that this information became available in the West. In 1891 another step in the discovery process occurred when another Russian, Dimitri Romanowsky, discovered the staining properties of a methylene blue—eosin stain. Romanowsky's stains and subsequent modifications by William Leishman, Gustav Giemsa, and James Wright could clearly demonstrate the presence of malaria parasites within red blood cells with a characteristic, blue-stained cytoplasm and a red nucleus. It was abundantly clear that malaria was caused by a protozoan, not a bacterium.

In 1877, Patrick Manson, a Scottish physician working in China, demonstrated that a microfilarial (small worm-like) parasite that causes a disease called elephantiasis was transmitted by the *Culex* mosquito. In 1881, Carlos Finlay, a Cuban physician, presented a paper entitled "The Mosquito Hypothetically Considered as the Agent of Transmission of Yellow Fever." Initially, his theory that the *Culex* mosquito served as the vector of yellow fever was overwhelmingly rejected, but he was later vindicated by his work with Walter Reed and the Yellow Fever Commission (see Chapter 11).

Dr. Albert King, Dean of the National Medical College at Columbia University, proposed in 1883 that, due to the correlation between mosquitoes and malaria, protection from mosquitoes would reduce the risk of contracting malaria if the mosquito origin of malaria was correct. In 1893 a physician and a veterinarian, Theobald Smith and Frederick Kilborne, respectively, working for the United States Department of Agriculture, showed that Texas cattle fever, a devastating disease of cattle, was caused by a protozoan that was transmitted from cow to cow by ticks (*Boophilus bovis*, now named *Rhipicephalus annulatus*). Their findings established for the first time that an arthropod vector, the tick, could spread a disease from animal to animal.

It took longer to determine the route of transmission and life cycle of malaria [8]. Building upon these earlier works, a British military physician working in India, Ronald Ross, studied avian malaria and by 1897 had demonstrated that mosquitoes were the vector of the disease in birds. (He transmitted the parasite between sparrows by means of mosquitoes and found the parasite in the mosquitoes' midgut and salivary glands. He eventually worked out the complete life cycle in birds.) Ross suspected that mosquitoes also transmitted human malaria. Before he could prove it, however, he was scooped by the Italians.

Between 1898 and 1900, Italian scientists Giovanni Grassi, Amico Bignami, Guiseppi Bastianelli, Angelo Celli, Camillo Golgi, and Ettore Marchiafava demonstrated that they could transmit malaria from infected patients to uninfected ones by means of mosquitoes. They demonstrated that only female *Anopheles* mosquitoes could do this and worked out the complete life cycles of *P. vivax*, *P. falciparum*, and *P. malariae*. It took until 1947 for the tissue forms of the parasite to be identified in the blood and as late as 1962 for the persistence of the parasites within liver cells to be confirmed.

Treatment

The discovery that material extracted from the bark of the cinchona tree (*Cinchona pubescens*) could cure malaria was made by an Augustinian Monk, Father Calanda, in Peru in 1633. The bark was shipped to Europe, where it was distributed by the Spanish Jesuits. As a result, it was also known as Jesuit's bark. For many years after, it was the only available cure for malaria, but it was very effective. The great Swedish botanist Carl Linnaeus named the tree "Cinchona" after the Countess of Chinchón, the wife of the Spanish Viceroy in Peru. She reportedly recovered from a tertian fever after drinking a bark infusion. As a result, it is claimed that she was the first to import the bark into Spain. In 1820, two French chemists, Pierre Pelletier and Joseph Caventou, were studying plant toxins. They isolated a gummy substance from cinchona bark and named the substance quinine. (This is a botanical mistake! The quinquinia tree was actually a different tree species that provided Balsam of Peru, a perfume and antiseptic.) Quinine was widely prescribed as both a treatment and preventative (Fig. 7.3), but it

the swamps and adding oil to any remaining water. The oil prevented mosquito larvae from breathing at the water surface. Gorgas also introduced the practice of burning pyrethrum, an insecticide derived from chrysanthemum flowers inside sealed buildings. He ensured that mosquito-proof screens were installed on windows and encouraged the use of quinine for malaria prophylaxis. Similar campaigns also occurred in New Orleans and during the construction of the Panama Canal. Within 3 years, Gorgas and his colleagues had reduced the prevalence of malaria in Panama from 800 cases per 1000 workers to 16. As Sir William Osler, the famous physician and one of the founding professors of the Johns Hopkins Hospital, said, "There is nothing to match the work of Gorgas in the history of human achievement."

Yellow fever was rapidly brought under control in the continental United States, malaria was not so fast. Yellow fever was an urban disease, and aggressive mosquito elimination measures could bring localized yellow fever outbreaks under control rapidly. Malaria, on the other hand, was widespread and by no means restricted to cities. Mosquito control in rural areas was a more difficult procedure. It was found to be more efficient to promptly and effectively isolate malaria victims through the use of bed nets while reducing mosquito populations by the use of potent insecticides.

As the country developed after the Civil War, greater efforts were made to provide proper drainage for agricultural land. This made better land available. Increased animal stocking densities provided the mosquitoes with alternative feeding sources. Housing quality also improved, thus isolating humans from mosquitoes. As a result, mosquito breeding sites diminished and the prevalence of malaria gradually declined. Outbreaks of the disease tended to occur in agricultural areas where poor housing was common. Malaria (and mosquitoes) were especially common in the swampy areas of south Florida and coastal Texas [9].

The early history of the University of Florida, Gainesville, attests to the prevalence of malaria and *Anopheles* mosquitoes. In the early 1900s, the campus was surrounded by sloughs, marshes, and water-filled sinkholes. Rutted dirt roads, open barrels, and other water containers provided additional mosquito breeding sites when it rained. The university's resident physician wrote that malaria was the most common complaint treated, with malaria diagnosed in more than a quarter of the 210 students in 1910. It was so pervasive that school athletes thought nothing of competing when sick. During a 1907 football game, Florida's halfback, William F. Gibbs, played the entire game with a malarial fever—he was bedridden for weeks afterward. Florida won 9—4. And the 1910 football team had 8 out of its 15 players diagnosed with malaria during their college career. During the latter part of the 19th century and up to World War II, the most southerly states were classified as a hyperendemic area, with the disease so prevalent that most people living there had at some time been infected (Box 7.1).

From 1933 to 1935 there were approximately 280,000 cases of malaria in the United States, resulting in 500 deaths annually. A survey of southern schoolchildren in 1933 found that 5.8% of the children had malaria. As a result, projects were instituted to prevent the disease by reducing the mosquito vector. In the second decade of the 20th century, The Rockefeller Foundation and the US Public Health Service joined forces to work on malaria eradication projects. Educational programs and mosquito control

BOX 7.1 Florida tourism.

The United States purchased Florida from Spain in 1821. At the time, it was almost totally underdeveloped, and in the following years its population grew very slowly. One reason for this slow development was that it was a very unhealthy place. The hot, humid, tropical climate together with its extensive network of rivers, lakes, and swamps ensured that it was heavily infested by mosquitoes! In fact, the central east coast of the state around Cape Canaveral was called "Los Musquitos" by the Spaniards. It was not considered a desirable place in which to live. The major cities from Florida's north Atlantic coast to the panhandle—Jacksonville, St. Augustine, Tallahassee, and Pensacola—were considered the "malaria belt." Those who could, moved north during the humid late summer months. When Orange County in Florida was first established in 1824, it was called "Mosquito County." The county was split in two in 1844 with the northern half becoming Orange County and the southern half eventually becoming Brevard County.

Along with the mosquitoes came malaria and yellow fever. Epidemics of both diseases were a continuing feature of Florida life throughout the 19th century. For example, there were major yellow fever outbreaks in Jacksonville in 1877 and 1888. These outbreaks forced the state to establish a Board of Health. Once the role of mosquitoes in transmitting malaria and yellow fever were understood, things changed. At the time of World War I, massive mosquito control and drainage projects reduced the disease burden while the climate increasingly attracted northerners escaping cold winters. In effect, the development of Florida tourism could not have happened without mosquito control and the consequent elimination of malaria and yellow fever.

helped reduce malaria in the South. State and county health departments expanded and devoted resources to mosquito control, such as using Paris green, a larvicide, and pyrethrum insecticides. Most cases occurred in August and September, and malaria was commonly classified as an "autumnal fever."

There was an increase in the number of malaria cases at the onset of World War II as a result of infected soldiers returning from overseas, especially the South Pacific. Not only were soldiers, sailors, and marines sent to fight in malarial regions, but thousands of troops were sent to be trained in the southern states. (John Kennedy was infected with malaria while a PT boat captain in the Solomon Islands.)

Because malaria persisted within much of the southeastern United States, in 1942 the US Public Health Service established the "Office of Malaria Control in War Areas (MCWA)" in collaboration with state and local health departments. The office was located in Atlanta, close to the malaria-endemic regions. Its mission was malaria prevention and the protection of military and civilian personnel in and around military training bases and essential war industrial areas. It utilized adult mosquito destruction projects, larvicidal programs, and drainage to reduce potential breeding areas and prevent malaria transmission. Educational material and training programs helped to inform the public and ensure their cooperation in eliminating mosquitos. These programs included "moving pictures" on malaria and mosquito control. In 1943, Walt Disney Productions produced a movie called *The Winged Scourge,* featuring the Seven Dwarfs (https://www.youtube.com/watch?v=y68F8YwLWdg)! These activities were successful in keeping malaria cases to a minimum among the troops, but also among the surrounding civilian population. In fact, they were so successful that within the areas of military activity, the infection rate for children dropped to only 0.1%. Outside MCWA areas of activity, however, malaria persisted.

At the war's conclusion, the MCWA was phased out, eventually being renamed the Centers for Disease Control and Prevention (CDC). But the Director of the US Public Health Service had two concerns: first, the continued presence of malaria in the southern states, and second, the potential for returning troops to reintroduce malaria. In fact, it is estimated that about 5% of troops that returned during the period from 1944 to 1946 were infected with *P. vivax.* Thus among the first tasks facing the CDC was the eradication of malaria from the United States. It oversaw the National Malaria Eradication Program (NMEP) and provided technical support to the states. Consequently, malaria control programs were extended and continued, but with the addition of a potent new weapon—dichlorodiphenyltrichloroethane, commonly known as DDT. During the war years, all DDT had been allocated for military use overseas. But with the war's end, it became available for civilian use. It acted as both a larvicide and an adulticide, providing mosquito protection when applied inside and outside of residences only a few times a year at critical intervals. Acceptance by homeowners was enthusiastic, as it not only killed the malaria mosquito but, more importantly, was very effective in providing lasting relief from flies, fleas, bedbugs, and roaches. The use of DDT as a residential treatment and in breeding environments was found to be very effective as well as economical. Major credit is given to DDT use in reducing *Anopheles* mosquito populations and thus being a key factor in controlling malaria.

Beginning in 1947, the CDC in cooperation with the local health agencies in the 13 endemic states began the eradication process. That year there were 15,000 malaria cases reported in the southern United States. In addition to the proven strategies of source reduction by wetland drainage and screening, a new component of their strategy was to apply DDT to the interior walls in homes and businesses in counties where malaria had been identified. By the end of 1949, more than 4,650,000 homes had been so treated. Aerial spraying was also used in some locations. By 1950, the number of reported malaria cases had dropped to 2000. By 1951, it was gone. The NMEP had succeeded in eliminating endemic malaria from the United States. The CDC was able to change its priorities to focus on surveillance of other infectious diseases.

It was fortunate that the scheme worked so well because DDT proved to be an especially persistent environmental poison that massively reduced many wild bird populations by interfering with their egg production. The harmful effects of the indiscriminate and massive use of DDT and other pesticides were addressed in the 1962 book *Silent Spring* by Rachel Carson. The book helped to heighten public awareness and concerns about pesticides and the environment. Such were the concerns over the detrimental effects of pesticides and pollutants on the environment and health that, in 1970, President Richard Nixon established the Environmental Protection Agency (EPA) to protect the American public from the effects of pollutants and to address "some pressing issues like DDT, which required immediate attention."

Malaria still occurs in the United States. Cases are sporadic and result from disease imported by immigrants and travelers. Most such cases originate in Africa (70%) with many fewer from South and Central America (6%).

These imported cases pose an ongoing threat because *Anopheles* mosquitoes are not uncommon in the United States and retain the potential to transmit malaria from returning travelers to their friends who stayed at home.

Ninety percent of global malaria deaths currently occur in West Africa. Unfortunately, the COVID-19 pandemic has forced the reallocation of scarce resources and lockdowns, leading to a reduction in disease control efforts in Africa. Ongoing climate change will likely permit the spread of both mosquito and malaria.

References

[1]   Carter R, Mendis KN. Evolutionary and historical aspects of the burden of malaria. Clin Microbiol Rev 2002;15(4):564—94. https://doi.org/10.1128/CMR.15.4.564-594.2002.

[2]   Esposito E. The side-effects of immunity: Malaria and African Slavery in the United States. American Economic Journal:Applied Economics, 2022 14(3):290—328.

[3]   Yalcindag Erhan, Elguero Eric, Arnathau Céline, et al. Multiple independent introductions of plasmodium falciparum in South America. Proc Natl Acad Sci 2012;109(2): 511—6. https://doi.org/10.1073/pnas.1119058109.

[4]   McCandless P. Revolutionary fever: disease and war in the lower South, 1776-1783. Trans Am Clin Climatol Assoc 2007;118:225—49.

[5]   Boyd RT. Another look at the "fever and ague" of Western Oregon. Ethnohistory (Columbus, Ohio) 1975;22:135—54. https://doi.org/10.2307/481642.

[6]   Winsor Roger A. Environmental imagery of the wet prairie of East Central Illinois, 1820—1920. J Hist Geogr 1987;13(4):375—97. https://doi.org/10.1016/s0305-7488(87)80047-6.

[7]   Ockenhouse CF, Magill A, Smith D, Milhous W. History of U.S. military contributions to the study of malaria. Mil Med 2005;170(4S):12. https://doi.org/10.7205/MILMED.

[8]   Cox FEG. History of the discovery of the malaria parasites and their vectors. Parasit Vectors 2010;3:5—13.

[9]   Carrasco P, Shandera WX. The past and present of Malaria in Houston. Tex Med 2012; 108(7):e1.

Smallpox—1518

Contents

Smallpox, "The Red Death," is caused by variola virus. This is a DNA virus belonging to the genus *Orthopoxvirus*. It is one of the largest and most complex viruses known, with 187 genes. Its closest known relative is the "Taterapox" virus that was isolated from a wild African gerbil (*Tatera kempi*) in Benin in 1975 [1]. The few differences between the DNA sequences of the two viruses indicate that they probably diverged about 3000 to 4000 years ago in East Africa [2]. It has been suggested that climate change, agriculture, and the establishment of towns and cities set up the conditions for the emergence of smallpox. The mummy of Ramses V, who died in Egypt in 1157 BCE, has skin lesions that may be due to smallpox or some other pox virus. Early Indian medical texts suggest that it may have been present in India between the 4th and 1st centuries BCE. There is also evidence suggesting that smallpox was present in China around the 3rd century BCE. Smallpox almost certainly caused the Antonine Plague in Rome in 170. Thus variola is a recently evolved virus and smallpox a recent disease. It

Great American Diseases
ISBN: 978-0-443-31404-9
https://doi.org/10.1016/B978-0-443-31404-9.00008-2

only infects humans. Each animal species has its own specific poxvirus; cowpox, horsepox, and fowl pox are examples (chicken pox is not caused by a poxvirus but by the herpesvirus *Herpes zoster*). The term pox is an old English term for spotty skin lesions. For example, the "Great pox" was the original English name for syphilis.

Smallpox: The disease

Smallpox virus is acquired by inhalation. The virus then grows in the cells of the upper respiratory tract and can be transmitted by coughing or in saliva. The incubation period of smallpox is 10 to 14 days. During this time, the virus moves between cells throughout the body. These cells release virus into the bloodstream to cause a viremia. The circulating virus accumulates in the spleen and bone marrow and results in the classic signs of sickness: fever, fatigue, malaise, aches, and pains for 2 to 4 days. The virus also moves to the skin and grows in epidermal cells. As a result, the patient develops a rash that develops into the classic feature of the disease—hundreds of small red pocks. These small red spots eventually grow into pus-filled blisters. They develop inside the mouth, tongue, eyes, throat, esophagus, and vagina. When the throat lesions rupture, the mouth can fill with pus that may then be coughed up. Victims cannot drink and may become severely dehydrated. The pustules rupture within 24 to 48 hours but are very painful and sting. Patients claim that they are "on fire." The pustules eventually dry up, form scabs, and shed highly infectious virus into the air. Victims are generally infectious for several days prior to developing clinical disease and infectivity lasts for about 24 days. The rupturing of the lesions in the mouth and respiratory system releases the virus, allowing it to be spread to others through the air when the infected person breaths, coughs, or sneezes. Survivors were often pockmarked and blind. Indeed, smallpox was the leading cause of blindness in Europe for hundreds of years. The scabs contain lots of highly infectious virus and protect the virus from the environment—virus found in the fluid from pustules can survive for weeks, but when the virus is incorporated into a scab, it can survive for months! Blankets or clothes from smallpox victims can contain infectious dander and dust from the shed scabs and pustule fluid. In Europe during the 17th and 18th centuries, it is estimated that smallpox may have killed 400,000 people annually. In 1660, out of a population of about 500,000 in London, 57,000 died of smallpox. The R_0 of smallpox is, as always, dependent on population density but ranges from 3.5 to 6. Thus it is a very highly infectious disease.

Smallpox had variable clinical presentations. One variant, *variola minor* or alastrim, had a relatively low mortality of 1% to 2% (*Alastrim* is a Portuguese word meaning uncountable or to spread out). The most serious variant was *variola major,* with a mortality of 10% to 35% among Europeans. Hemorrhagic smallpox was the predominant variant that developed in Native Americans. It was also known as the "Black pox." In these cases, bleeding occurred into the pocks, perhaps because they were more deeply located within the skin. The blood within the pustules made them turn black. Bleeding also occurred in the eyes so that they turned red. The virus also attacked the internal organs leading to abdominal bleeding and excruciating pain. Death invariably occurred within 8 days. Malignant smallpox or flat smallpox was a variant that occurred mainly in children. The lesions remained level with the skin. Internal damage was severe, and mortality was also close to 90%. In Native Americans, many victims developed confluent pocks. These spread and joined together. As a result, sheets of skin peeled off. The pain would have been agonizing. This loss of skin would have resulted in massive fluid loss, secondary infections, shock, and death.

History

When smallpox invaded a population, people either died or survived and became immune. As a result, the disease disappeared from a region for a while. However, the virus persisted elsewhere and would then reappear when sufficient numbers of susceptible children had been born. In Europe, this resulted in a 20-year epidemic cycle in specific locations. Natural selection would ensure that only those children with the greatest resistance to the virus would survive. Over many generations the population of Europe would therefore have developed significant resistance to smallpox. Nevertheless, it still killed 10% to 35% of affected Europeans in the 18th century.

The invasion of Mexico

In early 1518, a smallpox epidemic developed in southwestern Spain. It was carried by infected sailors to the Spanish colony of Hispaniola later that year. The infection spread and the few surviving natives on that island were soon wiped out. By November, smallpox had also spread to Puerto Rico and Cuba, and again, their native populations were largely destroyed. Native Americans had never encountered smallpox and never had an opportunity to evolve resistance. They provided "virgin soil" for the most catastrophic pandemic in human history.

It must, however, be pointed out that smallpox and European diseases in general were not the only cause of this demographic catastrophe. Wars, brutal slavery, and massacres contributed to social disruption. Likewise, the relative importance of each of these factors differed among different regions of the New World. It is also important to point out that the numbers of dead are in most cases no more than guesses and should not necessarily be relied upon.

Hernán Cortés, accompanied by 500 men, left Cuba in November 1518 to seek gold in what is now Mexico. However, he left without royal permission. This was deemed a crime because the king of Spain was trying to bring the freebooting conquistadors under control by regulating their explorations. As a result, the Governor of Cuba sent an expedition to capture Cortés and bring him back in chains. This second expedition, led by Panfilo de Narvaez, left Cuba on March 5, 1520. Narvaez sailed across the Straits of Yucatan and first arrived at the island of Cozumel. He left some men there to establish a colony. He also left smallpox [3]. The native Mayans began to die in large numbers (Fig. 8.1). Smallpox then gained a second foothold in

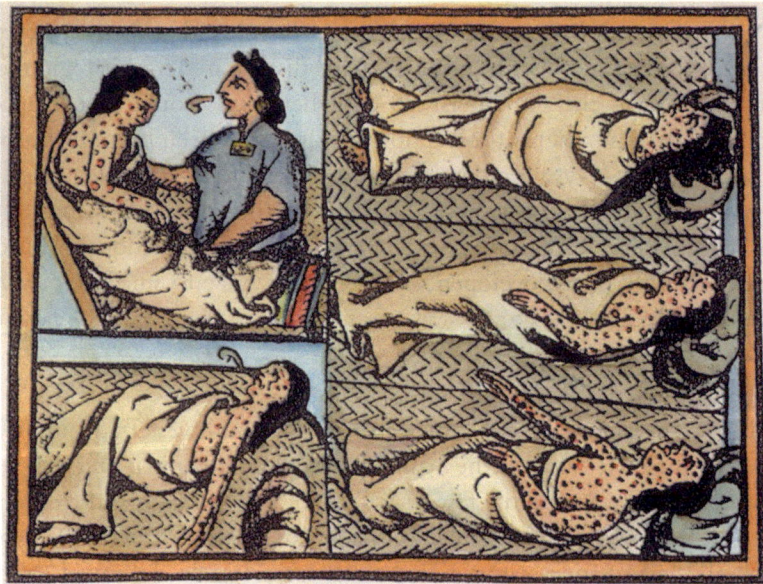

Fig. 8.1 A 16th century drawing of Mayans dying from smallpox. *Courtesy the National Library of Medicine.*

Cempoala near modern Veracruz. It probably arrived with Francisco de Eguia, a porter who worked for Narvaez. It spread to his neighbors, to his neighborhood, and to nearby towns inhabited by the Totonac tribe. When Narvaez caught up with Cortés that May, rather than arresting him, they joined forces and together moved inland to Tenochtitlan (Mexico City).

Meanwhile the smallpox also spread inland, traveling much faster than the Spaniards. It was not a mild strain. The natives had no immunity. None of the old tribal remedies worked. One after another, native villages were depopulated. The dead remained unburied; there were too many of them. The smell of dead bodies was overpowering. The natives despaired. As in Europe, any survivors were badly scarred or permanently blinded by the pox lesions. Mortality in this first epidemic reportedly killed half the population. Famine followed because the women who ground the corn died too, and sick individuals could not get out to search for food. The disease reached the valley of Mexico in September 1520. Nobody was spared. It killed the Aztec nobility as well as the poorest peasants. By late October it had reached Tenochtitlan. One of these deaths was that of Montezuma's successor as Aztec Emperor, Cuitlahuac, but many other leaders died and left the tribes seriously weakened in the face of the Spanish invasion. The first wave of the epidemic raged for 60 days and ended in December 1520. It may have killed as many as 120,000 people. Cortés' forces allied with local anti-Aztec tribes began the siege of city in May 1521, and effective resistance ended that August when the emperor Cuauhtemoc was captured. When they entered the city, the Spaniards found the canals and homes choked by dead bodies. The Spanish Franciscan Toribio Motolinia described how "they died in heaps like bedbugs." The Spaniards could not walk on the streets of Tenochtitlan without treading on the bodies of those killed in the fighting as well as smallpox victims. Because everyone fell sick at once, nobody could harvest or cook food, and as a result many died of starvation. The survivors couldn't bury them all. Sometimes they just pulled the adobe houses down over the bodies, burying them in the ruins. People fled from the infected city and scattered across Central America. But because the incubation period of smallpox is 10 to 14 days, apparently healthy individuals could carry the infection a long way before succumbing to the disease. Originating among the Aztecs and their neighbors in the valley of Mexico, it spread south to the Maya and north to the plains Indians. Eventually it spread across both North and South America. It has been estimated that half the Aztec nation died in this first epidemic.

Cortés replaced the dead chiefs as their chief ruler and became responsible for appointing their replacements. A feature of the disease that reinforced his authority was that the Spanish were somewhat resistant to the virus. Many had likely survived previous smallpox epidemics back home and were immune. This selectivity was noticed by the natives and was seen as a sign of divine favor for the Spaniards as well as being psychologically devastating. Clearly the God of the Spaniards was much more powerful than the Gods of the Aztecs. The Spaniards, in turn, although observing the epidemic, did not initially recognize its severity. Smallpox epidemics swept through Mexico again in lethal waves in 1531, 1545, 1564, and 1576.

The magnitude of the population loss that occurred in Mexico as a result of the introduction of smallpox and other European diseases such as measles and influenza as well as the mysterious cocoliztli is difficult to estimate with accuracy [4]. The Spaniards conducted a census in Mexico in 1560 and determined that the population then was 2.6 million. What, however, was the population in 1518? It was likely around 20 million. Thus it is estimated that about 90% of the population of Mexico and surroundings may have been wiped out within 50 years after the arrival of the Spaniards.

The massive drop in population due to epidemic diseases in the Americas after 1492 resulted in some significant global changes [5]. Prior to 1492, irrigated agriculture and terracing were widely used. The die-off of Native Americans, especially in Central and South America, led to a decline in this land use by about 60 million hectares. Fields were left uncultivated, and forests began to invade formerly agricultural lands. This secondary growth resulted in increased carbon uptake and a drop in atmospheric CO_2 by 7 to 10 ppm as measured in Antarctic ice cores. This resulted in a decline in average global temperature by 0.15°C. Thus the mass deaths of indigenous peoples in the New World had a measurable impact on the global climate.

The Inka in Peru

Smallpox traveled south from Mexico along the Pacific coast of South America, probably carried by trading vessels. By 1525, the chief Inka, Wayna Qhapaq (Huayna Capac), contracted smallpox and died at his home in Quito (modern Ecuador) [6]. The succession was confusing and violent. His designated successor, Ninan Cuyuchi, died of smallpox at the same time. The next in line, named Washkar Inka (Huascar), eventually succeeded but was immediately confronted by rivals. A brutal civil war followed in which Washkar was eventually captured and killed. In 1531, the victor, a general called Atawalla (Atahualpa), was on his way to the capital city of Cusco to celebrate his victory when he heard about the arrival of the Spaniards. Francisco Pizarro, with 168 Spaniards and 69 horses, had arrived on the coast of Peru. Within a few months, Pizarro moved inland and succeeded in overthrowing the Inka Empire by a mixture of aggression, treachery, and smallpox. The Spaniards murdered Atawalla in July 1533. When smallpox first struck in 1525, it was not only Wayna Qhapaq who had died but most of his immediate family, his senior generals, as well as most of the army's leaders. He had appointed regents to govern the southern part of the Inka empire—but they died too. With the political elite wiped out, the social bonds that maintained the empire collapsed and the Spaniards took over. Smallpox epidemics returned to Peru in 1533, 1535, 1558, and 1565. Measles, typhus, influenza, and diphtheria added to the mayhem. In Peru, the Inka were reduced from 8 million to 1 million between 1553 and 1791. It must, however, be pointed out that all these numbers are largely speculative and a matter of dispute.

Virginia colony

By 1585 and 1586, the settlers at the first English colony, Roanoke in North Carolina, had begun to notice that their Indian neighbors died in large numbers soon after the settlers visited their villages. As in other locations, this was probably due to many different infections introduced by the Europeans. Smallpox was not mentioned, but measles, mumps, rubella, and influenza would have been almost as lethal [7]. The second English colony founded at the James Fort (Jamestown) in Virginia in 1607 was more successful, although the settlers died in large numbers from starvation, typhoid, and dysentery. In April 1613, Matoaka, the daughter of a Powhatan chief and the rescuer of John Smith, was abducted and taken to England from Jamestown. Her nickname was Pocahontas (meaning "playful one"). She married a tobacco planter, John Rolfe. She was received at the court of King James but died while returning to Virginia, probably as a result of smallpox in 1617. She was only 20 or 21 years old.

The first major smallpox epidemic in Virginia began in 1667 when a sick sailor arriving in Northampton County (at the tip of the Delmarva peninsula) transmitted the disease to the local Indians and the usual mass mortality ensued. Another smallpox outbreak began in Jamestown in 1696 and eventually spread to the Carolinas and the lower Mississippi valley. Between the arrival of the English in 1607 and 1700, it is estimated that 75% of the native peoples of Virginia died, many as a result of measles and smallpox.

The Pilgrims in New England

The Pilgrims arrived in what is now Massachusetts in late November 1620, after a 65-day voyage in the Mayflower. They had intended to go to the Hudson River area but couldn't reach it because of shoals. Eventually they waded ashore and began to explore the area. The English encountered many abandoned native villages and overgrown cornfields. The countryside was depopulated. The pilgrims eventually chose a site for their settlement in the region of Plymouth. The land they chose had already been cleared and farmed for generations by the Wampanoag tribe but had been abandoned about 3 years previously following an outbreak of a "pestilential putrid fever." It is unclear just what this disease was. Some suggest that the disease was hemorrhagic smallpox. However, a century and a half later, Noah Webster reported that their "bodies all over were exceeding yellow" [8]. This suggests that they may have suffered from some form of liver disease. There are two possible candidates, a viral hepatitis or leptospirosis [9]. Leptospira are bacteria whose natural hosts are rats. It is suggested that the infection was introduced to the New World by shipborne rats. Infected rats may have "jumped ship" from European vessels engaged in fishing, whaling, and the fur trade in the area of New England and New France (the US northeast and the Canadian maritime provinces). European traders were known to have visited Patuxent in 1614 where they seized some of the Wampanoag as slaves. Another reported source of the disease was a French fishing vessel wrecked in that area in 1616. The Wampanoag killed most of the crew but let four live as slaves. When the fishing vessel sank, some rats could have made it ashore where they spread the infection via their urine to other rodents. At least one of the sailors (or their rats) carried disease. Although numbers are necessarily only informed guesses, it is estimated that of about 8000 Wampanoags living in the area in 1600, less than 2000 survived until 1620. On the site of modern Boston, so many natives had died that the surrounding woods were full of bones and skulls. The pestilence continued to kill the local tribespeople until at least 1622.

One local tribe, the Wampanoag, had been very much weakened by the disease and were under threat of attack by their neighbors, the Narragansett tribe, with whom they were not on good terms. In an effort to seek a defensive alliance, they chose to cooperate with the newcomers. Over the spring and summer of 1621, the Wampanoag helped feed the "pilgrims" and taught the interlopers how to fish and hunt and what and when to plant their crops. In consequence, about 90 Wampanoags participated in the Pilgrim's Thanksgiving feast that fall.

The Pilgrims believed in predestination where everything that happened was in accordance with a divine plan. Thus this disease outbreak among the natives was clearly an example of divine intervention. King James I (the Bible guy) certainly thought so. In the Royal Charter issued in 1620 authorizing a second colony in Massachusetts the document states:

> And also for that We have been further given certainly to knowe, that within these late Yeares there hath by God's Visitation reigned a wonderfull plague, together with many horrible Slaugthers, and Murthers, committed amoungst the Sauages and brutish People there, heretofore inhabiting, in a Manner to the utter Destruction, Deuastation, and Depopulacion of that whole Territorye, ...

John Winthrop, the governor of the colony wrote, "They are all dead of the smallpox so the lord clearath our title to what we possess." In general, the English were quite indifferent to, or even pleased at, the deaths of so many native Americans because in their minds it helped make their case that the lands were available for them to take. A major outbreak of smallpox in 1633–1634 broke out further west among the Mohegan and Pequots living along the Connecticut River. As elsewhere, the victims died agonizing deaths as their skin sloughed off. William Bradford wrote that "such a mortality that of a thousand, above nine and a half hundred of them died." From 1636 until 1698 there were at least six smallpox epidemics in the New England area.

The French in Canada

Jacques Cartier sailed up the St. Lawrence in 1534 and claimed "New France" for the French crown. He wintered at the Iroquoian village of Stadacona, near present-day Quebec City. Cartier noted before he left that about 150 of the natives had died from a disease that did not affect his men. Fur traders and missionaries soon followed. In 1603 when the next French explorer, Samuel de Champlain, visited Stadacona the community was deserted. Smallpox subsequently appeared in the community of Tadoussac on the St. Lawrence in 1616. It spread rapidly along the St. Lawrence River and north to James Bay, west to the Great Lakes, and east to the Maritime provinces. As elsewhere, the natives died in huge numbers.

The Iroquois Confederacy (The Five Nations) in New York and Ontario also lived in large communities and relied on farming for much of their food. They traded extensively with other nearby tribes. This population also began to suffer disease losses as soon as the French arrived in 1609. Their crowded communities had the usual problems with sewage disposal. They traditionally cared for sick relatives—a good way to contract smallpox. As village societies collapsed, the survivors fled to nearby villages that the new diseases had yet to reach and thus spread the infection (Box 8.1).

From 1634 to 1635 a smallpox epidemic broke out among members of the Five Nations, especially Mohawks, in what is now Upper New York State. It is estimated that the Mohawk population dropped from 7740 prior to the epidemic to 2835 thereafter, a drop of two-thirds. The disease is believed to have been introduced by Dutch settlers. Smallpox continued to spread north and west. By 1638 French Jesuit missionaries had introduced the disease to the Georgian Bay area, and by 1640 it had wiped out half the Huron nation. The Hurons blamed the European missionaries for their woes. They noted that victims often died after being sprinkled with holy water. Eventually the Hurons were so afraid of the Jesuits that they banned them from entering villages.

The French and Indian War, the American theater of the Seven Years War, lasted from 1756 until 1763. Much of the fighting took place in the Ohio region when the British and French fought for control of the Ohio River Valley; the French were soundly defeated. However, their native American allies were left very unhappy because they had sought to keep the British east of the Alleghenies. The British refused to leave the area after

---

BOX 8.1 Saint Kateri Tekakwitha.

Kateri Tekakwitha was a member of the Mohawk tribe in Upper New York State born around 1656. During a smallpox epidemic in 1661–1663, her family contracted the disease. As a result, her parents died, and she was left very badly scarred and partially blinded. She was raised by her father's sister. This was a period of intense French missionary activity among the tribes of the St. Lawrence region; Tekakwitha converted to Catholicism and was baptized when she was 19. She spent the rest of her short life ministering to the poor, praying, and steadfastly refused to marry. When she died at age 24 in April 1680, it was said that her smallpox scars vanished within minutes and her face appeared pale and beautiful. She was beatified in 1980 and canonized in 2012. She is the first Native American to be declared a saint. Numerous Catholic institutions, especially churches and schools, have been named after her.

the French had left, as they had agreed. They occupied the forts constructed by the French. The tribes in the Ohio region tried to force them out in what is known as Pontiac's rebellion. In May 1763 the Indians attacked Fort Detroit. The British were both irritated and frustrated that the Indians continued to fight. The Indians mopped up many small British outposts and eventually began to threaten Fort Pitt—modern Pittsburgh. The officer in charge of the fort, Captain Simeon Ecuyer, had his problems, because smallpox had broken out in the garrison and he was also sheltering many civilian refugees. He reported this situation to his superior officer, Colonel Henry Bouquet, who passed the word on to the Commander-in-Chief, Jeffery Amherst in New York. Amherst had learned about the Pontiac's revolt at the beginning of June and expressed his bewilderment and frustration with the natives. The smallpox at Ft. Pitt provided an opportunity. He replied to Bouquet and suggested: "Could it not be contrived to send the small pox among the disaffected tribes of Indians? We must on this occasion use every stratagem in our power to reduce them." Bouquet replied to Amherst on 13 July agreeing to use the blankets.

Ecuyer had, however, anticipated this response. In late June 1763 he and a colleague, William Trent, had been approached by some chiefs from the Delaware tribe encouraging him to surrender. Ecuyer declined but Trent gave the chiefs some gifts including food and a linen bundle containing two blankets and a handkerchief from the smallpox hospital. The blankets and handkerchief were unwashed, and Trent is quoted as saying, "I hope it will have the desired effect." There was a subsequent outbreak of smallpox among the local tribes, the Delaware and Shawnee, that may have originated with the blankets. About a hundred natives died. Bouquet and his troops eventually relieved the garrison at Ft. Pitt in early August. Trent billed the government for the blankets and handkerchief, and payment was approved. Amherst said, regarding Ecuyer, "I approve of everything he mentions to have done." Amherst's reputation has never recovered and in 2016, Amherst University in Massachusetts got rid of their unofficial campus mascot "Lord Jeff."

Variolation

Sometime around 1000 BCE the Chinese observed that individuals only got smallpox once. If they survived, they never got it a second time. As a result, they began the practice of variolation, in other words, deliberately infecting an individual with smallpox in such a way that they recovered and developed solid immunity. One way this was accomplished was by inserting a small amount of smallpox scab or pus into an incision on the upper arm. This worked; the individual generally developed localized smallpox lesions around the inoculation site and got quite sick, but in most cases recovered and was immune thereafter. The news of this technique gradually spread westward across Central Asia until it reached Constantinople where it was widely employed. Dr. Emanuel Timoni, the physician of the British Ambassador in that city, wrote a description of the process and sent it to England. The letter was forwarded to the Royal Society in London who duly published it in their Transactions. Given the innovative nature of variolation it was ignored for several years.

As early as 1647, the Puritans had established quarantine regulations to prevent the spread of infection by ships originating in the West Indies. Thus they prevented passengers and crew from coming ashore until the danger of disease had passed. By the late 1600s almost all the other colonies had established quarantine laws as well.

Despite these regulations, beginning in April 1721 when a British ship brought it from Barbados, a major smallpox outbreak developed in Boston and began to kill large numbers of victims. Out of a population of 11,000, 5759 smallpox cases were recorded, causing 844 deaths. The city had not had an outbreak for 19 years, so most young people were highly susceptible. Physicians were at a loss as to how to treat it. The Reverend Cotton Mather, an eminent clergyman, pushed hard for variolation. He had read Timoni's letter to the Royal Society and he had also heard about it from his African slave Oneismus. In a 1716 letter to the Royal Society of London, Mather proposed:

> "ye Method of Inoculation" as the best means of preventing smallpox and noted that he had learned of this process from "my Negro-Man Onesimus, who is a pretty Intelligent Fellow".

> Onesimus explained that he had undergone an Operation, "which had given him something of ye Small-Pox, and would forever preserve him from it, adding, that it was often used among [Africans] and whoever had ye Courage to use it, was forever free from ye Fear of the Contagion. He described ye Operation to me, and showed me in his Arm ye Scar." Onesimus told Mather: "People take the Juice of the Small Pox, and Cut the Skin, and put in a drop; then by 'nd by a little sick; then few small pox; and nobody dye of it; no body have small pox anymore."

Mather had some concept of infectious disease because he also wrote the following, "It begins now to be vehemently suspected that the Smallpox may be more of an animalculated business than we have been generally aware of. The millions of which the microscopes discover in the pustules, have confirmed the suspicion." (This was, of course, pure speculation. Neither Mather nor anyone else could have seen the smallpox virus in samples from the pustules, as the virus is too small to be seen by a light microscope—the first virus was seen using an electron microscope in 1938 (Chapter 1)).

Mather's proposal received a very mixed reception from the Boston medical community. Most physicians rejected the concept of variolation as hazardous and unproven. The only physician with a medical degree in the city, William Douglass, was bitterly opposed. (It must be remembered that physicians at that time had no concepts of viruses or immunity. It would have seemed to them as foolish folklore). The fact that Mather had learned of the concept from Oneismus, an African slave, also provoked racist fears as well as the idea that this was exotic sorcery.

However, Zabdiel Boylston, a local physician, agreed to try the procedure. With Mather's encouragement, on June 24th, Boylston variolated his 6-year-old son Thomas and two slaves. They all survived without complications. Boston society responded in two opposite ways. Most other physicians and clergy were horrified. This was unethical and "unscientific" and, as a result, the city council members banned the process.

> "for a man to infect a family in the morning with smallpox and to pray to God in the evening against the disease is blasphemy; that the smallpox is 'a judgment of God on the sins of the people,' and that 'to avert it is but to provoke him more;' that inoculation is 'an encroachment on the prerogatives of Jehovah, whose right it is to wound and smite."

> "The Select Men in duty bound to take Cognizance of the Matter, desire a Meeting of all the Practitioners in Town, to have their Opinion whether the Practice ought to be allowed or not; they Unanimously agreed that it was rash and dubious, being entirely new, not in the least vouched or recommended (being merely published, in the Philosophick Transactions by way of Amusement) from Britain, tho' it came to us via London from the Turk, and by a strong viva voce Evidence, was proved to be of fatal & dangerous Consequence. B—n is desired by the Select Men to desist." The New-England Courant, Aug 7, 1721.

As others were promoting variolation during the 1721 Boston smallpox epidemic, the Reverend John Williams wrote, in his 1721 treatise, that "Inoculation is unrighteous and unholy" and should not be practiced nor encouraged. He may have been one of the first instigators of the nonvaccination movement in America, but not the last.

The mob threatened to hang Boylston and he had to go into hiding for a while. Notwithstanding this, many parents came quietly to Boylston to have their children variolated. It continued to work well but was by no means free of risk. By the time the epidemic ended in 1722, Boylston himself had variolated 287 individuals and only 6 of them had died (2.4%). This was a great improvement over the 15% mortality in the uninoculated smallpox victims during this outbreak. In fairness to William Douglass, once he saw these results, he reversed his position and eventually performed inoculations himself.

As more people were variolated, the prevalence of smallpox declined. Across the American colonies, different jurisdictions had different regulations. Some encouraged variolation, whereas others banned it. As a result, immunity to smallpox varied greatly between the colonies.

The ambivalence in the American colonies towards variolation was not matched in England where it was introduced around the same time. The English Royal family, the future King George II, espoused it with enthusiasm and it was widely and largely uncritically used in England as well as in the British army.

Ben Franklin had lost his youngest son to smallpox.

In 1736 I lost one of my sons, a fine boy of 4 years old, taken by the small-pox in the common way. I long regretted that I had not given it to him by inoculation, which I mention for the sake of parents, who omit that operation on the supposition that they should never forgive themselves if a child died under it; my example showing that the regret may be the same either way, and that therefore the safer should be chosen.

Franklin bitterly regretted not having the boy variolated and wrote a pamphlet vigorously promoting the procedure.

A small Pamphlet wrote in plain language by some skillful Physician, and publish'd, directing what preparations of the body should be used before the Inoculation of children, what precautions to avoid giving the infection at the same time in the common way, and how the operation is to be performed, the incisions dress'd, the patient treated, and on the appearance of what symptoms a Physician is to be called, &c. might by encouraging parents to inoculate their own children, be a means of removing that objection of the expence, render the practice much more general, and thereby save the lives of thousands.

If the chance were only as two to one in favour of the practice [of inoculation] among children, would it not be sufficient to induce a tender parent to lay hold of the advantage? But when it's so much greater, as it appears to be by these accounts (in some even as thirty to one) surely parents will no longer refuse to accept and thankfully use a discovery GOD in his mercy has been pleased to bless mankind with.

The Revolutionary War

When the Revolution broke out and the newly formed Continental Army began to besiege Boston in 1775, other Americans decided to "liberate" Canada and so form a 14th colony. An expedition under General Richard Montgomery was therefore sent to Quebec in the fall of that year—a bad time to visit Canada weatherwise. Montgomery's troops easily occupied Montreal in November and then moved on to besiege the British stronghold of Quebec City where he joined another American force of 1100 commanded by Benedict Arnold. Montgomery was killed and Arnold seriously injured while attacking the city in late December. To make matters worse, smallpox-infected refugees carried the disease to the American forces (some accounts say they were prostitutes!). The defending British troops had been routinely variolated and were immune. By the spring of 1776, about half the American troops were sick and new recruits had no immunity. Without adequate shelter, thousands died and had to be buried in mass graves. When the troops retreated after losing about 3000 men, they brought smallpox back to Boston with them. Montgomery's successor, Major General John Thomas, was among the dead. Smallpox had saved Canada!

In June 1776 John Adams wrote, "the small-pox is ten times more terrible than Britons, Canadians, and Indians together." As a direct consequence, he and his wife Abigail decided to have their four children variolated. (Adams himself had been variolated in 1764.) The children got very sick as a result but eventually recovered. One son, the future president, John Quincy Adams, evidently had the least severe response. Abigail eventually died of typhoid fever.

George Washington was acutely aware of the damage that smallpox could do because he had suffered from the disease as a young man of 19. In 1751 he had traveled to the Island of Barbados with his brother Lawrence. (Lawrence had tuberculosis, and it was hoped that the island climate would help cure him—it didn't.) On arrival, George was sick for 25 days with the usual fever and chills, headache, and backache, as well as the typical smallpox rash and foul-smelling pustules. He survived, but the disease left him with a scarred nose for life.

During the siege of Boston in 1775–1776, a smallpox epidemic raged within the city. Washington made sure that refugees from the city stayed away from his troops, quarantined any of his troops showing signs of the disease, and set up a special smallpox hospital. When the British eventually evacuated Boston in March 1776, Washington suspected a trap so he first occupied the city with troops who had recovered from smallpox and were therefore immune. Washington also wanted very much to liberate the other major American cities that the British subsequently occupied, especially New York and Providence. With the experience of Quebec in his mind, he realized that British soldiers had been variolated and hence were immune while his own soldiers were not. His surgeon general, John Morgan, encouraged him to have his troops variolated. Washington resisted for two reasons. First, it was illegal in many areas, and he was not prepared to break the law. Secondly, variolation sickened patients for 10 to 14 days, during which time they could not fight. He worried that if he variolated all his men, the English would attack when all his soldiers were sick. As a result, Washington wrote to Congress thus......

The General presents his Compliments to the Honorable The Provincial Congress, and General Committee, is much obliged to them, for their Care, in endeavoring to prevent the spreading of the Small-pox (by Inoculation or any other way) in this City, or in the Continental Army, which might prove fatal to the army, if allowed of, at this critical time, when there is reason to expect that may soon be called to action; and orders that the Officers take the strictest care, to examine into the state of their respective Corps, and thereby prevent Inoculation amongst them; which, if any Soldier should presume upon, he must expect the severest punishment. Any Officer in the Continental Army, who shall suffer himself to be inoculated, will be cashiered and turned out of the army, and have his name published in the Newspapers throughout the Continent, as an Enemy and Traitor to his Country.

After the resignation of Morgan, however, Washington completely reversed himself. Initially he required that all new recruits be variolated immediately after enlistment. Hopefully they would be fully immune by the time they reached the army. In February 1777 he wrote to inform congress.

Finding the smallpox to be spreading much and fearing that no precaution can prevent it from running thro' the whole of our Army, I have determined that the Troops shall be inoculated. This Expedient may be attended with some inconveniences and some disadvantages, but yet I trust, in its consequences will have the most happy effects. [This was technically illegal.]

I have directed the Doctr. Bond to prepare immediately for the inoculating this Quarter, keeping the matter as secret as possible and request, that you will without delay, inoculate all the continental troops that are in Philadelphia and those that come in, as fast as they shall arrive....

I would fain hope that they will soon be fit for duty, and that in a short space of time we shall have an Army not subject to this, the greatest of all calamities that can befall it, when taken in the natural way.

In this first mass inoculation, 4988 men were variolated and 18 died (0.36%)!

While wintering near Trenton in January 1777, Washington had his men variolated and quarantined over a period of months. As he, himself said, "Necessity not only authorizes but seems to require the measure, for should the disorder affect the army …. we should have more to dread from it, than from the sword of the enemy." His men were indeed sickened by the procedure but the mass variation was kept a strict secret and the British did not learn about it. He informed Congress of his plan a month later. His troops in Morristown and Philadelphia were variolated en masse. The prevalence of smallpox in the Continental Army dropped from 17% to 1%. Congress subsequently legalized variolation. Thereafter, smallpox was no longer a major factor in the war.

In Virginia in the meantime, the British had gathered a large number of slaves, offering them freedom if they were to fight against the colonists. This "Ethiopia Regiment" posed a major threat to the Americans until epidemic diseases including a smallpox outbreak killed most of them.

Native Americans

While the Revolutionary war was underway in the East, smallpox raged unabated across the west of the continent, killing untold numbers of Native Americans [10]. The disease continued to wreak havoc with Native American populations until vaccination was introduced in the early 1800s. For example, the Cherokee suffered multiple outbreaks in the early 1700s. As a result, their population dropped by 75%. In the outbreaks in the 1770s, 30% of the natives in the Pacific Northwest died [11]. It killed thousands in New Orleans in 1779. It spread south to Mexico City and crossed the great plains to reach the Pacific coast and north to Alaska and Hudson's Bay [12]. This epidemic is estimated to have killed between 20% and 50% of native Americans and affected nearly every tribe on the continent (Box 8.2).

Vaccination

In 1795, Edward Jenner, an English rural physician, investigated a widespread local belief that an attack of cowpox would prevent smallpox. He demonstrated that cowpox scabs (vaccinia) (and sometimes horsepox) could replace variola in the variolation process, thus inventing the process of vaccination, a very much safer procedure. Cowpox is a virus closely

BOX 8.2 The underwater village.

The Kiowa and Apache tell this story. A lone hunter meets an old man emanating an awful stench and covered with oozing sores. As the hunter questions him, the man explains that he is Smallpox, that he comes from the east and from the white people, and he is traveling this way to find the Indians and make them sick because "it is what was willed." Because the hunter replies that his people have done nothing to deserve this, Smallpox lets him return home and flee with as many people as he wants. But no one will believe the hunter; therefore only his family escapes, guided by a "spirit" to hide under a lake. The rest of the tribe are killed by Smallpox, and later Kiowa-Apache people report seeing and hearing the signs of those that escaped the epidemic, still living under the lake where no one can reach them.

Story told by Eagle Plume (Frank Givens), translated by Ioleta MacElhaney, and recorded by Alice Marriott in Alice Marriott and Carol K. Rachlin, American Indian Mythology *(Thomas Y. Crowell Company, 1968), 49–50.*

related to variola, but the lesions it induces are restricted to hairless skin in cattle, especially the udder. Jenner published his observations in 1798 in a small book called *An Inquiry into the Causes and Effects of the Variolae Vaccinae, a Disease discovered in some of the Western Counties of England, particularly Gloucestershire, and known by the name of the Cow Pox.*

Vaccination was probably introduced into North America around 1798 by Dr. John Clinch, a classmate and friend of Jenner who worked in Newfoundland—the precise date is uncertain [13]. The next year, in 1799, a physician, Dr. Valentine Seaman in New York City, vaccinated his children with material sent to him by Jenner—unfortunately the virus had not survived the transatlantic voyage. Subsequently, Dr. Benjamin Waterhouse, a cofounder of the Harvard Medical School and Professor of Theory and Practice of Physic, received a copy of Jenner's booklet. After reading it, Waterhouse became highly enthusiastic and made several attempts to obtain Jenner's vaccine. He eventually succeeded in obtaining a glass vial containing threads soaked in cowpox lymph from England and vaccinated his 5-year-old son, Daniel on July 8, 1800. He went on to vaccinate three more of his children, other members of his family, and friends. Two months later he arranged for his children to be variolated with fresh material from a smallpox patient and showed that they were immune—a well-controlled experiment! After overcoming initial resistance from the medical profession, both Seaman and Waterhouse were enthusiastic proponents and encouraged public vaccination. Waterhouse wrote a pamphlet entitled "A Prospect of Eliminating Smallpox" and sent a copy to Thomas Jefferson, then Vice-President of the United States. For this, Jenner earned a nice thank-you letter from Jefferson. Vaccination soon replaced variolation. Once he became president, Thomas Jefferson had his entire family and his slaves vaccinated and moreover became an enthusiastic backer of vaccination. He ensured that supplies of the vaccine were sent to other cities in Virginia as well as Pennsylvania, New York, and Washington. Waterhouse, on the other hand, found himself and vaccination opposed by many in the medical profession who accused him of profiteering. He also ran afoul of conservative religious leaders and was forced to resign from Harvard (Box 8.3).

As more and more people got vaccinated, the prevalence of smallpox dropped progressively. Eventually James Madison signed "An Act to Encourage Vaccination" in 1813. This was the first federal law regulating vaccine production, distribution, and safety. The Act appointed a federal agent who was charged with preserving the genuine vaccine (fraudulent versions were circulating). The agent had authority to distribute the vaccine to any US citizen and do so postage free. The Act was repealed in 1822

BOX 8.3 The Milton experiment.

By 1809, the success of Jenner's vaccination method had been widely reported. As a result, the town of Milton in Massachusetts (now an affluent Boston suburb) arranged for all its inhabitants to receive Jenner's vaccine [19]. That July, over 300 individuals were vaccinated. Not content, however, with this achievement, a committee from the council decided to prove that the method worked and set up a public demonstration. On October 10, 1809, 12 previously vaccinated children were inoculated with fresh material from a case of smallpox! This was witnessed by 18 town members. The children were confined to a home. None developed any evidence of smallpox infection and were released 15 days later. The town formally declared that "He [meaning smallpox] is slain." The councilmen published a detailed account of the experiment and sent it to every town in Massachusetts. As a result, in March 1810, the State Legislature passed the Cow Pox Act (an act to diffuse the benefit of inoculation for the cowpox). This directed every community within the Commonwealth to provide for the vaccination of their inhabitants. For good measure, they also published the report from Milton at the same time!

following a smallpox outbreak in Tarboro, North Carolina in which 10 people died. It seems that the federal agent in charge of vaccines, Dr. James Smith, had accidentally supplied some live smallpox samples instead! Smith was dismissed and, in May 1832, Congress passed an act to encourage vaccination of Native Americans. The Army was directed to appoint physicians to go to the Indian villages and offer them vaccination [14].

The Royal Philanthropic Expedition

Around the end of the 18th century, during the reign of King Carlos IV, the Spanish royal family lost several prominent members to smallpox, most notably the king's daughter. That got the king's attention, and in 1798 he ordered that the whole population of Spain be variolated! A year later he read Edward Jenner's book on vaccination. Accordingly, he obtained the smallpox vaccine in 1800 and began to promote vaccination enthusiastically.

In 1802 a large outbreak of smallpox occurred in the Spanish colony of New Granada (now Colombia). Carlos called a meeting of the Council of the Indies to discuss the problem. They decided to send smallpox vaccine to the New World. Francisco Javier de Balmis was appointed to lead the vaccination expedition [15]. The corvette *Maria Pita* was selected for the task. The ship loaded 22 orphaned children aged between 8 and 10 at La Coruna on November 1803. Refrigeration was unknown and there was a question as to how to get the vaccine across the Atlantic. The vaccine was carried in the vesicle (blister) fluid of each child. As the vesicles matured and began to exude fluid it would then be transmitted to the next child by direct arm-to-arm inoculation. They first arrived in Venezuela. Some of the group moved on to Cartagena and Bogota, and everywhere they stopped they began vaccinating. In order to maintain the vaccine supply, they also infected large numbers of cows. By December 1805 the vaccinators had reached Ecuador and Peru.

de Balmis' team went first to Caracas and then on to Cuba [6]. They arrived in Mexico in June 1804 and moved across the country vaccinating as they went. It was a very organized campaign. The vaccination campaigns were regulated by the Catholic church, and priests were required to maintain immunization records. Vaccination boards met weekly to review the progress of the vaccination teams. By 1805, it is estimated that over 100,000 people in Mexico had been vaccinated.

Carlos IV then decreed that de Balmis continue onto the Spanish Colony of the Philippines. As a result, de Balmis left Acapulco in February 1805 accompanied by 25 Mexican orphans. The ship arrived in Manila in April 1805, and de Balmis' team vaccinated close to 20,000 individuals. The local Catholic church was at first reluctant to help but were persuaded when the governor had all five of his children vaccinated. After setting up vaccine production centers, the team moved on to Macau and Canton in China. de Balmis returned to Spain in July 1806.

In New Spain, the colonial government began to actively encourage vaccine production and use. A center was set up in Chihuahua. In the spring of 1805, the Spaniards began to vaccinate children from Chihuahua to Albuquerque and Santa Fe. Similar attempts in Texas initially failed, probably as a result of vaccine inactivation during the long overland journey. A second attempt, a year later in March 1806, succeeded in delivering the active vaccine to San Antonio de Bexar. Vaccination of the local population started on April 23 of that year.

The Russians

The Spanish authorities never succeeded in delivering viable smallpox vaccine to their settlements along the California coast. It was not until 1817 when it was introduced by the Russians! The Russian American Company was a trading company that dealt primarily in furs from Alaska, then a Russian territory, to California. Their most southerly outpost was Fort Ross, just north of San Francisco. A Russian trading vessel, the *Kutuzov*, picked up vaccine from Peru and delivered it to Monterey, San Francisco, and Fort Ross, where they vaccinated the natives. In 1835 the Russian-American Company in Alaska began vaccinating natives living near Russian settlements. Vaccination was not introduced into the Puget Sound area until many years later, in 1837.

Despite the availability of vaccination, even in the 1850s the Inuit in Alaska lost 50% of their population to smallpox. An outbreak in British Columbia and Washington state in 1862 killed 75% of the Native population. In 1868, the Mapuche tribe in Southern Chile and Argentina was effectively resisting the incursions of the central government until they were almost wiped out by a deliberately induced smallpox outbreak.

The High Plains Epidemic

In May 1837, as spring approached, the Indians of the Mandan and Hidatsa tribes gathered at Fort Clark beside the Missouri River in North Dakota to await the arrival of the first river steamer that year [16]. They planned to sell their furs in exchange for supplies. White fur traders had built a trade by exchanging trade goods for pelts. The side-wheeler, the *St. Peters,* accordingly docked at Fort Clark on June 19th and unloaded its cargo. The vessel left the next day going northward to Fort Union on the Yellowstone River. The captain was unaware that one of his passengers, Jacob Halsey, was suffering from smallpox and had infected several of the passengers and crew. As the ship proceeded northwards it had spread the infection to the local Indian tribe, the Lakota. At Fort Clark some passengers disembarked but were still contagious. None of the local Indians had been vaccinated.

The first Mandan death from smallpox occurred on July 14, 1831. The death rate accelerated, until 8 to 10 Mandan were dying daily. As in all such cases, their deaths were horrifying, with agonizing pain, hemorrhagic smallpox, bleeding from body orifices, and the agony of skin peeling off. Some tried to ward off the evil by clinging to their ceremonies and dances. Some attacked Whites as the bringers of the disease. Many committed suicide out of fear or despair. Some killed their families by mutual suicide. The deaths overwhelmed those who survived. Dead bodies accumulated. By mid-September 900 had died.

Upstream at Fort Union things were even worse. A bungled attempt at variolation resulted in many more deaths among Indians and Whites alike. The outbreak was largely met with indifference from the government in Washington and among the White fur traders.

Over a period of 7 months, the Mandan were reduced from 1600 people to 31. The 31 survivors were enslaved by other tribes who settled on their land. Half of the Hidatsa, Assiniboine, and Arikara had died. The Cree had also suffered enormous losses. The disease was carried further west by keelboat and began spreading among the Blackfoot (Siksika). The Blackfoot lost as many as 6000 people. The Great Plains were depopulated, and resistance to western expansion was effectively neutralized. The Lakota, while severely affected, remained powerful and moved into the depopulated regions. This was where they were when the Indian wars of 1860–1870 broke out. Perhaps as many as 20,000 Native Americans died in this, the last major smallpox epidemic in North America. It is worth noting that the Arikaras had also suffered grievously in the smallpox epidemics of the 1780s and 1803–1804 and were reduced from about from 30,000 members in 1776 to 2000 by 1804. In 1853 smallpox arrived in Honolulu on the ship *Charles Mallory* and killed more than 5000 Hawaiians.

The last century of smallpox

Abraham Lincoln felt weak and miserable on his way to Gettysburg in November 1863. On his return to Washington, he developed a high fever, a headache, backache, extreme fatigue, and small red blisters on his skin. Dr. Van Bibber was called in by the president's physician for consultation. "Mr. President, if I was to give a name to your malady, I should say that you have a touch of the varioloid." During his illness, Lincoln was quarantined in the White House. He resisted his doctors and is quoted as saying, "Finally I have something I can give everybody." The term varioloid was used to describe a very mild case of smallpox that occurred in someone who had been previously vaccinated or was otherwise immune. (There is no evidence that Lincoln had been vaccinated previously.) It took about 3 weeks before his skin lesions healed, and he fully recovered. There was a smallpox epidemic ongoing in Washington at that time, and Lincoln's servant, William Johnson, died as a result of smallpox in January 1864.

In Montreal, the largest and most prosperous Canadian city at that time, a major smallpox epidemic occurred in 1885, killing 5864 people and disfiguring about 13,000. It appeared to have arrived by rail from Chicago. The infected individual was railroad conductor George Longley. He was a Protestant, so he was taken to the Protestant, Montreal General Hospital; smallpox was diagnosed, but they refused to admit him! As a result, he went to the Catholic hospital, the Hotel-Dieu, where the nuns admitted him. Smallpox spread through the hospital, so the health department decided to release all patients who did not appear to be sick. Big mistake! The authorities believed that most of the inhabitants had been vaccinated and the disease eradicated. However, the French Canadians did not trust the British. They resisted vaccination and were encouraged by antivaccination campaigners. The disease spread across the city. Bodies were soon being carried in a continuous stream to the cemetery. Attempts to force vaccination on the unwilling citizens were resisted. Rioting followed in September, and a health center was wrecked. The smallpox epidemic finally ended in November. Ninety percent of the victims were French Canadian.

One ongoing problem with vaccination in the United States was the debate on whether to make it compulsory. The debate erupted with variolation and continues to this day. Sometimes the debate could get violent. For example, riots broke out in Milwaukee in 1894 (Fig. 8.2). The city was enforcing a strict quarantine. Infected individuals were taken from their

Fig. 8.2 The Milwaukee Vaccination Riots. *Courtesy the Library of Congress.*

homes and moved to the city's smallpox hospital. This removal was opposed by many in the city's immigrant community.

They believed that they were being discriminated against. They actively opposed any ambulances that tried to collect a child and drove them away. As in many other outbreaks, some victims were hidden from the authorities. An angry mob prevented health officers from taking a smallpox-affected child to the hospital. The riots continued throughout the month of August. The epidemic ended in early 1895 (Box 8.4).

BOX 8.4 Vaccine farms.

When Edward Jenner first developed vaccination, he transmitted the virus by taking the material from a scab on the arm of a vaccinated person and placing it in a small cut in the arm of the person to be vaccinated. This arm-to-arm transfer method was in use for many years afterward. It had, however, significant disadvantages. The supply was erratic, the quality uneven, and other diseases such as syphilis, leprosy, or erysipelas could be accidentally transmitted between individuals. As a result, beginning around 1860, smallpox vaccine began to be made by infecting animals.

BOX 8.4 Vaccine farms. (*cont'd*)

The animals (mainly calves, but occasionally donkeys and horses) were tied down on a specially designed vaccinating table, their bellies were shaved, and then cleaned thoroughly. About 100 to 120 small cuts, each about an inch long, were made in their skin, and a drop of vaccinia lymph was rubbed into each of them. About 6 days later, the local response resulted in the development of vesicles loaded with virus-laden fluid called lymph as well as extensive scab formation. The calf was killed, the vesicles scraped, and the mixture of skin, lymph, and blood—"lymph pulp"—collected, ground up, and suspended in a diluent with glycerine. This "animal vaccine" had a higher yield and ensured that a consistent supply of the vaccine was available while human disease transmission was minimized [20]. Despite some opposition, the method was rapidly adopted worldwide whereas arm-to-arm vaccination was eventually prohibited.

The method was introduced into the United States in 1870 by yet another Boston physician, Henry Austin Martin. Many "vaccine farms" were rapidly established by physicians who saw an opportunity for profit by producing "animal vaccine." It was often unclear just where their vaccine scabs originated. If they needed to be replenished, they were mainly imported from Europe. However, they were imported from multiple sources in different countries at different times and many would have been unlikely to be "pure stocks." Some companies simply purchased a vial of vaccine from one of their competitors. As a result, the vaccines produced in the United States appear to have been somewhat diverse. Modern analysis suggests that they probably contained viruses of cow and horse origin and even recombinants between the two.

As the science of microbiology developed, the vaccine farms eventually expanded into making other biologicals such as tetanus and diphtheria antitoxins as well as vaccines. This production was initially unregulated, but a series of unfortunate accidents eventually forced Congress to pass The Biologics Control Act in 1902 to regulate vaccine production.

Esparza J, Lederman S, Nitsche A, Damaso CR. Early smallpox vaccine manufacturing in the United States: Introduction of the "animal vaccine" in 1870, establishment of "vaccine farms" and the beginnings of the vaccine industry. Vaccine 2020;38:4773—79.

In Boston in 1901—1902, there was yet another smallpox outbreak. The City Board of Health mandated that everyone in the city should be vaccinated or revaccinated. Anyone who refused vaccination was fined $5 or 15 days in jail (worth about $150 today). The Anti-Compulsory Vaccination League was founded and took the city to court. The plaintiff was a pastor, Henning Jacobson, who had been vaccinated previously and claimed that it had caused him "great and extreme suffering." He maintained that the state was restricting his liberty. He was fined and appealed. As in so many cases, the legal arguments ended up in the US Supreme Court (*Jacobson v. Massachusetts*, 1905) where Jacobson lost. The justices voted 7 to 2 in favor of the state. In the majority opinion, they declared that the state could, at times, require vaccination in order to protect the public health and safety from a dangerous communicable disease. Freedom of the individual must sometimes be subordinated to the common welfare. The court also asserted that the police could enforce it. In a subsequent case, *Zucht v. King* in 1922, the Supreme Court ruled unanimously that a school district had the power to exclude unvaccinated students.

In 1947, three cases of smallpox occurred in New York City in patients recently arrived from Mexico. In response, the authorities recommended that everyone who had not been vaccinated within the previous 10 years should be revaccinated. People came to the clinics and lined up for their vaccination. The demand was such that the city's vaccine supply ran out! The *Times* reported "hundreds of eager men, women, and children queued up at Bellevue hospital at dawn, although vaccinations were not scheduled to begin until 10 am. At some stations the crowds did not take kindly to the news that the doctors had run out of vaccine and the police had a little difficulty dispersing a crowd of several hundred" outside one vaccine station. According to the city, between 5 and 6 million people were vaccinated within a few days; however, calculations based on newspaper accounts suggested that the number was closer to 2.5 million. Either way, science won, again [21]. In the end, there were only 12 cases and 2 deaths.

Despite arguments from those who opposed government involvement in such a personal decision, vaccination eliminated smallpox from the Western Hemisphere in April 1971 when the last cases occurred in Brazil. The last US victim to die was Lilian Barber (43) in an outbreak in Hidalgo County in the Rio Grande Valley on March 12, 1949 [17]. It is unclear where she contracted the disease, but she may have acquired it in a local hospital. Her husband also developed smallpox but survived. His son recalled his father covered in purple-black lesions. "I would change his sheets and each time I would burn a double handful of scabs that came off his body...." A mass vaccination campaign was introduced immediately in south Texas, and within a few days, 50,000 doses of vaccine had been administered.

Eradication!

In 1956, at the United Nations, it was proposed by the Health Minister of the Soviet Union, Victor Zhadnov, that smallpox be eradicated by a worldwide vaccination campaign. The UN initially allocated $100,000 for its budget! Ten years later the World Health Organization (WHO) proposed a budget of $2.4 million, but many developed countries objected to such a huge budget! The program got under way under the leadership of Dr. Donald Henderson at the WHO. It was a resounding success. This was facilitated by a very simple vaccination procedure using a double-tined (bifurcated) needle. Dr. Benjamin Rubin, an American researcher working for Wyeth Laboratories, invented the technique, which used less vaccine fluid to effectively vaccinate a person, at a time when vaccine supplies were limited—especially in poorer countries. The needle was dipped in the vaccine and then simply used to scratch the patient. The vaccinia established a mild infection and a local lesion at the scratch site and the patient developed immunity. Given the lack of animal reservoirs, the eradication proceeded swiftly so that the last natural case of the disease was recorded in Somalia in October 1977. An intensive search failed to find any others (Fig. 8.3). Smallpox was declared eradicated from the globe in April 1980 (Fig. 8.4) [18].

The last smallpox death occurred in 1978 when a scientist working at the University of Birmingham in England accidentally contracted the disease. Although smallpox is officially eliminated, samples are stored in the United States and Russia. This is a matter of concern because of its potential as a bioterrorism weapon. The complete genome of variola has been determined, and it would take relatively little effort to synthesize the virus. It is, however, illegal for US investigators to synthesize variola. Routine vaccination ended in the United States in the early 1970s.

Fig. 8.3 As smallpox cases dwindled, an intensive search was mounted for any remaining victims. This dramatic poster was used in an effort to find any remaining cases. *Courtesy US National Library of Medicine.*

Fig. 8.4 Directors of the Global Smallpox Eradication Program displaying the official magazine of the World Health Organization (May 1980) announcing the eradication of smallpox virus in the wild. (Dr. J. Donald Millar, *left*; Dr. William H. Foege, *center*; Dr. Michael Lane, *right*) *Courtesy Centers for Disease Control and Prevention's Public Health Image Library.*

References

[1] Parker S, Crump R, Hartzler H, Buller RM. Evaluation of taterapox virus in small animals. Viruses 2017;9:203. https://doi.org/10.3390/v9080203.

[2] Babkin IV, Babkina IN. The origins of variola virus. Viruses 2015;7:1100−12.

[3] Thomas H. Conquest: montezuma, cortes, and the fall of old Mexico. New York, New York: Touchstone books; 1993.

[4] Livi-Bacci M. The depopulation of Hispanic America after the conquest. Pop Dev Rev 2006;32(2):199−232.

[5] Koch A, Brierley C, Maslin MM, Lewis SL. Earth system impacts of the European arrival and great dying in the Americas after 1492. Quart Sci Rev 2019;207:13−36.

[6] MacQuarrie K. The Last days of the Incas. New York, New York: Simon and Schuster; 2007.

[7] Mintz S, McNeil S. *Death in early America*. Digital History. Retrieved from http://www.Digitalhistory.uh.edu.

[8] Webster, Noah. *A brief history of epidemic and pestilential diseases; with the principal phenomena of the physical world, which precede and accompany them, and the observations deduced from the facts Stated*. In Two Volumes. Hudson and Goodwin, Hartford. 1799.

[9] Marr JS, Cathey JT. New hypothesis for cause of epidemic among native Americans, New England, 1616-1619. Emerg Inf Dis 2010;16:281−6.

[10] Duffy J. Smallpox and the Indians in the American Colonies. Bull Hist Med 1951; 25(4):324−41.

[11] Fen EA. Pox Americana: the great smallpox epidemic of 1775-82. New York: Hill and Wang; 2001.

[12] Carlos AM, Lewis FD. Smallpox and native American mortality: the 1780s epidemic in the Hudson Bay Region. Expl Econ Hist; 2012. p. 277—90.

[13] Esparza J. Three different paths to introduce the smallpox vaccine in early 19th century United States. Vaccine 2020;38:2741—5.

[14] Riley JC. Smallpox and American Indians revisited. J Hist Med Allied Sci 2010;65(4): 445—77.

[15] Franco-Paredes C, Lammoglia L, Santos-Preciado JI. The Spanish royal philanthropic expedition to bring vaccination to the New World and Asia in the 19th century. Clin Inf Dis 2005;41:1285—9.

[16] Reinhiller J. Holding on to culture: the effects of the 1837 smallpox epidemic on mandan and Hidatsa. Butler Journal of Undergraduate Research 2018;4. Article 12. 198-213.

[17] Irons JV, Sullivan TD, Cook EB, et al. Outbreak of smallpox in the lower rio grande valley of Texas in 1949. Am J Publ Health 1953;43:25—9.

[18] Hopkins DR. Smallpox: ten years gone. Amer Jour Publ Hlth 1988;78:1589—95.

[19] Woodward SD. The story of smallpox in Massachusetts. Massachusetts. Annual Oration; 1932. http://www.massmed.org.

[20] Schrick L, Tausch SH, Dabrowski PW, et al. An early American smallpox vaccine based on horsepox. N Engl J Med 2017;377:15—6.

[21] Sepkowitz KA. The 1947 smallpox vaccination campaign in New York City revisited. Emerging Infect Dis 2004;10(5):960—1

Measles—1537

Contents

Many infectious diseases cause the development of spots on the skin. These include, most obviously, smallpox, typhoid, and measles. The term measles comes from Middle English, *masel,* meaning "little spot." Its Latin name is rubeola, referring to the reddish color of the rash. It was not until 1676 that Thomas Sydenham definitely showed to the Western medical establishment that smallpox and measles were in fact different diseases [1]. Although smallpox was extraordinarily lethal, measles was, in most cases, a fairly innocuous childhood infection. However, measles is extraordinarily infectious, and whereas only a small proportion of infected persons develop serious complications, these may add up to a lot of individuals.

The virus

Measles is caused by a single-stranded, negative-sense, nonsegmented RNA virus. It is classified as a member of the *Paramyxovirus* family and a member of the *Morbillivirus* genus (measles is also called *morbilli* in Italian). The morbilliviruses include the viruses that cause rinderpest in cattle, peste de petite ruminants in sheep and goats, and canine distemper in dogs. Although the measles virus has caused devastating human epidemics for centuries, it is in fact a relative newcomer to the world of pathogens.

Great American Diseases
ISBN: 978-0-443-31404-9
https://doi.org/10.1016/B978-0-443-31404-9.00009-4

The measles virus is believed to have arisen as a result of mutations in the bovine rinderpest virus. These genetic changes altered the virus' host preference so it was able to infect humans instead of cattle. It has been possible by examining the RNA sequences of viral genomes to determine the nature of their relationship. It is clear that measles and rinderpest viruses are very closely related indeed. Using estimates of the rates at which their viral RNA mutates, it has been calculated that the two viruses only diverged sometime between 600 and 1200. During that period, humans and domestic livestock in Europe and the Middle East lived in close contact, much more so than at present. As a result of this close contact and the naivety of the human immune system to this new virus, the mutated virus was able to infect humans and then spread from human to human. It eventually became endemic in Europe and Asia. Cities and towns, areas with dense human populations, provided the host availability that the virus needed for its spread and persistence. Thus cities and towns became endemic reservoirs of measles. In areas with sufficiently large populations, there are always some susceptible people who can become infected and so perpetuate the chain of infection. In situations where susceptible persons are scattered and at a lower population density, such as in rural areas, the disease develops a periodic epidemic pattern, infecting the nonimmune and causing an outbreak, then disappearing as most of the population becomes resistant only to return when the number of newly susceptible individuals once again reach critical numbers. Thus measles became a childhood disease that causes periodic outbreaks in groups of susceptible young people, such as in schools. Those who recover develop lifelong immunity.

The disease

Measles is transmitted and acquired via the respiratory route. Aerosolized droplets containing the virus can remain suspended in the air for up to 2 hours and are spread through coughing and sneezing. Measles virus is easily inactivated, so surface contamination is not a major source of infection. People infected with measles shed such large amounts of virus and the virus is so contagious that 90% of nonimmune people exposed to the virus will become infected. It has an R_0 that may reach as high as 15 to 20. Herd immunity is therefore difficult to achieve, and as a result, outbreaks of measles may occur even when fewer than 10% of the population are susceptible.

On inhalation, measles virus first infects the cells of the respiratory tract. After growing initially in the lungs, the host develops a viremia, that is, the virus circulates in the bloodstream. This spreads the virus throughout the body to other organs and tissues, such as lymphoid tissue, the gastrointestinal tract, kidneys, liver, and skin.

Measles has a relatively long incubation period of about 10 days to the onset of fever and 14 days to the onset of the rash. The infection begins with nonspecific symptoms. Thus it starts with a mild to moderate fever and malaise associated with persistent cough, runny nose, inflamed eyes (conjunctivitis), and sore throat. Two to 3 days after the onset of the fever, clinical measles starts with the appearance of Koplik spots—tiny white spots surrounded by a red ring, inside the mouth. A few days after the appearance of Koplik spots, a skin rash with slightly raised red spots develops behind the ears and spreads to the face. Over the next few days, the fever can reach 104°F to 106°F with the rash spreading to the trunk and extremities (Fig. 9.1). The skin rash is the result of virus-infected skin cells being destroyed by the host's cell-mediated immune response.

An infected person can transmit the virus from approximately 4 days before clinical symptoms occur to 4 days after the skin rash is gone. They can thus infect others long before anyone knows they have the disease.

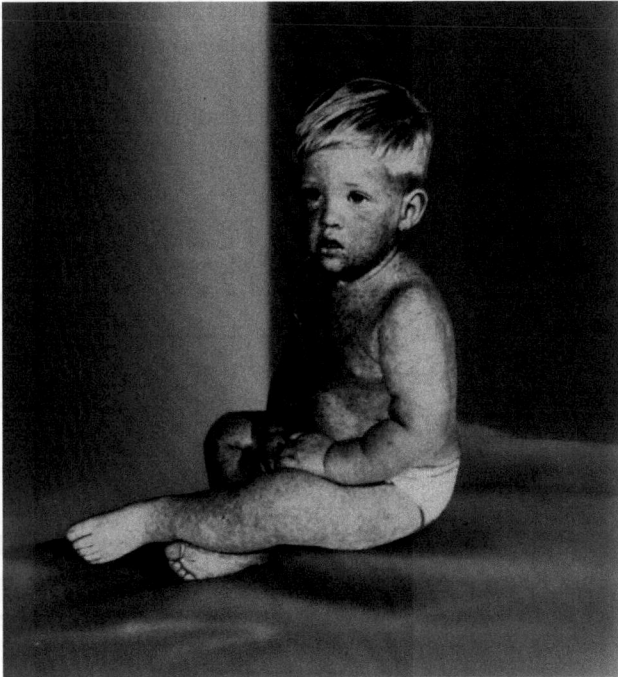

Fig. 9.1 *A case of measles in a child.* Courtesy National Library of Medicine.

Measles virus also causes a transient but overwhelming suppression of the immune system. This often results in secondary bacterial infections of the respiratory and gastrointestinal tracts. These secondary infections are the major cause of mortality in the disease. Complications include generalized bacterial infections, diarrhea, pneumonia, blindness, encephalitis, seizures, and death. Hearing loss from measles-related ear infections is common.

In developed countries, the case fatality rate of measles is usually below 0.01%, but in developing countries, it may exceed 5% as a result of lethal secondary infections. Infection with measles virus induces a lifelong protective immunity; people usually do not get measles a second time. Thus measles virus can only infect immunologically naive people who have not previously encountered the virus. The need for continual supply of susceptible victims is the reason that endemic infections tend to occur mainly in young children and adults living in densely populated areas such as cities, towns, army camps, and recruit depots.

History

Given its relatively recent appearance, it is unsurprising that measles only became endemic in Europe during the 15th to 16th centuries when the population density had grown sufficiently large to provide the virus with a steady supply of nonimmune little children to infect. As a result, measles was a common childhood disease with relatively low mortality. Measles was not present in either Native Americans or Pacific Islanders prior to their first encounters with Europeans. It likely never spread across the Bering Strait, and for most of the human history of the Americas, the population density would have been quite insufficient to maintain the virus.

At the time of the European invasions after 1492, although many Native Americans still lived dispersed in scattered bands, large, crowded cities had developed in both Mexico and Peru. These were large enough and dense enough to support the development of major epidemics when measles first arrived.

The Spaniards certainly were familiar with measles. The Spaniards knew about smallpox and gave measles a specific name—"sarampion" [1]. They described the first epidemic of sarampion in Mexico and Central America in 1531–1532. It was reported to have killed great numbers of the natives but not the Spaniards. Likewise, there was another massive outbreak of sarampion in Peru in 1537. The first Canadian measles epidemic was recorded in 1635 among the Indians along the St. Lawrence in French Canada [2].

In 1659 it was reported that a measles epidemic probably killed about 10,000 natives including Timuacans, Apalachee, Pensacola, Tocobaga, Calusa, Tequesta, Jeaga, Jobe, and Ais in Florida [3]. Measles eventually reached and almost wiped out the Yagan tribe on Tierra del Fuego in 1884.

> In a few days they were dying at such a rate that it was impossible to dig graves fast enough. In outlying districts, the dead were merely put outside the wigwams or, when the other occupants had the strength, carried or dragged to the nearest bushes.
>
> **—Lucas Bridges, a missionary in Patagonia [4]**

A major measles outbreak occurred among the natives in the Fiji Islands in 1875 when a British warship, HMS *Dido*, allowed sick sailors to go ashore. The outbreak that followed lasted for several months and killed almost a quarter of the population. Eventually the disease spread to other islands in the Fiji group and 20,000 to 40,000 persons died. In many of these cases, death resulted not directly from the disease but as a result of starvation because they were too sick to gather food. (This is a common feature of these serious debilitating disease outbreaks in societies where food needs to be harvested and cooked daily.)

Measles in North America

Although the Arab physician Rhazes in 900 distinguished measles from smallpox, it was not until 1676 that the English physician Thomas Sydenham clearly distinguished the two diseases [1]. However, even Sir William Osler, at the end of the 19th century (1892), acknowledged the difficulty of differentiating the smallpox rash from the early stages of measles. As with many of the other infectious diseases discussed in this book, it is important to remember that disease nomenclature was very much a matter of guesswork prior to the introduction of modern diagnostic methods. One example is Thomas Thatcher's 1678 guide entitled *Brief Rule To Guide the Common People of New-England how to Order themselves and theirs in the Small-pocks or Measels*. Thatcher appears to have considered measles simply as an early manifestation of smallpox. Noah Webster, in his comprehensive survey of epidemic diseases in the Americas, first mentions a spotted pestilent fever as occurring on a ship sailing from Cork to America in 1607 [5]. He recorded several other outbreaks of spotted fever and measles, so he apparently considered them to be separate entities.

Given the extraordinarily high R_0 of measles, it seems that it would have difficulty crossing the Atlantic when a good, fast voyage could easily take

90 days. Once it got on a ship, the virus would have rapidly infected everyone on board. As a result, the passengers and crew would have died or recovered long before they reached the New World. For example, John Winthrop reported that on one of the ships in the large fleet bringing Puritan refugees from England to Massachusetts in 1630, just prior to departure, there was a "Childe or 2 sicke of the Measells but like to doe well." The measles does not appear to have survived the Atlantic crossing on that occasion. An occasional, unusually short voyage may have allowed it to persist and reach the New World. Likewise, the virus could not have established itself in the English settlements until there was a population of sufficient size to allow it to become endemic. Prior to that, any outbreaks would have been self-limiting and died out rapidly.

The first record of measles occurring in colonial North America was in September 1657 when an outbreak occurred in Boston. Although morbidity was high, it appears that mortality was low; "only through the goodness of God, scarce any died of it" (John Hull) [6]. Then it appears to have disappeared for 30 years presumably dying out. It broke out again in Boston in 1687; "many people in Boston are sick of the measells but it is not mortall as yet." This time, it appears to have attacked both adults and children. This makes sense because none of the adults under 30 would have had any immunity. Measles outbreaks thereafter were fairly regular events in Boston occurring in 1693, 1713–1716, 1729, 1739–1740, 1747–1749, and 1769. The Puritan minister Cotton Mather (the smallpox vaccination guy) left a detailed description in his diary of the outbreak that affected his wife, their nine children, and a maidservant during October–November 1713 [7]. The outbreak in Mather's home was especially severe because he lost five members of his immediate family, including his wife, three children, and the maid. Mather also noted that "pleuritick fever" (pneumonia) was a serious complication. Mather went on to publish an open letter "for the benefit of the poor" in December 1713. In it he explained the clinical course of the disease, as well as some simple remedies. These were designed, as was expected at that time, to correct the "imbalance of the humours." Among them he suggested "hot beer and rum, hot cyder and hot honey."

A remarkably severe epidemic of measles began in 1772. Noah Webster claimed to have found records of it in all parts of North America [5]. It caused many deaths in New England. It spread to Rhode Island, Connecticut, New York, and Vermont. Benjamin Rush reported that it had caused a large number of deaths in Philadelphia as well. From New York it spread to Charleston in South Carolina. Once in Charleston the disease proved especially severe. It was reported that 800 to 900 children died out of an estimated population of about 14,000. This is an unusually high number and suggests that there were probably other diseases occurring at the same time. Universities appeared to be especially hard hit. For example, Harvard students suffered outbreaks in 1714 and 1759 (when the university had to close temporarily). Not to be outdone, Yale suffered major outbreaks in 1741 and 1783.

In 1783, Dr. Benjamin Waterhouse (the vaccination guy) gave a lecture on "Epidemic Diseases" at the Boston Academy of Arts and Sciences. He proposed keeping a record of such disease outbreaks, beginning with the measles currently raging in Boston. (It is remarkable how many times Boston appears to be the starting point for all these measles outbreaks.)

In 1847, a measles outbreak struck Hannibal, Missouri, the boyhood home of Mark Twain. Reflecting on the community's response, Twain wrote that

for a time a child died almost every day. ... Children that were not smitten with the disease were imprisoned in their homes to save them from the infection.

Fear, despair, and distress embraced the town, with the normal happy functions of the town ceasing. Twain, who was 12 and a half at that time, did contract measles and was extremely ill, but recovered with no complications.

Native Americans

Hawaii

In 1824, King Kamehameha II of the then-independent country of Hawaii decided to travel to London to meet King George IV. Kamehameha was fascinated by the newly arrived foreign missionaries. As a result, ignoring the advice of his courtiers, he and his favorite wife (he had five), Queen Kamamalu, sailed for England in 1824. Their visit was unannounced and unexpected, but as exotic visitors the royal couple were a sensation. They were warmly welcomed and taken to see the sights. They visited orphaned children at the Royal Military Asylum, where childhood diseases were rife. Within a month both had died as a result of measles.

In late 1848, a series of epidemics, including not only measles but also influenza, whooping cough, and dysentery, struck the Hawaiian Islands [8]. The measles had been brought from Mazatlan, Mexico, by an American frigate, the *Independence,* while around the same time a vessel from California brought the whooping cough. Although numbers are unreliable, one estimate suggests that it killed about a quarter to a third of the native population. The epidemics died out later in 1849.

West Coast

The first recorded measles epidemic in California occurred in 1806 and extended from San Francisco to Santa Barbara. A second epidemic occurred in 1827. As always, children suffered the most, and about 1600 died in Santa Barbara.

Measles reached the Pacific Northwest in 1847 and wreaked havoc among the Native American tribes [9]. The infection, although blamed on wagon trains carrying immigrating settlers, was probably introduced to Oregon from California. A party of Indians had traveled to the region of Sacramento to purchase cattle. When they returned, they brought measles with them. As they returned to their villages, they spread the disease. As a result, it is estimated that around 10% of the native population died.

This outbreak of measles began just before the first White settlers arrived along the Oregon Trail. The local tribes were unhappy because of ongoing disputes regarding land ownership. The missionaries were seen to be siding with the White settlers. The deadly measles epidemic added to their unhappiness. The local natives became convinced that the White settlers had brought the measles with them. Suspicion fell on a Calvinist missionary, Marcus Whitman, who was based with his wife Narcissa at Waiilatmu in the Oregon Territory near present-day Walla Walla. The Indians believed that shamans had the power to remove the spiritual causes of disease. However, they also had the power to summon the evil spirits through their magic. Marcus Whitman was regarded as a powerful shaman who could readily cause disease and therefore had to be removed.

As a result, on the morning of November 29, 1847, two members of the local Cayuse tribe, Tomahas and Tiloukait, attacked and killed Whitman and his wife Narcissa together with 12 other members of the mission in an effort to remove the cause of the pestilence [10]. It didn't work; about 40% of the Cayuse died as a result of measles. The disease spread north by land through what is now Oregon and Washington to British Columbia. It also spread by sea via the Hudson Bay Company steamer, the *Beaver*—and so it continued: An outbreak of measles killed 219 Cheyenne and Arapaho children in 1877.

In the summer of 1900, a massive measles epidemic broke out in western Alaska affecting Inuit, Yup'ik, Aleuts, and Athabaskan Indians alike [11]. Depending on the community, it killed from 13% to 74% of the native population. On average about a quarter died—they called it "The Great Sickness." Most deaths occurred as a result of gastrointestinal or respiratory complications, suggesting that they may have been a result of secondary bacterial infections and possibly influenza as well. The disease also affected the native peoples across much of Eastern Siberia. In contrast, the nonnative populations in Alaska, most of which likely had previous exposure to the diseases, experienced only mild symptoms and few deaths. Measles had reached other parts of Alaska previously, including the Aleutians and the Panhandle in 1848 and Kodiak Island in 1875 (Box 9.1).

BOX 9.1 Canine distemper—an American disease.

Canine distemper is a lethal disease in dogs caused by a morbillivirus. The first recorded outbreak of distemper occurred in Europe around 1760. This raises the question, where did it originate? It is now believed to have originated from measles virus in the Americas! The first records of canine distemper actually originated in Ecuador and Peru. The Spaniards had imported mastiffs and other large dog breeds into their colonies in order to intimidate the natives. These "war dogs" were especially effective in detecting native ambushes. Don Antonio de Ulloa provided the first credible report of the disease in 1735. He reported that the disease was especially lethal for young dogs. "However, if they recovered, they did not get it a second time." Distemper is not difficult to diagnose, and it is unlikely that it would have passed unnoticed had it occurred in Europe prior to that time. In fact, the first definitive reports of canine distemper occurred in Spain in the 1760s. As with any newly introduced disease, the initial outbreaks were highly lethal and killed hundreds of dogs, but over time its virulence has decreased.

Sequence analysis of the modern viruses indicates that canine distemper virus is in fact a derivative of measles virus. Thus analysis of the RNA of modern distemper virus isolates indicates that it shared a most recent common ancestor with measles virus in the United States in the 19th to 20th century. This is obviously incorrect because distemper was recognized and described centuries earlier. Nevertheless, it supports the theory that distemper virus first appeared in the New World relatively recently.

Given that extensive measles outbreaks followed the arrival of the Europeans, it appears highly likely that distemper is a mutant strain of an American measles virus that eventually spread to dogs in Europe and the rest of the world.

Uhl EW, Kelderhouse C, Buikstra J, Blick JP, Bolon B, Hogan RJ. New World origin of canine distemper: interdisciplinary insights. Int Jour Paleopathol 2019;24:266—78.

US Army

Measles was a major problem in both armies during the Civil War (Chapter 14). And it didn't go away. It came back when the United States entered World War I. Less than a year before the great influenza pandemic of 1918 killed thousands of US soldiers in their camps, there was an outbreak of a disease with an even greater case fatality rate—measles. (The case fatality rate is the proportion of deaths compared to the total number of diagnosed cases.) Until that time, measles was regarded as a disease that cycled through cities on a 2- to 3-year cycle. It mainly affected and killed children under 10. As described in Chapter 16, the massive US Army mobilization in 1917 resulted in over 400 million men being housed in large, hastily constructed camps while undergoing training. Although recruits were screened, measles cases appeared in the camps that summer and culminated in two peaks corresponding to the intakes of new recruits in November 1917 and in the late summer of 1918. Subsequent analysis counted 95,843 cases. Of these, about 23,000 were admitted to the hospital as a result of complications, and there were 3206 deaths. In some camps the measles killed more soldiers than did the 1918 influenza [12]. The case fatality rates varied greatly between camps, but the highest mortality occurred in troops from predominantly rural states such as Mississippi and Vermont. Recovery took many months and left many soldiers weakened and often permanently incapacitated.

The outbreak provided an enormous amount of useful data on the pathogenesis of measles and its complications. For example, about half of the infected soldiers also developed bacterial pneumonia, and half of these cases were severe and life-threatening. We now know that this was due to virus-mediated immunosuppression. The predominant cause of the pneumonia was the bacterium now called *Streptococcus pyogenes*. Soldiers with measles were 10 times more likely to die of secondary bacterial pneumonia than those without measles. Subsequent studies also revealed that *S. pyogenes* was carried in the noses of otherwise healthy soldiers. On arrival in camp, about 5% of recruits carried the streptococcus, but within a few days, this climbed to over 80%. (The prevalence of other streptococcal infections, such as scarlet fever and tonsillitis, also climbed at the same time.)

The Army, in response, initiated social distancing, prevention of soldiers gathering in large groups, spraying with disinfectants, and mounted massive hygienic efforts to clean up the camps—this may have helped, although this was never proven. Eventually a Military Pneumonia Commission documented the outbreak in detail and confirmed that the two outbreaks, measles and bacterial pneumonia, occurred at the same time.

Current status

Prior to the 1960s in the United States, almost all people contracted measles, usually in childhood. By the late 1950s and up to 1962, the year before the introduction of measles vaccination, there were approximately 500,000 cases reported annually in the United States, with 48,000 people hospitalized. Of these, 4000 developed encephalitis, a life-threatening inflammation of the brain. Measles-related complications such as pneumonia and encephalitis resulted in about 500 deaths annually; the decrease in measles-related deaths compared to earlier in the century was presumably a result of improved medical care, antibiotic therapy of secondary bacterial infections, and better overall nutrition.

As described in Chapter 13, Dr. John Enders won the Nobel Prize for medicine in 1954 by showing how to grow large quantities of the polio virus in monkey kidney cells. In that same year at Boston Children's Hospital, Enders and Thomas Peebles switched to studying measles. By collecting blood and throat-washing samples from children with measles and growing the samples on cultured cells, Peebles was able to isolate the virus. One sample was taken from a 13-year-old student, David Edmonston. This cultured measles virus was named the Edmonston strain and became the basis for the development of the measles vaccine.

During the late 1950s and early 1960s, continuing research in the United States and a large trial that vaccinated over 700,000 children in the West African country of Upper Volta (present-day Burkina Faso) culminated in the development and licensure in 1963 of two measles vaccines: a killed virus vaccine and an attenuated virus vaccine. The killed virus vaccine, composed of the whole virus that had been inactivated (killed), was produced and administered between 1963 and 1967. Use of the killed vaccine was, however, soon discontinued because it produced a relatively short-lasting immunity. An attenuated virus was produced by repeatedly growing and transferring the Edmonston virus through different cell cultures in the laboratory. This prolonged serial passage in cell culture eventually resulted in changes in the genetic makeup of the virus. It was able to infect cells but could no longer cause disease in humans—it was effectively attenuated.

Additional efforts further attenuated the vaccine virus and, as a result, improved its safety profile. This improved vaccine was released in 1968. One dose is about 93% effective in preventing infection, whereas two doses are 97% effective. Beginning in 1971, measles vaccines were combined with two other childhood vaccines, mumps and rubella, to produce a vaccine now called MMR. Before the MMR vaccine became widely available, in the 1960s, measles affected about 3000 per million. As a result of widespread vaccination, this figure dropped in the 1980s to 13 cases per million and to 1 case per million in 2000. They fell from 3 to 4 million cases annually to several thousand in the 1980s and to just a few hundred annually since the mid-1990s.

Vaccination, herd immunity, and vaccination hesitancy

The nationwide immunization program for measles was adopted in 1963, and in 1966, the government began a program to eliminate measles from the United States (Fig. 9.2). This program was based upon the use of the attenuated virus vaccine for routine vaccination of infants, vaccination of unvaccinated children upon entry to public school, and disease surveillance and control of outbreaks. By 1968, only 22,200 cases of measles were reported, a drop of nearly 95% compared to prevaccination levels.

However, vaccination coverage was not universal and well below the calculated 90% to 95% level that was needed to protect and ultimately eradicate measles from the US population—the herd immunity effect. Thus measles continued to occur in the population, cycling between years of low measles incidence followed by years with large outbreaks.

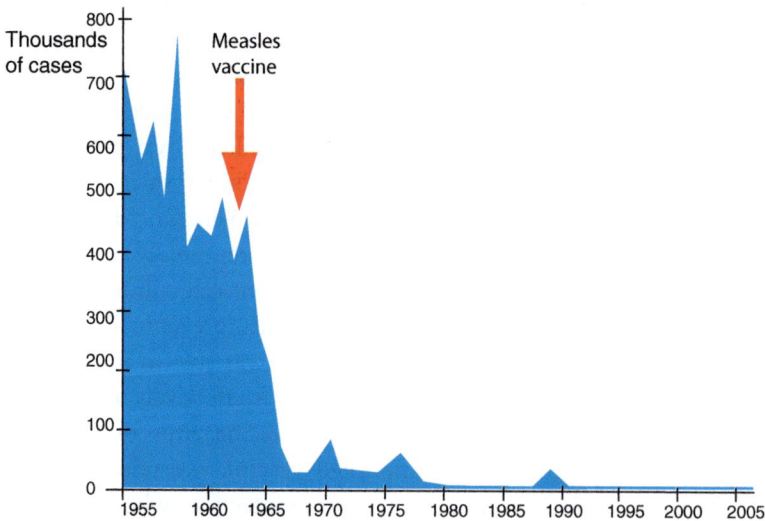

Fig. 9.2 *The number of cases of measles in the United States dropped sharply following the decision to attempt eradication in 1963. Unfortunately, vaccine hesitancy has prevented its complete eradication. Courtesy CDC.*

The effectiveness of the vaccine in protecting infants and school-age children was demonstrated during a disease outbreak in Texarkana in 1970–1971. The City of Texarkana straddles the Arkansas and Texas border, with the former requiring measles vaccine for school entry and the latter not requiring vaccination. Of the 633 cases of measles in the outbreak, 606 (96%) occurred on the Texas side of the border, whereas only 27 occurred on the Arkansas side. Texas communities, where no measles vaccination campaign had taken place, had an estimated 57% vaccine coverage, whereas in Arkansas there was a 95% vaccine coverage of children between 1 and 9 years old.

With the continued national immunization program, despite some outbreaks, measles was declared eradicated in the United States in 2000. This feat was accomplished through the use of an effective vaccine and employing a strategy that maximized population immunity through vaccination by the timely immunization of preschool children and re-vaccination of school-aged children.

However, the risk of viral reintroduction and spread continued to be a constant threat. Reintroduction could be caused by immigration from or travel to areas with continued high measles activity (which is most of the rest of the world). This threat is exacerbated as sections of the US population actively resist vaccination because of social, religious, psychological, philosophical, and technological issues; these are the vaccination-hesitant contingent. Resistance and fear of vaccination is nothing new, dating back to the very first vaccine, smallpox (see Chapter 8).

Today, the antivaccination movement relies on opinions and anecdotal evidence. A debunked and retracted publication in *The Lancet* in 1998 suggested that vaccination with the MMR vaccine was associated with the development of autism in children. Though the study was subsequently found to be based on flawed and manipulated methodology and data, and funded by organizations opposed to vaccine manufacturers, it resulted in a decline in vaccination rates and provided fodder for the antivaccination movement. Additionally, numerous perceived fears of vaccination are perpetuated by "expert" celebrities and politicians who provide wide exposure for their views on the internet. These people have proven influential in espousing the false notion that vaccines are unreasonably dangerous and thus persuading people not to vaccinate their children.

But the history and evidence of vaccination as an essential and beneficial tool for public health is overwhelming. Measles outbreak investigations, such as the aforementioned 1970 Texarkana outbreak, are excellent examples that demonstrate how unvaccinated people can serve as a means of introducing the virus and leave their communities vulnerable to viral spread.

The continued sporadic measles outbreaks that have occurred since the declared elimination of measles in the United States demonstrates the ongoing risks and consequences resulting from travel and the presence of an unvaccinated population. Some religious communities, due to issues pertaining to research, the presence of forbidden animal proteins or fats, purity of the body, inappropriate interference in God's work or plan, predestined fate, etc., have high rates of underimmunized children. This sets up an environment for disease resurgence. Two recent incidents in the past decade are illustrative.

During a 4-month period in the spring of 2014, 383 cases of measles were reported in Ohio. The source of the outbreak was traced to four people from the Amish community who had returned from a visit to the Philippines, a country that was in the midst of a measles outbreak at that time. The spread of measles was limited almost exclusively (99% of the cases) to the Amish community, a population that is vaccination hesitant. In another example, in the latter part of 2018, a nonvaccinated person with a history of recent travel to Israel was identified as the source of an outbreak in New York and New Jersey. During a 6-month period, 275 cases of measles were identified, with the majority of the cases being members of the Orthodox Jewish community, a group that also has low vaccination compliance.

By early 2020, the United States was in the midst of the worst measles outbreak since the "elimination" of the disease in 2000. During 2019, there were 1282 cases of measles in 31 states, with the highest concentration of cases occurring in New York and Washington. Of these cases, 128 patients were hospitalized, and measles-related complications such as pneumonia and encephalitis occurred in 61 cases. The majority of the cases were associated with New York Orthodox Jewish communities. Many other factors contributed to the disease's spread, but vaccination rates of school-aged children were paramount. About 90% of the cases occurred in people not vaccinated or with unknown vaccination history. Vaccination has been and continues to be the backbone of measles control and eradication.

Following this decline in vaccine usage there has been a massive rise in measles cases. In the United States there were 1200 cases reported in 2019. There were 121 cases in 2022 and 58 in 2023 (CDC). In Europe there was a 45-fold increase in measles cases from 2022 to 2023. Globally, the number of measles cases increased by 18% between 2021 and 2022, and deaths from measles increased by 43% (WHO). Following this drop in vaccine usage there has been a massive rise in measles cases. These cases were largely a result of importation by travelers infected in other countries. Unfortunately, they have the potential to spread widely in unvaccinated populations.

References

[1] Chuna BA. Smallpox and measles: historical aspects and clinical differentiation. Inf Dis Clin N Amer 2004;18(1):79–100.

[2] Caulfield E. Early measles outbreaks in America. Yale J Biol Med 1943;15(4):531–56.

[3] Black FL, Hierholzer W, Woodall JP, Pinhiero F. Intensified reactions to measles vaccine in unexposed populations of American Indians. J Inf Dis 1971;124:306–17.

[4] Bridges EL. Uttermost part of the earth. Indians of tierra del Fuego. New York: Dover Publications Inc.; 1949.

[5] Webster, Noah. *A brief history of epidemic and pestilential diseases; with the principal phenomena of the physical world, which precede and accompany them, and the observations deduced from the facts stated.* In Two Volumes. Hartford: Hudson and Goodwin; 1799.

[6] Clarke HF. John Hull Colonial merchant. Trans Collections Amer Antiq Soc. 1936;3: 197–218.

[7] Morens DM. The past is never dead—measles epidemic, Boston Massachusetts, 1713. Emerg Inf Dis 2015;21:1257–60.

[8] Schmitt RC, Nordyke EC. Death in Hawaii: the epidemics of 1848–1849. Hawaiian J History 2001;35:1–13.

[9] Boyd RT. Pacific Northwest measles epidemic of 1847–1848. Anthropology Faculty Publications and Presentations; 1994. p. 147. https://pdxscholar.library.pdx.edu/anth_fac/147.

[10] Norton MJ, Booss J. Missionaries, measles and manuscripts: revisiting the Whitman tragedy. J Med Lib Assn 2019;107. https://doi.org/10.5195/jmla.2019.538.

[11] Wolfe RJ. Alaska's great sickness, 1900: an epidemic of measles and influenza in a virgin soil population. Proc Am Philos Soc 1982;126:91–121.

[12] Morens DM, Taubenberger JK. A forgotten epidemic that changed medicine: measles in the US Army, 1917–1918. Lancet Infect Dis 2015;15(7):852–61.

Typhoid and typhus—1607

Contents

Prior to the identification of bacteria as the cause of infectious diseases, clinical diagnosis was a very inexact procedure. In practice, three major disease states were recognized: fevers, diarrhea, and rashes or pox. Therefore, when we read about these diseases of the premicrobiology era, we must acknowledge that the diagnosis of a specific infection before the late 19th century was, in many cases, no more than an educated guess. This is especially a problem when discussing two diseases with very similar clinical signs, such as typhus and typhoid. The name is derived from the Greek *typhos,* which refers to smoke or bad smells. Presumably this refers to the bad smells derived from putrefaction and diarrhea. For many years, they were often considered two variants of the same disease. Thus typhus was simply considered a more severe form of typhoid.

In 1837 an American physician, William Wood Gerhard, was working at the University of Pennsylvania Medical School. An outbreak of disease broke out in the poor, crowded neighborhoods on the southern edge of Philadelphia. The clinical disease closely resembled typhoid—it was a confident diagnosis. Gerhard was very familiar with typhoid. He had seen and

Great American Diseases
ISBN: 978-0-443-31404-9
https://doi.org/10.1016/B978-0-443-31404-9.00010-0

autopsied many typhoid cases while studying in Paris. He also knew that typhoid patients had extensive gastrointestinal lesions, especially ulcers. However, when Gerhard autopsied his Philadelphia patients, he found that their intestinal tracts were apparently normal; they showed none of the lesions associated with typhoid. He reported his observations in 1837 in the *American Journal of Health Sciences*, where he concluded that typhoid and typhus were two distinctly different diseases. Typhoid was a gastrointestinal disease, whereas typhus was muscular and neurological. Gerhard also concluded that although some patients caught the disease from environmental factors such as miasmas, others caught it by close contact with sick patients.

Typhoid and typhus were not reliably distinguished clinically until 1850 when an English physician, William Jenner, published *On the identity or non-identity of typhoid and typhus fevers*. We now know that typhoid fever is caused by the gram-negative bacterium *Salmonella* Typhi, whereas typhus is caused by a different bacterium, *Rickettsia prowazekii*. It was subsequently demonstrated that typhoid is generally acquired from drinking contaminated water (the fecal-oral route), whereas typhus is acquired from the feces of infected body lice. In the early 19th century, both diseases were common among the poor and starving. The word typhoid means typhus-like due to the similarity of its symptoms to that louse-borne disease. It is difficult to distinguish them clinically [1]. Some still call typhoid "abdominal typhus." Similar considerations apply to diarrheal diseases and their identification as cholera or dysentery. Now we know they are caused by different organisms; back then they were not so sure.

Typhoid fever
The bacterium

The gram-negative bacillus suspected of causing typhoid was first described in 1880 by Karl Eberth, who observed the bacteria in the lymph nodes and spleen of victims. This was confirmed 4 years later, and the organism was called *Eberthella typhosa*. It was subsequently renamed *S. Typhi* after Daniel Salmon, a US government scientist who isolated *Salmonella choleraesuis* from pigs and was the first to produce a vaccine against Salmonella. Today, genetic studies have demonstrated that the most important Salmonellae actually belong to a single species, *Salmonella enterica*. This organism is classified into six subspecies and over 2600 different "serotypes" based on their antigenicity. The organism that causes typhoid is now called

S. enterica, subspecies *enterica*, serotype Typhi. For simplicity, we will call it *S. Typhi*. A related bacterium, serotype Paratyphi, causes paratyphoid fever, a somewhat milder disease [2]. Both species are found only in humans in contrast to many other Salmonella serotypes that originate in animals and birds and are thus zoonotic. These infections are almost always acquired orally through contaminated food or drink. In the case of typhoid, as few as 10^5 to 10^7 organisms may cause disease.

The disease

Typhoid affects the gastrointestinal tract. It is spread by the fecal-oral route as well as by contaminated water, contaminated surfaces, and even on the legs of house flies. It was thus associated with poor sanitation and inadequate hygiene. It has a case fatality rate of 10% to 20% when untreated. Its incubation period is about 2 weeks. Typhoid has three clinical stages; each stage lasts roughly a week. In the first week, the patient gradually develops a fever with malaise, headache, abdominal tenderness and distention, and a cough as well as constipation. In the second week, the fever may reach 105°F and may remain high for 2 weeks as cytokines continue to act on the brain. The patient may become delirious or simply dull. Characteristic skin lesions called "rose-colored spots" may develop on the abdomen and chest. (These are small spots 2 cm in diameter caused by bacteria clumping in small blood vessels. They last 2 to 4 days but come and go.) Cramping and constipation occur. (Vomiting and diarrhea generally do not.) In the third week, complications may arise. For example, in some victims, the lymphoid tissues (Peyer's patches) in the intestine are destroyed, leaving a hole. This results in intestinal ulceration, bleeding, and possibly perforation that requires immediate surgery. (Intestinal perforation was one of the major ways that patients with typhoid died.) The bacteria can also invade the bloodstream and be carried throughout the body. As a result, they may be found in the patient's urine. Some patients develop heart disease, encephalitis, and delirium. Bronchitis or pneumonia can also complicate matters. The fever drops, and recovery begins by the fourth week. The infection may be confirmed by culturing the patient's blood or stools.

Typhoid does not always produce acute disease. Some individuals develop a mild, short-lived fever and apparently recover fully. Unfortunately, the organism is resistant to killing by bile, so it may colonize the gall bladder, where it may persist indefinitely. Large numbers of the bacteria can then pass down the bile duct to the small intestine, where they are shed

in the stools. About 2% to 3% of typhoid cases become healthy chronic carriers in this way and shed *S. Typhi* in their stools and urine. They can therefore spread the infection to other vulnerable individuals. Until a patient has
been shown to have cleared the infection, they must not be permitted to
prepare food. There is a blood test designed to detect antibodies to the organism in exposed humans, the Widal test, named after its discoverer,
Georges-Fernand Widal (1896). (A bacterial suspension is mixed with blood
serum and, if positive, the antibodies make the bacteria clump.) Thus the
Widal test helps diagnose the infection and also screen for potential typhoid
carriers.

When the sources of typhoid were clarified, hygienic measures and water
purification massively reduced its prevalence in developed countries. Nevertheless, there was a period from the middle of the 19th to the beginning of
the 20th century where clean water supplies were unable to keep up with
urban growth, and typhoid outbreaks were common. Typhoid has effectively been eliminated in Western countries (Fig. 10.1), although it persists
in many third-world countries.

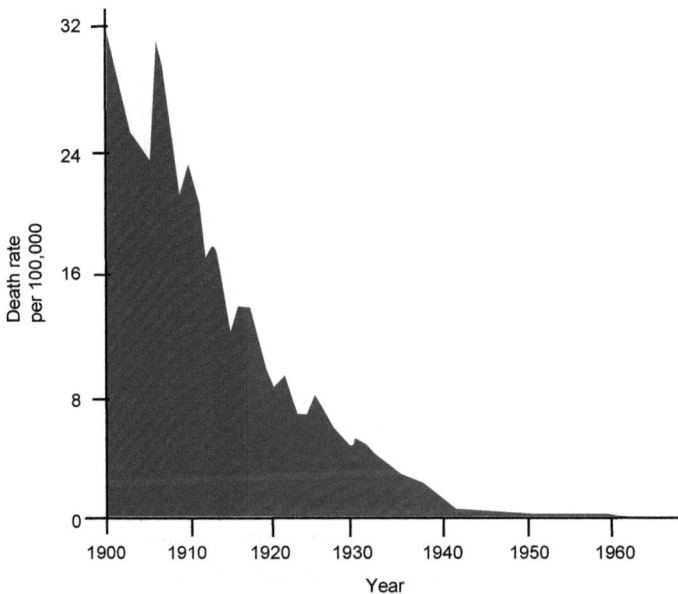

Fig. 10.1 *The progressive decline in deaths from typhoid fever since 1900 in the United
States. This primarily reflects a progressive improvement in the quality of the drinking
water supply. Data courtesy CDC.*

History

Typhoid is probably an ancient disease that often affected humans who lived beside rivers and drank contaminated water. The first known outbreak in the New World occurred with the foundation of Jamestown in May 1607. In order to protect themselves against hostile natives, the settlers selected a marshy site surrounded on three sides by the James River as the place to build their fort. Presumably the settlers believed that the flow of the river would flush their waste away and so used the river as a toilet. There was no point in digging a well. Unfortunately, it didn't work that way and their drinking water was contaminated [3]. The first settler died from the "bloody flix" within 6 weeks of their arrival. Of the first 104 settlers, half were dead within 1 year. It is estimated that in the first 17 years of the settlement, as many as 6000 settlers died as a result of enteric infections such as typhoid and dysentery. These diseases were only controlled when the Virginia Company relocated the settlement to a healthier site in 1624. Enteric diseases were a fact of life in both rural and urban societies in the 17th and 18th centuries and were rarely newsworthy.

President William Henry Harrison died in April 1841, 1 month after he became the ninth president of the United States. It was generally believed that he developed pneumonia after delivering his long-winded inaugural address on a cold, wet March day without a hat or coat. His physician reported that he had died of "pneumonia with congestion of the liver and derangement of the stomach and bowels." However, the written descriptions of his final illness are much more consistent with typhoid. Thus he was severely constipated, but after 4 days, developed diarrhea. He suffered severe stomach pains. He had an unproductive cough. His stomach became distended and he developed systemic illness before he died. His weak pulse and cold extremities suggest septic shock (Chapter 2). All these signs are consistent with typhoid or paratyphoid leading to a perforated small intestine.

Although Washington had many splendid buildings, little attention had been paid to providing clean water. Water was simply drawn from the Potomac into which sewage also flowed. The primitive sewage system was totally inadequate, so it is entirely plausible that typhoid could have killed Harrison. As late as 1861, Stephen A. Douglas, the Democratic candidate who opposed Lincoln in the 1860 presidential election, also died from typhoid fever.

Grosse Île

The typhus/typhoid epidemic that struck in 1847 killed upward of 20,000 people in Canada, mainly Irish refugees at the Grosse Île Quarantine Station. This facility was originally established on an island in the St. Lawrence River by the Government of Quebec in anticipation of the cholera epidemic of 1832 [4]. The quarantine station was designed to identify sick passengers, isolate them, and house them until they either recovered or died. It was also designed to handle the passengers from one arriving ship at a time. In theory, the next ship had to wait at anchor until all the passengers from previous ships had been released. However, the Grosse Île station became the primary port of entry for impoverished Irish emigrants into North America. This flow increased dramatically with the advent of the Irish potato famine in 1847. Hundreds of thousands of sick and starving Irish made their way across the Atlantic in grossly overcrowded and squalid "coffin" ships. Thus, the first such vessel, the *Syria*, arrived at Grosse Île on May 17, 1847. It carried 430 febrile immigrants. A few days later 8 more ships arrived, and another 17 came the next week. The island hospital had only about 300 beds, but the ships kept arriving. By May 29, 36 ships had arrived and were lined up along the St. Lawrence River for 2 miles. Every vessel was loaded with fever and dysentery victims. Any sick passengers who managed to make it to the shore were laid out in tents, huts, and the chapel. Some new arrivals were simply laid out in the open. They crawled ashore from the beach and died by the hundreds. Healthy and sick were mixed indiscriminately. The island was totally overwhelmed.

Because each ship had to take its turn, some passengers had to wait on board for many days. Under the quarantine rules, healthy passengers were required to remain on their ship for 15 days after the sick had been removed. Each new case that occurred extended that period. As a result, the disease continued to sicken and kill those remaining on the anchored ships. Eventually these quarantine regulations had to be abandoned because they could not be enforced. Doctors simply let passengers line up and walk past them. Only the obviously sick were stopped. The others went on their way and thus carried the disease to the ports upstream. In Montreal, between 3500 and 6000 Irish emigrants died in the fever sheds. Some reached Kingston, where 1400 died, and even as far as Toronto, where 863 died. Some even managed to make it to New York, where 147 cases were treated and about 15 died (Fig. 10.2). Most were in newly arrived Irish immigrants, but about a fifth resulted from secondary spread.

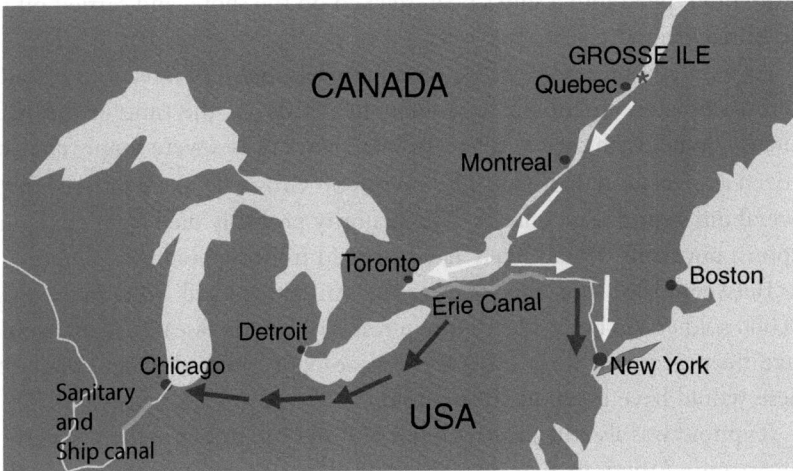

Fig. 10.2 *The spread of disease from the Grosse Île quarantine station in Quebec, Canada.* It was first overwhelmed with cholera victims in 1832. That disease was subsequently carried to major urban centers such as New York City by way of the newly constructed Erie Canal *(yellow arrows, light gray in print version)*. The station was overwhelmed for a second time by typhoid victims arriving as a result of the Irish potato famine in 1847. In this case, the disease primarily traveled by railroad *(red arrows, dark gray in print version)*.

Meanwhile, back on Grosse Île disaster reigned. Patients lay on bunks and feces dripped down from the upper ones onto those beneath. Two or three patients were placed in one bed irrespective of gender. There was no bread, just fluids, tea, broth, and gruel. Even water was limited. The sheds lacked ventilation and privies. The toilets were situated along the shoreline so that the feces dropped into the sea only at high tide. There were not even enough staff to remove the dead bodies. While medicine was primitive, there were insufficient doctors and nurses to care for the sick and dying. The doctors and nurses frequently contracted the diseases themselves. At least four doctors died of typhus. Priests were available to comfort the dying, but they too died in significant numbers. Case fatality rates reached as high as 20% to 45%. Nearly 90,000 people passed through the quarantine station in 1847. In one mass grave on Grosse Île, there are over 5424 dead, primarily from Ireland, and another 4000 were buried elsewhere on the island.

Although Grosse Île was the destination, many victims never made it that far and were simply buried at sea. It is estimated that over 5000 died en route. Once they arrived in St. Lawrence, the newly dead were stored in the ship's hold until burial could be arranged. Then they were pulled out of the hold, piled into a ship's boat, stacked on the shore, and carried off to the burial grounds.

As pointed out in Chapter 1, the microbial origins of infectious diseases were completely unknown until after the 1850s. At the time of the Irish potato famine, with the exception of smallpox, diseases were simply characterized as fevers and diarrheas. The victims on Grosse Île likely suffered from several different disease entities. The majority probably died from epidemic typhus, but many would likely have resulted from typhoid.

Between 1861 and 1865 during the American Civil War, more than 80,000 soldiers on both sides died as a result of enteric diseases. Most would have been acquired from drinking sewage-contaminated water. Many of these would have been due to typhoid.

Typhoid was also the primary killer of American troops in the first part of the Spanish-American War until superseded by yellow fever. By the 1890s, the concept of bacterial disease had been broadly accepted, as was the concept of filth diseases. Diagnosis was also much more precise. The high prevalence and mortality of this typhoid epidemic caused the War Department to establish a Typhoid Board in 1898 under the chairmanship of Major Walter Reed, the Army's expert on typhoid fever. The board members immediately began to visit the newly established Army training camps, identify unsanitary conditions, and issue recommendations regarding improving cleanliness. Initially their recommendations were ignored by both officers and men, and as a result, 90% of the regiments suffered losses from typhoid. Eventually, out of 170,000 troops, 20,700 contracted typhoid and over 1500 died. One thing the commission noted was that camps that had been thoroughly cleaned would often suffer recurrences of typhoid. They correctly drew the conclusion that some soldiers must be carriers of the bacterium. In 1896, Richard Pfeiffer in Germany and Almroth Wright in England produced the first effective vaccines against typhoid. (Pfeiffer published his results about 3 months before Wright, but Wright has tended to get the credit!) These vaccines were subsequently used in troops in both the Boer War and the World War I and, although crude by modern standards, largely eliminated the disease. Typhoid vaccines have been significantly improved since then.

Water quality

It was not until 1854 that the link between water quality and outbreaks of enteric disease was established. Prior to that time, almost no effort was made to separate human waste from drinking water. Air, not water, was considered the source of infection. Water was simply pumped from nearby rivers or lakes and, if these were not available, from shallow, hand-dug wells. Water was often sold by traveling water sellers carrying lake or river water in horse-drawn barrels. Sewage was disposed of in the most convenient manner, into rivers and lakes if available, and otherwise into nearby cesspools. If they overflowed or seeped into other sites, it was no big deal. As in the early colonial settlements in Jamestown, both water supplies and sewage disposal were simply placed most conveniently as near to homes as possible. The only apparent downside was that drinking water sometimes smelled bad.

As the cities of the United States began to grow, such ad hoc arrangements no longer sufficed. For example, Manhattan, although surrounded by brackish water, had to rely on shallow wells for drinking. As a result, by the late 1820s, New York City began construction of the Croton Aqueduct system to carry clean drinking water from inland lakes to the city. This, no doubt, helped considerably in reducing the impact of the 1832 cholera outbreak. (Cholera, discussed in Chapter 12, is another waterborne bacterial disease that causes lethal diarrhea.) Other cities also began to construct mechanisms to provide their inhabitants with clean, potable water. The identification of polluted water as the source of typhoid and cholera provided a further and very significant incentive for the development of public sanitation systems [5].

Following the Civil War, American cities expanded rapidly while drinking water and sewage systems struggled to keep up with their growth. For example, in 1885 a typhoid epidemic affected Plymouth, Pennsylvania. The problem was traced to the town's water supply. The source of the disease was eventually traced to a farmer who lived beside Coal Creek, a mountain stream that served as the source of the town's water. At Christmas 1884 the farmer had visited his brother in Philadelphia and returned home with typhoid. When he was sick and the ground around his house was covered in snow, it became highly contaminated with his stools. In March, the thaw and rain came, and the snow melted. The contaminated snow ended up in the creek and the creek flowed into the town's reservoir. From here it was drawn into the water supply. By the time the outbreak ended in September 1885, there were more than 1100 cases and 114 deaths recorded out of a population of 8000 (probably a low estimate).

On a much larger scale, typhoid was common in Chicago from 1850 on [6]. Major outbreaks occurred in 1852–1854, 1864–1866, 1881, as well as in 1890–1892. The 1891 outbreak of typhoid in Chicago killed 1997 individuals resulting in a death rate of 174 per 100,000 people. (This death rate was seven times that in New York!) Chicago obtained its drinking water from Lake Michigan while at the same time discharged all its raw sewage and refuse either directly into the lake or into the Chicago River that flowed into the lake. (They had clearly failed to see any relationship between sewage and enteric disease.) Heavy rains regularly flushed the river's contents out into the lake. The next year, 1892, the city of Chicago began the process of reversing the flow of the Chicago River by building the Sanitary and Ship Canal system designed to separate urban waste from Lake Michigan [7]. The canal connected the Chicago River to the Des Plaines River through the hills west of the city. As a result, the city's sewage flowed into the Illinois River and eventually into the Mississippi rather than into the lake (Fig. 10.2). The result was a 31-mile-long canal that eventually cost $70 million. It was completed in 1900. As a consequence, the prevalence of typhoid in Chicago dropped significantly. It is not clear just what the inhabitants living downstream thought of the project, but it can be imagined.

Similar outbreaks occurred in other towns, such as Ithaca, New York, as they sought to clean up and improve their supply of drinking water. In the case of Ithaca, in 1903, typhoid killed at least 82 individuals, of whom 29 were students at Cornell University.

In 1905, in response to threats of a typhoid epidemic, Washington, DC belatedly constructed a massive water filtration system. The water was filtered through concrete cylinders containing layers of sand and gravel before entering the city's water mains. The remains of these cylinders can still be seen today along Michigan Avenue.

In 1906 typhoid broke out in Philadelphia. It resulted in 9712 cases and killed 63 people. In May 1912 Wilbur Wright (the airplane guy) died in Dayton, Ohio, as a result of typhoid acquired from eating contaminated oysters. In 1924–1925 there was another typhoid outbreak in Chicago with 129 cases, the source of which was shown to be contaminated oysters originating in Long Island.

In February 1927, an outbreak of typhoid occurred in Montreal with 5353 cases and 538 deaths over 3 months [8]. The source was found to be milk produced by the Montreal Dairy Company. Much of their milk output appeared to be improperly pasteurized, and water used to wash equipment was pumped directly from an obviously polluted river. (It should be pointed out that the dairy cows were not infected with, nor the source of, the *S. Typhi*. *S. Typhi* and *S. Paratyphi* are found only in humans.)

The first attempts to sterilize drinking water by chlorination were made in Hamburg, Germany, in 1893. By 1897, the town of Maidstone in England was able to chlorinate its entire water supply. The first continuous use of chlorination in the United States occurred in 1908 when water from the Boonton Reservoir supplying Jersey City, New Jersey, was routinely chlorinated. Low concentrations of chlorine effectively kill bacteria. Dilute solutions of chloride of lime (calcium hypochlorite) were added to the drinking water at a level of 0.2 to 0.35 ppm. Chlorine is a potent oxidant that can damage bacterial proteins and lipids and so kill them; chlorination can also kill viruses and some waterborne parasites. Over the next few years, thousands of towns and cities across the country followed. In 1900, there were about 100 cases (and 32 deaths) of typhoid per 100,000 people in the United States. As a result of the widespread adoption of chlorination, outbreaks of typhoid and cholera dropped precipitously. By 1920 the prevalence of typhoid had dropped to 33.8 cases per 100,000, and by 2006 it had dropped to 0.1 cases per 100,000 (Fig. 10.1). (That was 353 cases, of which three-quarters were introduced by international travelers.)

Typhoid Mary

In 1883 a young woman called Mary Mallon migrated from Cookstown, County Tyrone, Ireland, to the United States. By 1906 she was employed by the wealthy Warren family in Long Island as a cook. She was evidently a good cook. One of her specialties was fresh peaches with ice cream. (A dish in which S. Typhi would have survived.) Not long after she began working for them, six members of the Warren family developed typhoid fever when renting a house for the summer at Oyster Bay on Long Island. The outbreak was investigated by a sanitary engineer named George Soper at the request of the house owner (he wanted to rent the house again). Soper interviewed all the family members, excluded other causes such as contaminated clams, but found that they had hired a new cook, Mary Mallon, 2 weeks prior to the outbreak [9]. Soper looked further into Mary's past. He interviewed her former employers and found out that following her 8 previous jobs, there had been 22 cases of typhoid, some of whom fell sick and some had died. Soper managed to obtain a court order to obtain fecal samples from Mary and had them examined. They contained huge numbers of S. Typhi. When he told Mary that she was the source of typhoid she attacked him with a carving fork!

Mary Mallon was an asymptomatic carrier of S. Typhi. She showed no clinical signs, but by shedding the bacterium and not rigorously washing her hands after going to the toilet, she effectively contaminated food. As a result, the New York City Health Department isolated Mary in a cottage on the grounds of Riverside Hospital on North Brother Island in the middle of the East River from 1907 until 1910. Her stools were tested repeatedly, and it was shown that she was an intermittent carrier with 120/163 samples testing positive for *S. Typhi*. It seems that nobody tried to explain the situation to her; she was never sick, and she never really understood why she was being confined. The authorities offered to remove her gall bladder (where the infection persisted), but she declined. She was, however, released in 1910 after giving a pledge that she would never work in any job that required the handling of food. Her confinement was seen by some as evidence of anti-Irish prejudice, and there was significant public pressure, especially from William Randolph Hearst's newspapers, for her release. Once released, she first worked as a laundry maid but, not enjoying the job, broke her promise and went to work as a hospital cook at the Sloane Maternity Hospital in Manhattan in 1915 under the name of Mary Brown. In the subsequent typhoid outbreak, 24 people developed the disease and at least 2 died. Investigation showed that the cook was Mary Mallon working under an assumed name. As a result, "Typhoid Mary" was confined for the last 23 years of her life on North Brother Island and died in 1938. She continued to shed *S. Typhi* until she died. Mary Mallon was not unique. It is estimated that about 3% of those who recover from typhoid can become chronic carriers. However, no other carriers have been forcibly confined in this way since Mary Mallon.

Current status

There are about 22 million cases of typhoid annually worldwide that result in at least 200,000 deaths. These primarily occur in developing countries. Fewer than 500 cases occur annually in the United States, with the vast majority acquired while traveling outside the country. A vaccine is available for those at high risk. Antibiotics such as ciprofloxacin have proven effective against typhoid, but multidrug resistance is a growing problem, especially in the Indian subcontinent and Southeast Asia.

Epidemic typhus

The history of typhus .... Is the history of human misery.
—**August Hirsch, Physician and Historian**

Between 1845 and 1849 more than a million people died in Ireland as a result of starvation and disease originating from the massive failure of the potato crop. The three great killers were fevers, dysentery, and smallpox. There were, of course, other infectious diseases, such as tuberculosis, measles, and pneumonia, that also contributed to this massive loss of life.

As pointed out in Chapter 2, fever reflects the innate immune response to microbial invasion. The elevated body temperature is accompanied by chills as the brain is misinformed about its temperature. Aches and pains are attributed to a reallocation of resources to the immune system as well as lethargy and lack of appetite. Physicians at the time did not know what caused these fevers, and some maintained that it was a direct result of famine and starvation. It was considered by many to be a consequence of crowding, dirt, malnutrition, and drunkenness as well as an inappropriate lifestyle and, as a result, punishment from the Almighty for manifold sins. With hindsight, epidemic typhus was a significant cause of these fevers and the resulting mass mortality among the immigrants. Epidemic typhus still occurs today in countries afflicted by war, malnutrition, and poverty.

The bacterium

Epidemic typhus is caused by a small, gram-negative, intracellular bacterium, R. prowazekii [10]. As an intracellular parasite, this organism does not need some of the metabolic pathways that extracellular bacteria need. As a result, it has a very much reduced genome containing only about 800 genes. R. prowazekii grows in the intestinal wall of the human body louse, Pediculus humanus. Body lice are blood-sucking parasites that live and multiply in clothing. They feed several times daily and in doing so cause severe itching. The victim naturally scratches these bites and so rubs louse fecal material into the wound. R. prowazekii is excreted in louse feces; it is not transmitted by the bite of the louse. Contaminated dust containing dried louse feces may also be inhaled or enter through the eyes. Infected lice die, but some survive to maintain disease transmission. The bacterium can survive in dry louse feces for many months. Disease develops after an incubation period of 10 to 14 days. Because lice thrive under crowded, unhygienic conditions, times of war and famine generally trigger outbreaks. Jails, ships, and hospitals were sites where it thrived. Epidemic typhus was one of the major disease killers in the German concentration camps in World War II.

The disease

Once it enters the body, *R. prowazekii* penetrates the cells lining blood vessels, where it grows until the cells rupture. It then invades other nearby cells. After an incubation period of about 14 days, symptoms develop. These symptoms include not only a prolonged high fever with chills, but also cough, extreme muscle pain and headaches, low blood pressure, delirium, coma, and death. A marked feature of the disease is the rash that develops on the chest around 5 days and eventually spreads from the trunk to the face and limbs. In fatal cases, death usually results in about 14 days. Mortality in untreated cases is under 5% in children but may reach 60% in middle-aged or older adults. The infection may persist as a subclinical infection for life and recur under stressful situations. When this happens, louse-infested victims can continue to transmit the bacteria to their lice.

History

Typhus may have been first recognized in Spain during the 11th century [11]. It was associated with many of the European wars in France and Italy in the 16th and 17th centuries. As of today, it persists in impoverished, starving, and destitute populations. However, the nonspecific symptoms of epidemic typhus are such that it is difficult to confirm with any degree of confidence that historical outbreaks of disease diagnosed as typhus were in fact the same disease. Many other infections display similar symptoms.

It is unclear just when typhus first appeared in the Americas. It was, however, well recognized as a disease of prisoners held in crowded and unsanitary jails where louse infestation was common. When English felons were shipped to the Americas it was common for "ship fever" to break out and cause significant mortality among both prisoners and sailors alike.

In 1746, the French made one last attempt to retake Canada from the British. They dispatched a large fleet under the Duc d'Anville to capture Nova Scotia and burn Halifax. It was a long and stormy voyage, and typhus broke out on board the French vessels. They arrived weakened and sick and were forced to return to France. When they departed, the French left behind a quantity of used blankets and discarded old clothes that were promptly scavenged and used by the local tribe, the Mi'kmaq. It is believed that either surviving lice or dried louse feces then triggered a massive epidemic of typhus that wiped out more than a third of the Mi'kmaq nation.

Ship fever was probably imported into many East Coast ports throughout the 18th century although it does not appear to have spread far. However, there was a general awareness that lice infestations were best avoided, and newly recruited sailors were often stripped, shaved, and washed before being issued with fresh clothes—it appeared to work and reduced the prevalence of the disease.

Typhus (or typhoid) may have also played an important role in the Revolutionary War. Thus in 1776 the disease broke out among the American troops defending New York. Among its victims was General Nathanael Greene. Greene was a first-rate general. Unfortunately, while he was sick with the fever, the Battle of Brooklyn Heights was fought and lost by troops under George Washington. Had Greene been healthy, it is speculated that the outcome of the battle may have been different. He recovered within a few weeks, but the American forces were obliged to evacuate New York.

The thousands of destitute Irish immigrants who fled the potato famine in 1847−1848 brought typhus with them, although the disease did not spread far from these groups. For example, in New York in May and June 1847 there were 138 cases of typhus of which 131 were in recent immigrants [12]. The seven Americans who contracted the disease had all been in close contact with Irish immigrants. In 1852, the Marine Hospital on Staten Island housed 3040 cases of typhus and 17% died. All were Irish immigrants. Some Irish immigrants even arrived via New Orleans. Many had to be hospitalized and died. However, the deaths were restricted to immigrants, doctors, and nurses. Subsequent waves of Russian emigrants also brought typhus with them in 1892. They were promptly quarantined on North Brother Island in New York Harbor. The disease got ashore but was restricted to eight boarding houses on the Lower East Side (Box 10.1).

Typhus was also an urban disease associated with squalid, crowded urban housing, filthy, unwashed clothes, and shared beds. It tended to be a winter disease, when victims crowded indoors and put on every piece of clothing they possessed in order to remain warm [13]. As described earlier, a major typhus epidemic was associated with the Irish potato famine in 1846−1849. It spread to England, where it was known as the "Irish Fever."

Despite its obvious infectivity and its introduction into North America during the Irish Potato Famine, epidemic typhus never established itself as a major disease in the United States (Fig. 10.3). For example, during the Civil War when most soldiers were heavily infested by lice, typhus was rare [14].

In 1917, a typhus epidemic in Mexico caused the US government to place a typhus quarantine along the US–Mexico border. Immigrants had

BOX 10.1 In defense of the miasma theory.

Microorganisms were highly unlikely to be considered the cause of disease until they were seen under the first microscopes. Once Antony van Leeuwenhoek reported their existence, however, thoughtful persons such as Cotton Mather and Ben Franklin suggested that these "animalcules" could cause disease but had no evidence to support this. Those who thought about such things recognized that diseases could spread between individuals in the absence of physical contact. (A notable exception to this was the sexually transmissible disease syphilis.) The obvious explanation was that the disease was caused by some form of poison. This could have been acquired from food or water, but it was considered equally likely that it could be airborne. In an environment much smellier than the present, it made sense to attribute disease to some form of airborne poison. Smells were often much more obvious and pervasive in homes and areas occupied by the poor, sick, and dying. The dead obviously smelled bad. Moving away from such smells whenever possible made sense, as did the custom of rapid, deep burial. As a result, the "miasmatic" theory of disease was widely accepted. It held that most, if not all, diseases were the result of inhaling air that had been polluted by rotting vegetable matter.

From 1858 to 1859, the River Thames in London smelled abominably as a result of massive fecal pollution. In consequence, the "Great Stink" was considered to be the cause of a typhoid outbreak in London at that time [6]. (Retrospective analysis showed, however, that sickness and death rates at that time were below average, probably because people avoided drinking the smelly river water.)

In a variation on this theme, it was suggested by Sir Francis Head, Lieutenant Governor of Upper Canada (modern Ontario) in 1835, that "some settlements in the Americas had been rendered dangerous by the ploughing of virgin soil, which had exposed decaying vegetable matter and the 'miasms' that arose from it."

to be inspected and bathed before entering the United States. Sanitation stations remained in operation for the next 40 years, long after the initial typhus epidemic had ended. It contributed significantly to antiimmigration sentiment (Box 10.2).

Current status

It was not until 1909 that typhus was shown to be transmitted by body lice. Charles Nicolle in France determined that typhus patients were no

Fig. 10.3 *A mobile typhus testing station in New York.* Courtesy US National Library of Medicine.

longer contagious after they had bathed and changed their clothes. *R. prowazekii* was only shown to be the cause of epidemic typhus in 1916. The bacterium was named after two bacteriologists who had contracted typhus and died, Howard Ricketts and Stanislaus von Prowazek. A vaccine against typhus is now available, and the disease can be treated with antibiotics such as tetracycline or chloramphenicol. In such cases, it is also essential to get rid of all the lice in the patient's clothes, bedsheets, and hair by the aggressive use of insecticides, hot water washing, and heated drying.

In 1975, *R. prowazekii* was isolated from flying squirrels living east of the Mississippi. Between then and 2001 there have been 39 human cases of epidemic typhus reported in the United States. All were in close contact with northern flying squirrels (*Glaucomys volans*) or their ectoparasites [15]. One case has even been associated with sand fly bites on South Padre Island, Texas! This squirrel disease appears to be milder than classic louse-borne epidemic typhus. The patients presented with an abrupt onset of headaches, fever, muscle pain, and skin rashes.

BOX 10.2 The El Paso bath house riots.

In contrast to its absence from the United States, epidemic typhus was not uncommon in Mexico. Between 1915 and 1917, typhus broke out in Mexico City
and spread to several other Mexican states. The mayor of El Paso, Texas, Thomas
Lea, demanded that a quarantine be put in place to prevent an influx of "dirty
lousy destitute Mexicans." A disinfecting station was therefore set up at the
foot of the Santa Fe bridge in El Paso. Mexicans crossing into the city from Juarez
were sent to separate buildings based on gender, undressed, their clothing and
valuables steamed, and the naked individuals examined by inspectors for nits
and lice. If lice were found, the men's hair was clipped. Women's hair was washed
in a mixture of vinegar and kerosene. Once they passed the test, everyone was
sprayed with a highly inflammatory mixture of soap and kerosene. Once they
dressed, the Mexicans were vaccinated and given a card certifying them as
lice free. Only then could they enter the immigration building for processing.
Sometimes, it was claimed, that the naked women were photographed and
the photographs circulated illicitly in local bars.

On the morning of January 28, 1917, inspectors attempted to remove several
Mexican women for bathing and fumigation from the cross-border trolley on
their way to work. One woman, Carmelita Torres, had heard about the photographs as well as the fire risk due to the kerosene. She refused to leave the trolley
and undergo the humiliating process, and she encouraged the other women to
protest as well. They began to hurl abuse, insults, and eventually bottles and
stones at the health and immigration officials. A crowd gathered and thousands
of people joined the process. Several trolley operators were injured. The crowd
was dispersed that afternoon by mounted soldiers. The next day the rioting
resumed, mainly by men this time. Again, the rioters were dispersed by Mexican
cavalry. Some compromises were made by the US authorities after negotiations
and complaints from employers employing immigrant labor. For example, certif-
icates were valid for 1 week, and the United States agreed to accept Mexican cer-
tificates. But the bathing and fumigations continued for another 40 years until
the 1950s. It did nothing to help cross-border relations.

Endemic typhus

There are two different diseases called typhus, endemic typhus and epidemic
typhus. Epidemic typhus was the great killer. Endemic typhus, also termed
murine or flea-borne typhus, is transmitted by the rat flea. It is caused by a
different bacterium, *Rickettsia typhi*, and is found worldwide. This disease
was almost eradicated from the United States in the 1940s following the
introduction of the potent insecticide DDT. By 1958 there were less than

100 cases reported annually. However, since then, the number of reported cases of murine typhus has progressively increased in California, Texas, and Hawaii. Between 2008 and 2018, 3507 cases of flea-borne typhus have been diagnosed in Texas, predominantly in counties bordering Mexico along the lower Rio Grande River and in counties surrounding the major metropolitan areas of Austin, Houston, and San Antonio. The infection appears to be maintained by a life cycle involving opossums and cat fleas. (Despite their name these fleas are not restricted to cats!)

References

[1] Cunha BA. Osler on typhoid fever: differentiating typhoid from typhus and malaria. Infect Dis Clin North Am 2004;18(1):111–25. https://doi.org/10.1016/S0891-5520(03)00094-1.

[2] Paterson JCS. Typhoid and the paratyphoids: a review. Postgrad Med J 1949;25(287): 413–20. https://doi.org/10.1136/pgmj.25.287.413.

[3] Earle Carville. Environment, disease and mortality in early Virginia. J Hist Geogr 1979; 5(4):365–90. https://doi.org/10.1016/0305-7488(79)90224-x.

[4] Forster MM. Quarantine at Grosse Ile. Can Fam Physician 1995;41:841–8.

[5] Head F. Report on the sanitary conditions of the labouring population of Great Britain. Q Rev 1842;71:1842.

[6] Jordan EO. Typhoid fever and water supply in Chicago. JAMA 1902;XXXIX(25): 1561–6. https://doi.org/10.1001/jama.1902.52480510001001.

[7] Spalding Heman, Bundesen Herman N. Control of typhoid fever in Chicago. Am J Public Health 1918;8(5):358–62. https://doi.org/10.2105/AJPH.8.5.358.

[8] Usher SJ. A study of one hundred and seventy five cases of typhoid fever in children, occurring during the recent epidemic in Montreal. Can Med Assoc J. 1927;17: 1486–92.

[9] Marineli F, Tsoucalas G, Karamanou M, Androutsos G. Mary Mallon (1869–1938) and the history of typhoid fever. Ann Gastroenterol 2013;26(2):132–4.

[10] Woodward TE. A historical account of the rickettsial diseases with a discussion of unsolved problems. J Infect Dis 1973;127(5):583–94. https://doi.org/10.1093/infdis/127.5.583.

[11] Zinsser H. Rats lice and history. Black Dog & Leventhal Publishers, New York, NY; 1996.

[12] Hardy A. Urban famine or urban crisis? Typhus in the Victorian City. Med Hist 1988; 32(4):401–25. https://doi.org/10.1017/s0025727300048523.

[13] Raoult D, Woodward T, Dumler JS. The history of epidemic typhus. Infect Dis Clin North Am 2004;18(1):127–40. https://doi.org/10.1016/s0891-5520(03)00093-x.

[14] Humphreys M. A Stranger to our camps: typhus in American history. Bull Hist Med 2006;80(2):269–90. https://doi.org/10.1353/bhm.2006.0058.

[15] Duma RJ, McGill TM, Sonenshine DE, et al. Epidemic typhus in the United States associated with flying squirrels. JAMA 1981;245(22):2318–23. https://doi.org/10.1001/jama.1981.03310470032022.

Yellow fever—1647

Contents

Yellow fever has its origin in the rainforests of West Africa, where the virus circulates among monkeys high in the treetops. It is transmitted between these monkeys by female mosquitos. Historically, the disease was largely restricted to people that lived in or ventured into the forests. As the slave trade developed, however, it was inevitable that some kidnapped individuals would have been infected with the yellow fever virus. Slave ships also required plenty of drinking water, so they filled their water barrels at any convenient lake or river, where they would also have picked up mosquito larvae at the same time. It is also likely that infected mosquitos could have traveled across the Atlantic in the dark spaces between decks. On arrival in the New World about 1640, these mosquitos rapidly established themselves in the forests of Central and South America, where the climate was ideal and there were plenty of monkeys to feed on. The virus also traveled with the mosquitos to American cities where the dense human populations enabled the mosquitos to find plenty of susceptible victims. As a result, yellow fever became a major killer from the United States to Brazil.

Great American Diseases
ISBN: 978-0-443-31404-9
https://doi.org/10.1016/B978-0-443-31404-9.00011-2

Yellow fever is caused by an enveloped RNA virus belonging to the genus *Flavivirus*. (This genus also includes Dengue virus and West Nile virus.) It has a positive-sense RNA genome of about 12,000 base pairs. It is transmitted by the bite of the day-flying *Aedes aegypti* mosquito. The virus is taken up in the female mosquito's blood meal. Once inside the mosquito, it invades the epithelial cells lining the mosquito intestine, passes through its hemocoel, and then to its salivary glands. This takes about 12 to 16 days, depending on temperature. The next time the mosquito feeds, it injects the infected saliva into the victim's skin.

Two major cycles are important for yellow fever transmission in the Americas. In nature, it causes a mosquito-borne infection of monkeys and thus cycles between monkeys in the jungles. This is called sylvatic (or jungle) yellow fever and is only encountered by those individuals working deep in tropical forests. The second type of transmission occurs when infected mosquitos enter cities where there are lots of susceptible individuals. This urban yellow fever does not involve monkeys because there are plenty of available humans and its R_0 is high.

Yellow fever

Should an infected female mosquito bite a human, the virus will grow in the victim's lymph nodes before spreading to the liver and causing acute hepatitis and hemorrhagic fever. The victim may develop jaundice, hence the term yellow fever, and bleed into the stomach, hence the Spanish term *vomito negro*. The incubation period of the disease is 3 to 6 days. The patient first develops an "acute phase" characterized by the nonspecific signs of sickness including fever, muscle pains, especially a backache or headache, nausea, and vomiting. After 3 to 4 days, most patients recover. However, about 15% of yellow fever victims, after a brief period of lucidity, the second phase, then enter a third "toxic phase." The patient becomes jaundiced and develops a yellow skin color accompanied by abdominal pain and vomiting. Bleeding from the mouth, nose, eyes, and stomach results in clotted blood appearing in their vomit and feces. Their pulse slows down. Kidney failure follows, perhaps leading to complete loss of kidney function. Half of these patients die of delirium, shock, and multiple organ failure within 10 to 14 days. Survivors remain seriously ill for several weeks, and recovery is slow. Even today there is no specific treatment or known "cure" for yellow fever. Disease management is symptomatic, and intensive nursing care is often necessary. However, there is a very effective vaccine available.

History

The first confirmed yellow fever epidemic in the New World (as opposed to single cases in slaves) likely occurred in 1647 on the island of Barbados. Noah Webster, who cataloged these outbreaks, reported, "There dies in Barbados and St. Kitts, 5 or 6,000 each" [1]. As a result, ships from the West Indies were not permitted to land in Boston without undergoing inspection. The disease was called the "Barbados distemper" by John Winthrop, the governor of Massachusetts. Winthrop required ships with infected passengers and crew to fly a yellow flag and undergo inspection before docking, hence the term "Yellow jack." The population in Barbados was sufficiently dense (about 200 persons per square mile) to sustain the epidemic and many were unacclimated refugees from the English Civil War. The Barbados outbreak probably killed between 30% and 50% of the population. Another outbreak occurred in Guadeloupe in 1648. From Barbados and Guadeloupe, the virus and the mosquito spread around the Caribbean. They reached Campeche and Merida in the Yucatan, as well as Cuba, Florida, and northern South America within a year. The Maya called the disease *xekik,* meaning "blood vomit." Yellow fever reached St. Augustine, Florida, in 1649, probably carried in water barrels from Havana. It killed the Spanish governor, several royal officials, numerous Franciscan friars, as well as many of the local natives.

The first northern outbreaks of yellow fever occurred in New York in 1668 and in Philadelphia in 1669 [2]. Webster reported, "In 1668, appeared a comet with a stupendous coma. This was attended by an excessively hot summer and malignant diseases in America. In New York the epidemic was so fatal, that a fast was appointed in September, on that account" [1]. There were outbreaks in Boston in 1691 and in Charleston in 1699 and 1706. Webster also reported, "In 1699 raged in Charleston South Carolina and in Philadelphia, the most deadly bilious plague that probably ever affected the people of this country." French colonists in Louisiana recorded widespread outbreaks in 1669. Yellow fever eventually became endemic in much of Central America; the Caribbean islands; the southern United States, especially New Orleans, Florida, and the Texas coast; and occasionally reached as far north as Charleston, Baltimore, New York, and Philadelphia, the cities that traded with the West Indies. It persisted in these urban areas because of the presence of readily available susceptible hosts and plentiful mosquitos [3].

The islands of the West Indies were treasured colonies of the European powers Spain, France, and England because they were the source of high-value crops, especially sugar. These three powers were almost constantly at war with each other during the 18th century, and they sent troops in very large numbers to the sugar islands. Actual fighting was fairly uncommon, but yellow fever killed European soldiers and sailors in huge numbers. Thus a British fleet sent to the West Indies to conquer the French Island of Martinique in 1693 lost at least half of its troops. The death rate for British soldiers in Jamaica was seven times that in Canada, mainly because of yellow fever.

Saint-Domingue

Epidemics of yellow fever were common in the Caribbean after 1690. They tended to occur as new colonies were developed and susceptible colonists moved in. The island of Hispaniola was divided in 1697 by the Treaty of Ryswick between a French half to the west called Saint-Domingue, which eventually became Haiti, and a Spanish half to the east called Santo Domingo, which became the Dominican Republic. Saint-Domingue, on the western side, was France's richest colony. It was a major sugar-producing region, producing about 40% of the world's sugar. It also produced 60% of the world's coffee, cocoa, and indigo. All this productivity required huge numbers of African slaves to work the plantations. In 1787 alone, the French imported about 20,000 slaves. Half of them died within a year, primarily from yellow fever. They were worked as hard as possible to get as much out of them before they died. Nevertheless, the slaves outnumbered the free population by about 10 to 1.

When the French Revolutionary Wars broke out in 1793, the British Royal Navy effectively cut Saint-Domingue off from France. Beginning in 1791 the slaves there revolted, killed their masters, and expelled the French. Once the war started, the British saw an opportunity and stepped in "at the request of the French colonists" to protect them and ensure that the rebellion did not get out of control. This also prevented the colony from being used by French buccaneers. The British landed 7500 men on the island in September 1793, but these were too few to conquer the colony and were totally susceptible to disease [4]. They remained until 1798 and suffered incredible losses as a result of yellow fever. (It should be pointed out that multiple other tropical fevers affected the troops, especially malaria and dengue, and the three were often confused.) Within 2 months the British had lost 40 officers and 600 men. Thereafter the disease spread rapidly, and by June 1794, 14% had died. Within another 3 months, losses in different regiments ranged from 24% to 44%. Case fatality rates reached

70%. The monthly death loss from August to December averaged 9.5%. The next year was worse. By 1795 some regiments had buried over 75% of their men. New replacements direct from Britain died as they arrived. The Army began to cover up the losses, so reported numbers then became unreliable. From May to September 1796 over 600 soldiers died each month. Mortality dropped in 1797 as the survivors became acclimatized. In 5 years, out of 20,200 British troops stationed in Saint-Dominique, about 12,700 died and about 1500 were sent home as invalids. Ninety-five percent of the deaths were a result of disease, primarily yellow fever.

The British withdrew from Haiti in 1798 having wasted large numbers of lives and huge amounts of money. The local colonists and slaves suffered much less as a result of acquired immunity. Saint-Domingue was not unique. In 1781 the British sent 1400 men up the Nicaragua River to occupy San Juan and lost 77% of them. In 1762, 40% of the British troops that occupied Havana died of disease.

In 1801 a brief truce between the British and French permitted Napoleon to attempt to reassert French sovereignty over Haiti. In February 1802 he sent 35,000 soldiers under the command of his brother-in-law, General Victor Leclerc, to retake the colony. The French fought a war against the rebels for several months and rapidly achieved military success, but by the end of March, 5000 French soldiers had died and another 5000 were hospitalized. Like the British on the same island, the number of deaths is unclear. Perhaps as many as 80% to 85% of the French troops died. Leclerc wrote to France reporting that he had 3600 men in hospital and was losing 30 to 50 a day. He eventually lost 25,000 men. Leclerc himself died late in October 1802, as did five other generals. An additional 20,000 troops were sent from France with Leclerc's successor General Rochambeau. In November 1803 the French were decisively defeated at the Battle of Vertières, in which Jean Jacques Dessalines, a former slave, led Haitian troops to victory over Napoleon's forces. This marked the last major battle of the Haitian Revolution, and the next month, Rochambeau surrendered to the British with 3500 survivors. Haiti declared its independence in 1804.

Napoleon had planned to revive the French colonies in the New World, but the first step was to recapture Saint-Domingue. He soon recognized that he could no longer afford to retain possession of his American colonies. He also recognized that Saint-Domingue was no longer generating the profits it once had. So rather than let the British capture the colonies, he abandoned any ideas of New World reconquest and decided to cut his losses by selling

Louisiana to the United States [5]. On April 11, 1803, Napoleon told his finance minister, Talleyrand, that he could not afford an American campaign and he worried that the British might seize Haiti. Besides, he needed the money to finance his proposed invasion of England. In the meantime, Thomas Jefferson was concerned that the French might choose to block American use of the Mississippi River. He therefore sent two senior diplomats to Paris to negotiate the continued use of the river by American vessels. They were authorized to offer to buy New Orleans for up to $10 million. To their surprise, the French offered to sell them the whole of the Louisiana Territory for $15 million. They immediately agreed! The Louisiana Purchase Treaty was signed in May 1803 in Paris. In return for the money, Napoleon renounced all claims to Louisiana. For less than 3¢/acre it effectively doubled the size of the United States and opened up the Mississippi to American trade and settlement. Given the importance of this to the United States, it is no exaggeration to say that yellow fever can be considered the most significant disease in American history. It was directly responsible for the Louisiana Purchase and indirectly responsible for the creation of Haiti. On January 1, 1804, Saint-Domingue, France's "pearl of the Antilles," became the country of Haiti and the second country in the Americas to gain its independence. The remaining refugee French planters with their slaves and diseases fled Haiti to New Orleans. Subsequently, yellow fever became endemic and major outbreaks occurred in later years in New Orleans, Florida, Memphis, and Texas.

The United States
Philadelphia

In the summer of 1793 ships from Saint-Domingue began to arrive in Philadelphia loaded with refugees fleeing the slave revolt and its accompanying massacres in Saint-Domingue. At this time, Philadelphia was the largest city in the newly established United States with a population of about 50,000. It was the interim capital of the country while the government waited for the construction of Washington DC to be completed. Many of the founding fathers lived in town, most notably George Washington, John Adams, and Thomas Jefferson.

On August 3 the first deaths from yellow fever occurred in Philadelphia. The first to die were two recent immigrants, one Irish and one recent arrival from Saint-Domingue staying in Denny's lodging house on North Water Street. More cases followed within weeks. Dr. Benjamin Rush, one of the city's most distinguished physicians (and politicians), recognized the disease as yellow fever and alerted his colleagues and the local government [6]. The government believed that the refugees were the source and established a quarantine of 2 to 3 weeks on immigrants and their goods, but they were too late; the next cases occurred in residents—the virus had landed (Box 11.1).

Philadelphia at that time clustered close to the Delaware River. The first yellow fever cases occurred in the wharf region, especially near the Arch Street wharf. There was an abandoned heap of rotting coffee beans piled near that wharf, and Benjamin Rush, adhering to the miasma theory, attributed the disease to its smell. The Mayor of Philadelphia, Matthew Clarkson, sought help from the city's College of Physicians. They wrote a report and published it in the city newspapers. They suggested some ways of remaining healthy: Avoid fatigue, hot sun, too much alcohol, and night air. Use vinegar and camphor to cleanse contaminated rooms. They advised the city

BOX 11.1 Dr. Benjamin Rush (1746–1813).

Dr. Benjamin Rush was a leading intellectual in the city of Philadelphia. He was active in the political life of the city and a signer of the Declaration of Independence. He served as a Surgeon General of the Medical Department of the Continental Army but resigned in protest over corruption and mismanagement. He was a Professor of Medical Theory and Clinical Practice at the University of Pennsylvania. He wrote the first American chemistry textbook. He was also a leading medical doctor of his day. He was somewhat old fashioned in his treatments and a great proponent of bleeding for almost any illness. He even insisted that he himself be bled as he lay dying. He worked diligently to save the sick and dying during the great yellow fever epidemic of 1793 by staying in Philadelphia and treating hundreds of disease victims. However, his use of copious bleeding and high doses of calomel (mercury chloride) as an emetic and laxative probably killed more victims than the bed rest recommended by experienced French physicians from the West Indies. Rush was a proponent of this "heroic medicine," arguing a serious disease needed a serious treatment to shock the body back to health. (Poor George Washington was subjected to the same "heroic" treatment on his deathbed. Benjamin Franklin, on the other hand, pointed out that in Barbados, the sick recovered "only after the doctors had run out of medicine.") Rush was the first to describe dengue fever in the United States when he wrote a report in 1789 on a case during the 1780 epidemic. He was probably the most famous physician in the United States when he died.

to make funerals private, to clean up the streets, and to burn gunpowder "to increase the oxygen supply."

Controversy arose as to the best way to treat the disease. Benjamin Rush adhered to the somewhat old-fashioned methods of bleeding (at least a pint of blood) and purging with high doses of mercury salts to cause profuse diarrhea—the heroic treatment. If these didn't work, he increased the dose and took more blood [7]. The basic theory was that disease was like a fire so the way to put it out was to remove excess fuel by bleeding and purging. Others, such as Alexander Hamilton, took the "Bark and Wine cure." This involved taking Chinchona bark containing quinine and diluted Madeira wine. This was the treatment recommended by experienced French doctors. Neither treatment would have cured the disease, but the bark and wine treatment was probably somewhat safer and much less stressful. Given that both Rush and Hamilton were active politically the argument extended into the political arena. There was debate as to whether the disease was contagious, but it was considered best to avoid the sick as much as possible. Regular citizens simply smoked cigars or chewed a garlic clove.

By late August the exodus from Philadelphia had begun. It is estimated that 10,000 to 20,000 people fled the city through September, including the leaders of the Federal government. Alexander Hamilton, the secretary of the Treasury, developed the disease. He tried to move to New York but was refused entry. He ended up in Albany under armed guard until he recovered. Many government workers refused to stay. They locked their offices and got out of town fast. Other citizens left the city in panic. However, many neighboring communities blocked the roads to prevent the refugees from entering. Both Baltimore and New York rejected refugees and quarantined goods arriving from Philadelphia. Stagecoaches and boats originating in Philadelphia were not allowed in many cities. (However, the yellow fever broke out in Baltimore between August and October 1794 and eventually reached New York in the summer of 1795 and persisted until 1804.) President Washington and his cabinet were meeting in Philadelphia and continued to do so until he left that city for Mount Vernon on September 10. He returned on November 11. The national capital moved to Washington shortly thereafter before Philadelphia was declared free of the disease. Congress adjourned and moved to the nearby village of Germantown. The state legislature carried on with its session until a dead body was found on the steps of the State House. However, banks, the post office, and other government employees bravely remained in the city throughout the epidemic. The city council fled with everyone else, but

Mayor Clarkson stayed and organized the city's responses. He set up committees to oversee the fever hospital at Bush Hill about 2 miles outside the city, arrange visits to the sick, and, if necessary, arrange for their transport to the hospital. The deaths of many adults resulted in large numbers of orphans, so orphanages had to be set up. They also arranged burial parties. However, without any effective treatment, mortality was around 50%.

Another eminent physician who stayed and ministered to the sick was Philip Physick. As a local practitioner, he became the physician at Bush Hill Hospital. Among his patients was Dolly Madison, the wife of the future president, James Madison. He also treated Benjamin Rush when he contracted the disease. Physick became the preeminent surgeon of his time.

Between August 1 and September 7, 456 people died in Philadelphia and another 42 died on September 8. In the week beginning October 7, 711 deaths were reported, and the daily death toll remained over 36 until October 26. The outbreak ended following the arrival of the first frosts in late October. The official death count was 4044 deaths between August 1 and November 9, but this was based on grave counts, so the total was probably closer to 5000—one-tenth of the city's population.

In response to the yellow fever outbreak, in 1799 the city of Philadelphia built the Lazaretto Quarantine station on the Delaware River. Its function was to act as a checkpoint where everyone and their cargo entering the city by ship could be inspected and the sick quarantined. It operated in this role until 1895. It is the oldest such facility in the United States. (The term "Lazaretto" may be derived from the Latin, Lazarus, the man brought back to life by Jesus, and which subsequently morphed into the Italian, Lazaro, meaning a beggar. Alternatively, it may be a corruption of Nazaretto, a quarantine island in the Venetian lagoon named Santa Maria di Nazareth that was established in 1423 during a plague outbreak.)

Although the 1793 outbreak primarily affected Philadelphia, in 1795 yellow fever reached New York, where it killed over 700 individuals that fall. The disease returned again in 1798 and killed another 3000 individuals before cold weather arrived. The city purchased Belle Vue farm several miles north of Lower Manhattan in 1798 to deal with the influx of the yellow fever victims. Located beside the East River, Bellevue Hospital is the oldest public hospital in the United States.

Contagion?
During most of the 19th century, when the cause of yellow fever was unknown, there was much debate as to whether the disease was introduced

by outsiders or arose locally. For example, the 1793 outbreak in Philadelphia divided physicians and politicians between those who believed that the disease had been introduced by the French refugees and those who believed it was due to miasma, especially that famous pile of rotting coffee. Those who believed that the disease was introduced strongly recommended quarantine as the prime method of control. Those who favored miasmas emphasized cleaning up the urban environment. Initially quarantine was the favored remedy, but it did not always appear to work, and it impeded commercial activity, which upset the merchants. It was also obvious that the disease was the same as that seen in the West Indies, and it was largely restricted to seaports. On the other hand, Benjamin Rush argued that waterfronts were so filthy that this was a place where epidemics would be expected to originate. The disease was seasonal, and the summer heat would increase the production of miasmas. The disease usually disappeared with the first frosts. Those who did not believe in contagion opposed quarantine but advocated aggressive methods to clean up the city with sewer construction and waste removal. It is interesting to note that both sides were partially correct. The disease was introduced by outsiders, but it was maintained by mosquitos thriving in polluted stagnant water.

The myth of Black resistance

It was widely believed, totally incorrectly, that African Americans were immune to yellow fever. As a result, during the Philadelphia outbreak, so many White people fled the city that an appeal was published in the newspapers seeking volunteers from the free African American community to clean and care for the sick and bury the dead. Led by the Free African Society and three leading pastors, Absalom Jones, Richard Allen, and Anne Saville, they coordinated the relief efforts by their congregations. They provided nursing care, picked up bodies, drove them to the cemeteries, and dug graves. The Black community responded positively even though subsequent analysis indicated that their mortality rate was similar to that of Whites and about 240 African Americans died. Likewise, their bravery and hard work were not appreciated at that time. In the winter of 1793, Mathew Carey published a pamphlet, *A Short Account of the Malignant Fever Lately Prevalent in Philadelphia*. In it Carey barely mentioned the efforts of Black citizens. Although Allen and Jones published a rebuttal pamphlet the next year, the damage had been done.

The myth of Black immunity persisted. Even 100 years later in the southern states, doctors promoted the idea that Black people had a natural immunity to yellow fever. Not only that, but this claim was also used to justify slavery. Thus it was argued, because Black people were resistant to this disease, that their slavery would protect Whites from work that would kill them. Black people could work outside in yellow fever areas without any risk. In other words, God had made Black people resistant in order to save White people from death. Not everyone believed in the theory of Black resistance, but it was widely asserted and clearly thought to be true by many. Nevertheless, an acclimated Black slave sold for up to 50% more than a nonacclimated slave (Box 11.2).

In northern cities, the cold winters killed the *Aedes* mosquitos so that outbreaks of yellow fever only occurred when the virus and the mosquito vector arrived about the same time. In southern cities, however, the mosquitos could survive year-round. The Louisiana Purchase and the availability of land in the Mississippi and Alabama Territories encouraged more and more American settlers to move to New Orleans and the Gulf Coast. As a result, during the 19th century yellow fever outbreaks were much more common in the southern states. They were especially common in the port cities along the Gulf Coast and along the southern Atlantic Coast. In 1821, within a few weeks of the transfer of Florida from Spain to the United States, yellow fever broke out in St. Augustine. About a third of the US troops that occupied the city died of yellow fever as did many bureaucrats who had arrived from Washington (Fig. 11.1). Yellow fever outbreaks were also common in ports that traded with the West Indies because infected sailors could spread the disease. Thus outbreaks occurred in Portsmouth, Providence, Newburyport, New London, New Haven, Baltimore, Norfolk, Wilmington (both Delaware and North Carolina), Charleston, Savannah, Pensacola, and Galveston, and most especially in New Orleans [2]. It is interesting to note, however, that New Orleans was free of yellow fever during the Civil War, presumably because it was unable to trade with the West Indies.

Memphis

After the Civil War the southern economy recovered, and with the establishment of efficient river transport, yellow fever returned to New Orleans in 1867 and 1873 and spread north along the Mississippi River. Yellow fever reached Memphis in both 1867 and 1873 and killed over 2000 people in the 1873 outbreak, at that time the greatest mortality recorded for a noncoastal city. Memphis had thrived during the reconstruction period following the Civil War, and by 1878 its population had reached 48,000. In June 1878

BOX 11.2 Another Lincoln assassination attempt!

As the Civil War was drawing to a close in 1864, a Confederate agent tried to use yellow fever to kill President Lincoln. In 1864, the agent, Dr. Luke Blackburn, traveled to Bermuda, where a yellow fever epidemic was raging. Dr. Blackburn had some experience with the disease and, like others at the time, believed that it was transmitted by fomites, such as contaminated blankets and clothing. He therefore began to collect such items from infected patients and put them in a trunk with some new coats and shirts, hoping to contaminate them as well. He then took his luggage with him when he returned to Nova Scotia but had the trunks shipped to Boston and eventually to Washington. The clothing was sold at auction in Washington and New Bern, North Carolina. Blackburn also expressly ordered that a valise of infected clothing be sent as a gift to President Lincoln. However, the valise was apparently never shipped. Blackburn returned to Bermuda to await events. Nothing happened!

He tried to repeat the process of shipping infected clothing to New York City, but he got caught after an informant tipped off the police. Blackburn was arrested in Canada but eventually acquitted there—there was no evidence that the trunks had ever been on Canadian soil! Blackburn had engaged a man called Hyams in Nova Scotia to ship his trunks to Washington but had failed to pay him! Hyams, in frustration, went to the United States consul's office in Toronto in April 1865 and offered to expose Confederate plots! Two days later Lincoln was assassinated, and Hyams went to Washington to serve as a witness in the trials of his assassins.

Blackburn returned to the United States in 1867 and eventually back to his native Kentucky. He ran for governor in 1878 and won. When questioned by the press about the Yellow Fever Plot he said that the accusations were "too preposterous for intelligent gentlemen to believe."

Yellow fever also played a role in the consequences of the successful assassination of Lincoln. Dr. Samuel Mudd received a life sentence for treating the broken leg of Lincoln's assassin, John Wilkes Booth. Mudd was imprisoned in remote Fort Jefferson, on an island in the Dry Tortugas in the Gulf of Mexico, 70 miles west of Key West. Having no natural source of drinking water on the island, rainwater was collected in open cisterns and barrels with the predictable result—lots of breeding mosquitos. During the late summer and fall of 1867, an outbreak of yellow fever occurred at the Fort. After the army physician, Dr. Joseph Smith, died from the disease, Dr. Mudd volunteered to care for the sick prisoners and soldiers alike. Dr. Mudd's actions were so appreciated by the prison staff that they petitioned President Andrew Johnson for his release. Recognizing Dr. Mudd's contribution in saving lives during the outbreak, as well as other issues, President Johnson proclaimed, "Thus, now, therefore be it known that I, Andrew Johnson, President of the United States of America, in consideration of the premises, divers other good and sufficient reasons me thereunto moving, do hereby grant to the said Dr. Samuel A. Mudd a full and unconditional pardon." Though yellow fever killed many people, it was instrumental in giving Dr. Mudd his life back.

Fig. 11.1 *Florida being dragged down by yellow fever and being rescued by the United States. Courtesy Library of Congress.*

news arrived in Memphis of another outbreak of yellow fever in New Orleans. The Memphis business community rejected suggestions to take precautionary measures deeming them bad for trade. The Board of Health started cleaning up the streets and, in late July, established quarantine stations downstream for any arriving refugees. The local newspaper advised, "Avoid patent medicines and bad whiskey!" But it was a hot, wet summer—ideal for mosquitoes. On July 27 the first case was reported in Vicksburg, Mississippi, downriver from Memphis. On August 13, the first yellow fever victim in Memphis, Mrs. Kate Bionda, died. The next day there were 55 more cases. In the first weeks 10 patients died daily, which soon climbed to 50 daily and by mid-September it had climbed to 200 deaths a day. As in other disease outbreaks throughout the world, those with means tried to flee the city. Twenty thousand citizens fled Memphis, and as a result, cities further north tried to protect themselves by banning riverboats and trains from Memphis. The first frosts arrived in mid-October and the outbreak was declared over at the end of that month. About 16,000 people died in the Memphis region by the time the epidemic finally ended in early 1879.

Due to the widespread disruption caused by introduced diseases, President Rutherford B. Hayes signed the Quarantine Act in April 1878. This gave the Marine Hospital Service responsibility for preventing the importation of disease by quarantining incoming vessels. However, the act was unfunded, so rather than stopping shipping, the Service usually just fumigated vessels!

Texas

Yellow fever spread along the Texas coast, so that Galveston, for example, had at least nine outbreaks between 1839 and 1867. The 1839 epidemic in Galveston was well documented by Ashbel Smith, the former Surgeon General of the Texas Army [8]. Smith had strong opinions regarding whether yellow fever was contagious. He pointed out that most of the cases occurred in the region of "The Strand," Galveston's main street, where there were "animal and vegetable matters abounding around the houses, and a marsh exposed to the heat of the sun ...." Patients were moved from the affected districts to "healthy" districts, and none of the attendants nor their new neighbors developed the disease. "Dr. Smith performed many dissections, examining everything closely, and immersing his hands freely in black vomit, &c., but with perfect impunity. He *tasted* the black vomit repeatedly when freshly ejected from the stomach of the living with equal impunity." He did not get sick and pronounced the disease "noncontagious"!

In the 1853 Galveston epidemic there were over 500 deaths. In 1858 there were 175 yellow fever deaths in Houston. In Galveston's 1867 epidemic there were over 700 deaths. Outbreaks continued to occur along the coast and up the Brazos River valley until 1903 when aggressive mosquito control began.

The Spanish-American War

For most of the 19th century, yellow fever continued to be a scourge throughout the Caribbean basin, and sporadic outbreaks occurred regularly in New Orleans, in northern Florida, and around the US Gulf Coast. Although it was still not known how the disease was transmitted, by the 1880s there was a growing recognition that some diseases such as malaria, for example, could be transmitted by mosquitos. In 1881, at a meeting of the Havana Academy of Sciences, a distinguished Cuban physician, Dr. Carlos Finlay, went so far as to propose that yellow fever was not contagious but mosquito borne, but he was unable to prove this. Finlay was ridiculed for his preposterous theory. (His theory was correct, but his experimental timing was wrong! He had left insufficient time for the virus to grow within the *Aedes* mosquito after it fed on a yellow fever victim before allowing it to bite a healthy human volunteer.)

In 1898, the Spanish-American War broke out as a result of Spanish attempts to suppress a revolt in its colony of Cuba. The United States resolved to expel the Spaniards and did so in a short campaign that ended at the beginning of July. The Spanish Government sued for peace, so most American troops left quickly. However, a force of 50,000 had to be left behind in Cuba to maintain order. It was planned that they would stay for 4 years. Brigadier General Leonard Wood was appointed governor.

The first task was to clean up the country. Cities such as Havana and Santiago were filthy. Under the command of George Sternberg, the Surgeon General of the Army, strict orders were given to improve sanitation and keep the army camps spotlessly clean. He appointed Major William Gorgas and his sanitation teams to remove sewage and garbage from the army camps and Havana, the capital—this had worked well for typhoid. The army leaders, recognizing the yellow fever threat, also sought to move their men away from the coast where the threat was greatest to healthier high ground. But in June 1899 yellow fever reappeared in the camps, and by mid-July there were over 150 cases. The sick were separated from the healthy. The sick troops were sent to a camp at Siboney. (This is a coastal city on the south coast, not far from Guantanamo Bay.) In the belief that African Americans were more resistant to yellow fever than Whites, the 24th Infantry Regiment, composed of African Americans, was also sent there to provide care. The regiment lost more than a third of its men to yellow fever within 40 days. A few cases also occurred in Hampton, Virginia, and these were traced back to Cuba. Despite all the efforts of Gorgas and his men to clean the place up, yellow fever was not going away. Seriously alarmed, in the spring of 1900 George Sternberg requested that the Army establish a "Yellow Fever Commission." Major Walter Reed was appointed chairman (Fig. 11.2). (Reed had previously chaired the successful Typhoid

Fig. 11.2 *Major Walter Reed.* *Courtesy US National Library of Medicine.*

Commission [Chapter 10].) Their task was to determine its cause and seek to prevent yellow fever. Reed had three assistants, James Carroll, Aristides Agramonte, and Jesse Lazear, and Agramonte and Lazear were civilians.

Walter Reed had graduated from the University of Virginia in 1869 and joined the United States Army Medical Corps in 1875. After initially serving in remote western outposts, Reed participated in a Pathology and Bacteriology course at Johns Hopkins Hospital. Reed so impressed Sternberg that he was appointed Professor of Clinical and Sanitary Microscopy at the Army Medical School in Washington. (In 1892, Sternberg himself had published the first comprehensive bacteriology textbook in the United States. His *Manual of Bacteriology* was the standard reference text for many years.)

James Carroll was an Englishman who had emigrated to the United States from Canada in 1874 and enlisted in the Army. While in the Army he pursued and obtained a medical degree from the University of Maryland. He then worked at Johns Hopkins where he acted as the laboratory assistant to Major Walter Reed. They worked together for several years.

In 1899, Sternberg had Reed and Carroll investigate an organism called *Bacillus icteroides* that the Italian bacteriologist Giuseppe Sanarelli had claimed was the cause of yellow fever—it was not, but just a contaminant. As a result, Sternberg appointed Carroll as acting Assistant Surgeon and sent him off to Cuba as Reed's deputy. The men were assisted by Aristides Agramonte, a Cuban American who was the laboratory director at the military hospital in Havana. Jesse Lazear was head of the clinical laboratories at the Johns Hopkins School of Medicine. An experienced bacteriologist, he also studied malaria and was regarded as an expert in tropical diseases. He wrote to Sternberg seeking a temporary assignment to study Cuban diseases. Sternberg promptly appointed him to the Yellow Fever Commission.

Although Sternberg firmly believed that yellow fever was a filth disease spread by contaminated material, others disagreed. Thus Dr. Henry Carter of the Marine Hospital service and the Havana quarantine officer had noted during the Memphis epidemic that 2 weeks generally elapsed between the first and second waves of yellow fever and wondered what was happening during that time. Carter happened to be in Cuba at the time Reed arrived and they had a discussion on the topic. Reed was impressed. As described earlier, in 1881, Dr. Carlos Finlay had speculated that *A. aegypti* was the carrier of yellow fever. Finlay had done some preliminary studies, but they were inconclusive and generally disbelieved by the medical community. When Reed and his commission arrived, Finlay offered his services to them.

Reed and his team met for the first time in Cuba in June 1900. It was commonly believed at that time that yellow fever was spread by "fomites," in other words, contaminated material such as clothing or bedding from yellow fever victims. Reed, however, noted in one army camp that nurses and orderlies who were in closest contact with contaminated material had not been infected. In another situation, a soldier confined to the guardhouse for 6 days contracted yellow fever despite not being in contact with fomites. And his cellmates remained healthy! In addition, the reports from London regarding Ronald Ross' finding that malaria was transmitted by mosquitos rekindled interest in the mosquito theory (Chapter 7). As a result, the committee visited Carlos Finlay that July and decided to test Finlay's mosquito theory once more. Finlay provided Lazear with *Aedes* mosquito eggs. Lazear hatched the eggs and let them feed on an infected patient and then on nine army volunteers (Fig. 11.3). None got sick! Carroll then agreed to be bitten by a mosquito that had fed on a yellow fever victim 12 days earlier, and he subsequently developed the disease. For a while his life was in the balance, but he recovered, to everyone's relief.

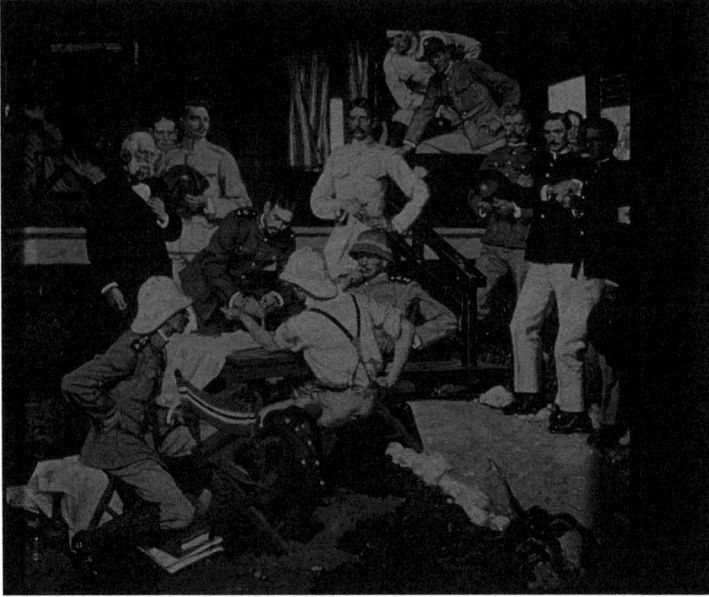

Fig. 11.3 *The yellow fever transmission study in Cuba in the presence of Walter Reed and Carlos Finlay.* *Courtesy US National Library of Medicine.*

Because Carroll had been in contact with a yellow fever case a few days previously, the experiment was repeated on two more volunteers who had not been in contact with such cases. One, Private William Dean, developed a mild case of yellow fever 2 days later. It appeared that the agent had to be in the mosquito for at least 10 days before it reached the salivary glands and could transmit the disease.

The other volunteer, Jesse Lazear, also developed yellow fever, suffered fever, delirium, and black vomit, and died on September 25, 1900, after a brief illness.

The Commission now conducted a series of elegant experiments that confirmed conclusively that yellow fever was transmitted by the *A. aegypti* mosquito. In November, Reed established a new experimental station outside Havana that he called Camp Lazear. Here they proved the mosquito theory and disproved the fomite theory. Volunteers who lived in a mosquito-proof room sleeping with the clothing and blankets of yellow fever victims never developed the disease. Carroll subsequently showed that yellow fever could be transmitted by a blood transfusion. Finally, he showed that if the blood was appropriately filtered to exclude bacteria it was still infectious. Yellow fever was caused by a "filterable agent." In other words, a virus.

Nurse Clara Maass

When seeking human volunteers, Reed and his colleagues recognized that their experiments might prove lethal, so they applied the principle of informed consent. They insisted that the volunteers fully understand just what they were volunteering for. All had to sign a consent form. They were paid $100 (the equivalent of about $3000 today) when they volunteered and an additional $100 if they became ill. The only woman to volunteer was Clara Maass, an army nurse who participated in the 1901 experiments (Fig. 11.4). Born to a family of German immigrants in New Jersey, Clara Maass volunteered as a nurse in the Army. She served in the Philippines in 1899—1900 and was posted to Cuba. In an attempt to show that patients who had a mild case of yellow fever would transmit a mild disease to others, Ms. Maass was bitten by an infected mosquito. Nothing happened the first time. In order to determine if she was immune, she was bitten a second

Nurse Clara Louise Maass, U.S.A.
Yellow Fever Heroine. Class of 1895, Lutheran Memorial Hospital, now Clara Maass Memorial Hospital, Newark, N. J.

Fig. 11.4 *Nurse Clara Maass, who gave her life in the studies on yellow fever. Courtesy US National Library of Medicine.*

time in August 1901. She developed a severe case of yellow fever and died within 2 weeks. The resulting public outcry resulted in the termination of experiments using human volunteers for yellow fever studies. Ms. Maass' sacrifice was commemorated with a US Postage Stamp in 1976.

The next step was to control or eliminate the disease. The Chief Sanitary Officer in Havana, William Gorgas, persuaded Governor-General Leonard Wood to focus the government's efforts on mosquito eradication.

A. aegypti is a small dark mosquito with white markings in the shape of a lyre on its upper thorax and white bands on its legs. It is common in areas lacking proper sanitation systems. It rarely leaves the house where it has fed. It lays its eggs in damp areas close to water. They can hatch within 2 to 7 days when damp, but they can survive for more than a year when dried and hatch when exposed to water. The rate at which the larvae develop depends upon the temperature, perhaps persisting in the larval stage for many months when it is cool.

Major William Gorgas, despite being skeptical of Reed's mosquito theory, organized an aggressive and very effective anti-mosquito campaign. The city was under martial law so Havana was divided into 20 districts and teams of sanitation officers assigned to each. They visited every home and building. They focused on mosquito breeding sites. By eliminating standing water from barrels, cisterns, and drainage ditches and covering all stagnant water with a layer of oil, the mosquito larvae suffocated. Mosquito traps were also set out. Sanitary officers went into homes, sealed off any cracks in the wooden buildings, and fumigated them by burning pyrethrum. Likewise, all yellow fever cases had to be reported to the authorities. Citizens were fined if mosquitos were found in their homes. As a result, the mosquito population was reduced to such an extent that the number of yellow fever cases in Havana fell from 1400 in 1900 to 20 in 1901 to none in 1902.

The last US epidemic

Between 1817 and 1900, yellow fever had been an almost annual visitor to the cities of the southeast and gulf coasts. New Orleans was often the center of these outbreaks with an epidemic on average every 3 years. This had important social consequences because employers tried not to hire anyone who had not recovered from the disease and was therefore not immune. Unacclimated individuals could not live in certain neighborhoods and were often not permitted to rent rooms. They were unwelcome in some social circles. (Yellow fever was nicknamed "The Stranger's Disease.") Why waste time and money to train someone who might soon be dead? The acclimated survivors, on the other hand, gradually accumulated significant social, political, and economic power.

The city's location just above sea level always caused problems in getting and keeping clean, fresh drinking water. Wells would not work. As a result, the residents collected rainwater in massive open-topped cisterns made of wood or iron. These cisterns attracted hordes of mosquitos who laid their eggs there. Applying Gorgas' mosquito control methods to these cities, the prevalence of yellow fever dropped dramatically. However, the methods had to be rigorously applied. That was not always the case, and as a result, New Orleans suffered the last yellow fever epidemic in 1905. The outbreak began in July 1905 in the lower French Quarter and eventually affected 3402 individuals, killing 452. It killed the archbishop. The orphanage in the French Quarter filled up, and a new one had to be built. The epidemic peaked on August 12 and the last case occurred in November. But by this time the source was known and a massive mosquito eradication campaign was mounted. Citizens were required to screen windows and cisterns. They had to sleep under mosquito nets. Oil was poured on any standing water, such as the cesspools and cisterns, to kill mosquito larvae. It worked; yellow fever was eradicated from N'awlins.

The Panama Canal

The French engineer Ferdinand de Lesseps completed the building of the Suez Canal in 1869. The Suez Canal is essentially a level ditch across the desert. His success, however, encouraged him to attempt to build a Panama Canal across a much more difficult, hilly, and heavily wooded country. The French began digging in 1882 but gave up after several years as a result of several factors, the most important of which were the ravages of yellow fever and malaria. The French, not knowing how these diseases were transmitted, permitted lots of standing water in the construction zone, and as a result, mosquitos thrived. In the first year of construction 400 workers died; in the second, 1300. It was considered a graveyard for French engineers; in just 2 months, March and April 1882, 37 of the fewer than 100 engineers sent to work on the canal died! So inescapable was yellow fever that Jules Dingler, Director General—who pompously had stated that yellow fever only affected drunkards and the lazy—saw his wife, son, daughter, and her fiancé succumb due to the fever. When construction was abandoned

7 years later, Gorgas estimated that more than 22,000 workers had died of disease, mainly from malaria and yellow fever. De Lesseps and his company went bankrupt in 1889, and the French sold their rights to the United States for $40 million 10 years later. The Americans took over, established Panama as an independent country, and obtained rights to the Panama Canal Zone in 1903. Theodore Roosevelt decided to complete the construction of the canal in 1907. William Gorgas (by now a colonel), the eliminator of yellow fever from Havana, arrived in 1904 and, after a hard fight, used the findings of the Reed Commission to persuade the authorities to convince the Canal Commission that mosquito control was necessary. With his colleague Henry Carter and chief engineer John Stevens, Gorgas pursued aggressive mosquito control policies in the canal zone. Removal of standing water and screening of workers' housing worked well. (Gorgas even insisted that the holy water in Panama Cathedral be changed daily!) The death rate from yellow fever was reduced to insignificance and the last case of yellow fever in Panama occurred in 1906. They also eliminated malaria at the same time. The canal was completed in August 1914 and Gorgas was promoted to Surgeon General of the Army.

Yellow fever vaccine

In the years following Carroll's demonstration that it was caused by a virus, yellow fever research continued. The leading institution in this subject was the Rockefeller Foundation in New York. Not all their scientists were convinced that the disease was caused by a virus. As a result, much effort was put into developing a vaccine against a bacterium found in some yellow fever patients. It turned out to be a spirochete called *Leptospira icterohemorrhagica*. Undeterred, the Rockefeller Foundation continued its efforts. Eventually Dr. Max Theiler, a South African scientist working for the foundation, showed that the virus would grow well in mice when inoculated into the brain. By passing the virus many times through these mice and then over 100 times through chick embryos, he was able to ensure that the virus lost its virulence for both mice and monkeys. (About the 30th mouse passage, Theiler himself contracted a mild case of yellow fever.) This attenuated live strain designated 17D could no longer induce disease in human volunteers but did trigger a protective immune response. It was first field tested in Brazil in 1937. The 17D strain has proved itself to be an incredibly safe and effective vaccine and has been used to protect millions of people [9] (Box 11.4).

BOX 11.3 The yellow fever vaccine problem of 1942.

As America mobilized to fight the Germans and Japanese, the US Army decided that it would be a good idea if all its troops were immunized against yellow fever. After all, nobody knew where the fighting might eventually extend to. The vaccine was provided by the Rockefeller Foundation without charge. The attenuated 17D vaccine at that time needed to be "stabilized" by the addition of normal human serum to prolong its shelf life. Unfortunately, seven vaccine lots were contaminated with serum from donors carrying hepatitis B virus although this was unknown at the time. Some batches were much more infectious than others. Thus a batch in one camp caused 1004 cases of hepatitis among 5000 troops. Overall, more than 300,000 troops were infected, there were 50,000 clinical cases, and 29,000 cases of jaundice developed. For a short time, it was believed that the vaccine itself may have caused yellow fever. Once the problem was recognized, the contaminated batches were promptly withdrawn and the use of the vaccine containing human serum was discontinued. This was the largest vaccine-related epidemic in US history.

BOX 11.4 Viruses and the slave trade.

It has proved possible to detect DNA viral genomes in ancient skeletal remains from Mexico. Thus dental root canal samples obtained from individuals buried in mass graves in a colonial hospital and a colonial chapel in Mexico City in the 16th century have been shown to contain DNA from human parvovirus B19 genomes and a hepatitis B genome. These victims were identified as first-generation African slaves and at least one indigenous person. These viral DNA sequences have proved to be identical to those in virus strains originating in Africa. Thus there is direct molecular evidence that ancient viruses were transmitted from Africa to New Spain via the transatlantic slave trade. Clearly yellow fever was not the first of these lethal diseases to be introduced.

(From Guznan-Solis AA, Villa-Islas V, Bravo-Lopez M, Sandoval-VelascoJulie M, WespJorge K, Gómez-Valdés. A, et al. Ancient viral genomes reveal introduction of human pathogenic viruses into Mexico during the transatlantic slave trade. eLife 2021; https://doi.org/10.7554/eLife.68612.)

A single dose provides immunity for at least 10 years. Max Theiler received the Nobel Prize for his discovery in 1951 (Box 11.3).

Despite the vaccine, yellow fever virus persists in the jungles of central Africa and South America. Major outbreaks of yellow fever have occurred in South America, especially Brazil, and in Africa, especially Angola and the Democratic Republic of the Congo [10]. In Brazil, it was preceded by an epizootic in brown howler monkeys that almost wiped that species out. But it then spread to humans. Between December 1, 2016 and May 2018 there were 2050 confirmed yellow fever cases and 681 deaths in Brazil, a case fatality rate of 33%. In early 2017 the World Health Organization launched a vaccination campaign in Brazil using 3.5 million doses of yellow fever vaccine. In March 2018 the Brazilian government announced that it would vaccinate all its citizens by April 2019. Unfortunately, this has yet to be achieved [11].

References

[1]   Webster N. *Two Volumes Hudson and Goodwin A brief history of epidemic and pestilential diseases; with the principal phenomena of the physical world, which precede and accompany them, and the observations deduced from the facts stated.*
[2]   Blake JB. Yellow fever in eighteenth century America. Bull N Y Acad Med 1968; 44(6):673–86.
[3]   Chippaux JP, Chippaux A. Yellow fever in Africa and the Americas: a historical and epidemiological perspective. J Venom Anim Toxins Incl Trop Dis 2018;24(1). https://doi.org/10.1186/s40409-018-0162-y. Available from: http://www.jvat.org/.
[4]   Geggus David. Yellow fever in the 1790s: the British army in occupied Saint Domingue. Med Hist 1979;23(1):38–58. https://doi.org/10.1017/s00257 27300051012.
[5]   Marr JS, Cathey JT. The 1802 Saint-Domingue yellow fever epidemic and the Louisiana Purchase. J Public Health Manag Pract 2013;19(1):77–82. https://doi.org/10.1097/ PHH.0b013e318252eea8.
[6]   Fried S. Rush: revolution, madness and the visionary doctor who became a founding father. 2018. p. 2018.
[7]   Benjamin North RL. Rush, MD: assassin or beloved healer? Proc Bayl Univ Med Cent 2000;13:45.
[8]   Bishop BP. Yellow fever in Galveston, Republic of Texas, 1839: an account of the great epidemic. J Am Med Assoc 1952;149(1). https://doi.org/10.1001/jama.1952. 039201800970371952.
[9]   Frierson JG. The yellow fever vaccine: a history. Yale J Biol Med 2010;83(2):77–85. Available from: http://www.ncbi.nlm.nih.gov/pmc/articles/PMC2892770/pdf/ yjbm_83_2_77.pdf. United States.
[10] Barrett ADT. The reemergence of yellow fever. Science 2018;361(6405):847–8. https://doi.org/10.1126/science.aau8225. Available from: http://science.science mag.org/content/361/6405/847/tab-pdf.
[11] Rogers DJ, Wilson AJ, Hay SI, Graham AJ. The global distribution of yellow fever and dengue. Adv Parasitol 2006;62:181–220. https://doi.org/10.1016/S0065-308X(05) 62006-4.

Cholera—1832

Contents

The term *"cholera morbus"* was first used in 1673 by the English physician Thomas Sydenham to describe an acute diarrheal disease with stomach cramps and vomiting and to distinguish it from plain *"cholera,"* which means bad temper! The modern usage of the term referring to a specific bacterial disease stems from much later in the 19th century.

Cholera as a specific infectious disease appears to have originated in India in the Ganges River delta. This is a densely populated riverine swampy area readily contaminated by human feces. The cause of cholera, *Vibrio cholerae,* is a waterborne bacterium and as a result, anyone who drinks contaminated water is a potential victim. In the absence of proper sewage treatment the bacterium can spread rapidly and widely. This is especially consequential in urban areas where the diarrhea from large numbers of individuals may overwhelm primitive waste treatment facilities and contaminate drinking water supplies. It is a singularly nasty disease, killing about 90% of those affected with the most severe form.

The bacterium

The German microbiologist Robert Koch searched for the cause of cholera during an outbreak in Cairo, Egypt in October 1883. When he examined the intestinal tissues of autopsied victims, he saw huge numbers

Great American Diseases
ISBN: 978-0-443-31404-9
https://doi.org/10.1016/B978-0-443-31404-9.00012-4

of a comma-shaped bacterium. He also found the same organism in the feces of cholera patients and in the bodies of those who died. He was even able to trace the organism to a contaminated water tank and was subsequently able to grow it in pure culture. What he found was a motile gram-negative, facultatively anaerobic bacillus that is characteristically curved or comma shaped and has single flagellum at one end. As it is propelled through liquid by its flagellum, it appears to vibrate, so Koch called it *V. cholerae*. The organisms may join end to end to form spiral or S-shaped chains. Although *V. cholerae* only affects humans, there are many other diverse vibrios found naturally in aquatic environments, especially in seawater. There are two biotypes of *V. cholerae* O1: the classical biotype, cholerae, and the current predominant biotype, El Tor. These are subdivided into serotypes Inaba, Ogawa, and Hikojima, based on the structure of their surface (O) antigens. The lethal strains of the bacteria secrete toxins that are responsible for their pathogenicity. Gastric acid may kill small doses of the cholera bacteria, but in large numbers some will get through the stomach to invade the intestinal tract and so cause disease.

The disease

V. cholerae does not invade the body but grows within the intestine where it releases its toxin. Cholera toxin binds to cell receptors and stimulates the intestinal lining cells to excrete large amounts of water, sodium, potassium, and bicarbonate. As a result, huge volumes of fluid, perhaps as much as 2 to 3 gallons a day, leave the patient in the form of watery, slightly cloudy diarrhea—so-called "rice water." Death results from extreme dehydration and the loss of essential salts. The diarrhea starts abruptly 2 to 5 days after infection, depending upon the dose of bacteria ingested. Humans have to ingest at least 1 billion vibrios in order to develop disease. (This is in marked contrast to Salmonellae that can infect victims with as few as 100 organisms.) The patient's blood pressure drops precipitously within hours of disease onset and death may occur within a few days. Flecks of mucus in the stools are responsible for the cloudiness. In addition to the profuse diarrhea, victims may also suffer from severe vomiting. The patient's complexion may become bluish-gray as a result of poor circulation and lack of oxygen. Sunken eyes, dry mouth, cold skin, and wrinkled extremities develop. Electrolyte loss results in severe leg cramps. About 75% of those infected by *V. cholerae* remain healthy, but the bacteria persist in their feces for 7 to 14 days and can still contaminate the environment and infect others. Of the 25% who develop symptoms, about 80% develop a relatively mild diarrheal disease while the remainder get the very severe form. As with other infections, malnourished and immunodeficient individuals, especially children, are most susceptible. In less developed countries, drinking contaminated water or eating foods that have been washed in that water are the primary sources of infection. Most cholera cases in developed countries now result from ingestion of contaminated seafood, especially insufficiently cooked crabs, shrimp, or oysters.

History

Cholera has been endemic in the Ganges River delta of India and Bangladesh for perhaps thousands of years. Indian manuscripts from 400 BCE describe what appears to be typical cholera—vomiting, diarrhea, and a blue face. Surviving in contaminated rivers, ponds, and water tanks, cholera was only recognized by the world at large when British military and commercial activities opened up the Ganges area of Bengal to a wider world. The first recognized cholera pandemic probably originated around 1817. It appears to have started in a village near Kolkata. It killed several thousand British soldiers in India before it spread by ship along the trade routes from Bengal to Sri Lanka, Burma, Thailand, Java, the Philippines, China, and Japan in 1822 to 1824. It spread westward to the Persian Gulf and Southern Russia in 1821 to 1824 before it died out. In India, the disease often accompanied British forces as they moved around the country. They carried it to Afghanistan and the Middle East.

The second pandemic

The second cholera pandemic also began in Bengal around 1826 and spread first to Afghanistan, Persia, and Russia. It reached Mecca in 1831 at the time of the Haj, and returning pilgrims took it home with them. Russia was at war with Persia and Turkey and dealing with a revolt in Poland at that time. The disease reached Moscow in 1830 where the mortality was over 50%. It spread westward to Poland and eventually reached the British Isles in November 1831. Martin Van Buren, the American Secretary of State, reported its arrival in England that year and told his government to prepare for the worst. The disease reached Belfast, Ireland in February 1832 and spread throughout the island. From there it was carried by immigrants on ships sailing to North America.

The government of Lower Canada (modern Quebec) heard of the disease outbreak in Britain and Ireland and as a result sought to prevent its entry by establishing a quarantine station at Grosse Île, an island in the St. Lawrence river downstream from Quebec City (Chapter 10). All vessels with more than 15 passengers had to stop there to be inspected. The number of arriving vessels could reach 40 to 50 a day, and 51,000 immigrants passed through the station in 1832. The number was so great and the system so disorganized that many ships sailed past Gross Île and simply landed all their passengers in Quebec City [1]. It has been claimed that the disease arrived sometime that June in a vessel from Belfast. Cholera cases soon broke out and then spread with the immigrants to Quebec City and Montreal. In the early 1830s, the most efficient form of transport in the United States was by boat. During June and July 1832, cases of cholera occurred in Syracuse and Detroit, presumably arriving by boat up the St. Lawrence. From there, the disease traveled south along the newly completed Erie Canal and the first cases appeared in Albany and New York City that July [2] (Fig. 12.1). The New York City Health Department had been established in 1793 in order to prevent the influx of yellow fever from Philadelphia (Chapter 11).

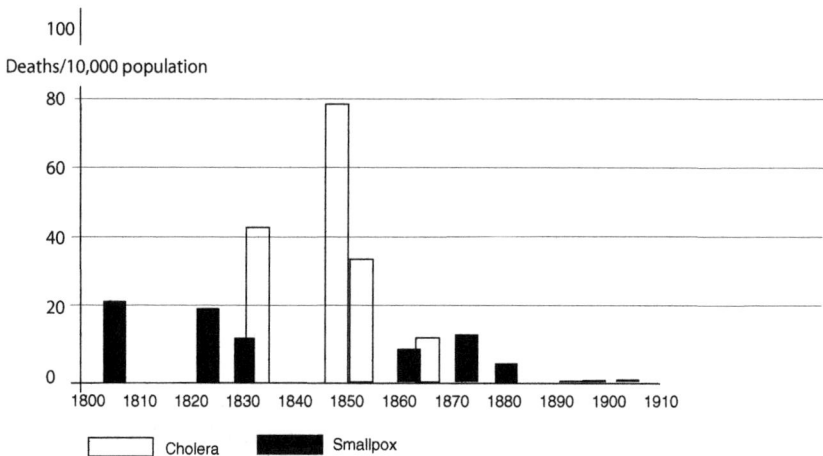

Fig. 12.1 *The lethality of two major epidemic diseases that have affected New York City since 1805. The sudden appearance of cholera in 1834 is obvious as is its rapid disappearance once clean water supplies were assured. Smallpox (red bars) gradually declined as the population was vaccinated so that the last such epidemic occurred in 1901–2. Data from New York Department of Health and the CDC.*

As in other epidemics, the belief that local toxic vapors from rotting vegetation, so-called "miasmas," were the cause ensured that those who could afford to left town by the first available stagecoach. The poor and impoverished were left behind. There was also a widespread belief, at least among the upper classes, that it was your own fault if you got the disease (Fig. 12.2). Merchant and civic leader John Pintard wrote that cholera "is almost exclusively confined to the lower classes of intemperate, dissolute & filthy people huddled together like swine in their polluted habitations." This was code for recent poor Irish Catholic immigrants (Box 12.1).

The New York City Health Department responded to the appearance of cholera by quarantining ships and cleaning up the filthy streets, especially in the poorer districts such as the Lower East Side where overcrowding and

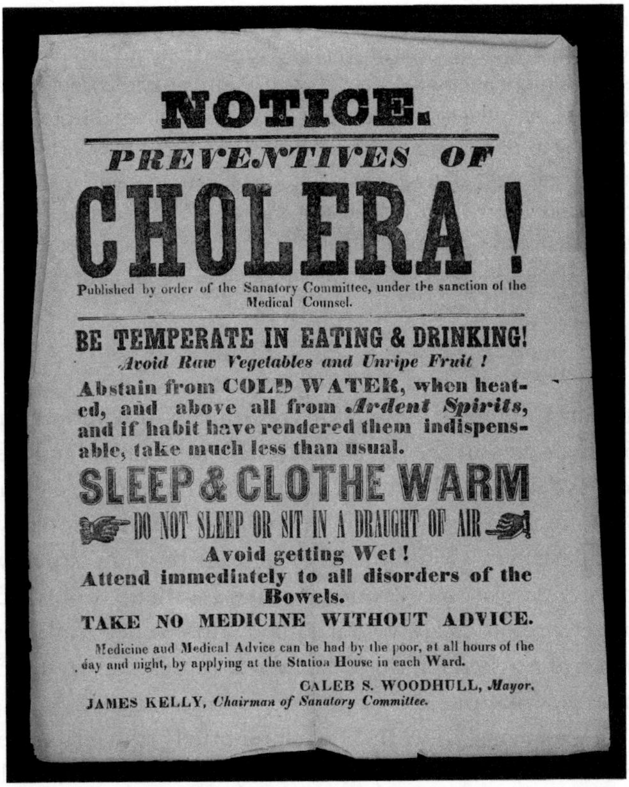

Fig. 12.2 *A poster from New York City recommending how to avoid getting cholera. Good behavior was essential. Courtesy Library of Congress.*

BOX 12.1 Religion and disease in America.

Prior to the dawn of microbiology, in the absence of an obvious physical cause for infectious diseases, people naturally turned to the alternative spiritual causation. In a highly religious age, disease could be attributed directly to God as a form of punishment or to Satan as an example of his wickedness. In general, the first was favored in Western society. The Old Testament contained several examples of God punishing the wicked by causing plagues. People were also intensely aware of their own sinfulness and many saw disease as a just punishment. This was especially the case when the disease affected others. Thus disease was also considered a manifestation of some sort of character defect, especially when it preferentially affected the poor, starving, and impoverished. Cholera was considered to affect primarily the "intemperate."

A logical corollary to this was the concept that attempts to prevent or treat disease were considered to thwart God's will and were thus inappropriate. One such example was the opposition elicited among many in Boston by Mather and Boylston's attempts to variolate children against smallpox (Chapter 9). Disease was Apollo's punishment for Syphilis failing to worship him and has long been regarded as appropriate punishment for sexual misbehavior. Major disease outbreaks such as the influenza epidemic of 1918 were claimed by preachers such as Billy Sunday to be God's punishment on a wicked country. Much of the early hesitancy in confronting the AIDS epidemic resulted from a similar attitude. Even today, recent surveys have indicated that 15% of the US population still regard the AIDS epidemic as God's punishment for homosexual behavior (Chapter 17). Some have suggested that COVID-19 is also a punishment from God.

inadequate sanitation were prevalent. Nevertheless, in 2 months more than 3500 individuals died in the city, mostly Irish immigrants and free Blacks. The scene was described thus by an assistant to the landscape painter Asher Durand: "There is no business doing here if I accept that done by Cholera Doctors, Undertakers, Coffin makers, &c. Our bustling city now wears a most gloomy & desolate aspect one may take a walk up & down Broadway & scarce meet a soul." By July cholera had also reached Philadelphia, and by August it had reached Chicago. The disease spread down the East Coast and reached Washington, DC in the late summer of 1832, killing an estimated 12 to 13 persons daily [3]. The winter briefly slowed its spread, but it reached New Orleans the next year either by the coastal route or by way of the Mississippi. Six thousand died in New Orleans, reflecting the inadequate sewage facilities and the difficulties in draining low-lying swamps in that region.

During 1833 and 1834, cholera arrived in Texas and Mexico, probably by way of New Orleans. It was not understood at the time that this was a waterborne disease. Some believed that it was spread by fresh fruit; others claimed that pork carried it, and some communities banned pork consumption. Houses were fumigated with smoke. As in all major disease pandemics, people fled the towns and cities. Some communities were hardly affected, whereas others suffered severely. For example, in Texas, the small town of Goliad lost 91 people over 3 weeks in June 1834 [4]. That was 20% of the residents. They blamed fruits and vegetables and banned their sale. The dead were hidden out of view. And water had to be filtered through burnt toast! San Antonio, in contrast, had few deaths. When a refugee from Goliad arrived with the news of the outbreak, the authorities took precautions to prevent panic; orders were issued that the church bells not be tolled for the dead; troops were posted on the roads out of town to exclude refugees from Goliad. Citizens were encouraged to remain in place, but most people decamped to the Texas Hill country, leaving the city almost deserted—something that probably helped considerably. Stephen F. Austin was in Mexico City when cholera reached that city. He asserted that there were 43,000 sick and about 18,000 dead. "I have never witnessed such a horrible scene of disease and death." Austin himself came down with a mild case of the disease. "I was taken about 3 o'clock pm with excessive purging of a white mucos character, great pain in the bowels, cold feet, legs, hands, etc., pains all over the body." Minutes later, Austin reported, "relieved by a fine perspiration which I think saved my life, for others have died in less than one hour whose simptons were similar to mine."

For the first part of the 19th century, cholera was considered a miasmatic disease, brought on by bad smells resulting from filth and decay. It was not considered contagious. The miasmas were believed to originate in rotting vegetation. Thus a combination of miasmas and immoral behavior was deemed seriously lethal. As with other infections, a multitude of ineffective home remedies were tried without success. There were variations of the miasma theory; it could be noxious fumes, it could be the wind. The occurrence of many diseases in marshes or swampy areas tended to support this theory. Even as late as 1885 the New York City Board of Health suggested that digging trenches for underground telegraph lines could release noxious fumes! A variant of this was the filth theory, specifically the smell of feces. The move to indoor toilets instead of outdoor privies reduced these smells. Certainly, unhygienic environmental conditions were associated with the disease. It was not until the work of Dr. John Snow in London identified sewage-contaminated water as its source, and Robert Koch identified the cholera vibrio, that rational control procedures could be established [5].

There were always a few individuals who considered possible other sources of the "poison." For example, Henry Boyd, a prominent Cincinnati furniture maker, proposed that water was the source of cholera in 1832. He was correct but ignored. Likewise, John Lea of Cincinnati claimed in the 1848 outbreak that cholera was spread by water. Lea noted that cholera occurred in homes whose water supply came from wells but not in homes that relied on rainwater. Lea was an amateur geologist and as a result attributed the disease to cholera poison derived from alkaline soils, specifically calcium and magnesium salts in the drinking water. (Magnesium sulfate [Epsom salt] is a potent laxative so John Lea was correct in some respects.)

The third pandemic

The third cholera pandemic began in India in 1846, reached England in 1848, and the United States in 1849. This time it was not only carried by waterways such as the Ohio and Mississippi but also by the growing and much faster railroad network [6]. The westward movement of settlers and the California Gold Rush in 1849 ensured that cholera spread through the Mississippi valley, up the Missouri, and along the wagon trails to California. The disease reached San Francisco by boat in early October 1850 (perhaps by the same boat that brought the news that California had been admitted to the Union!). Within a few days it had moved on to Sacramento [7]. It is estimated that 6000 to 12,000 migrants died from cholera on their way to California between 1849 and 1855. New Orleans experienced two disease outbreaks in 1853—yellow fever and cholera. About 8000 died of yellow fever and 129 from cholera. As always, it hit those who had no access to clean drinking water, impoverished recent immigrants, and free Blacks. In 1849 the cholera also reached Chicago where 678 people died. A second outbreak in that city in 1854 killed 1424 (about 5% of the population). The 1849 cholera epidemic killed 5071 in New York. It was during this pandemic in 1854 that Dr. John Snow in London demonstrated that cholera was acquired from contaminated drinking water. It took some time for the news to reach America.

Cholera most likely contributed to the death of President Zachary Taylor in 1850. He suffered 4 days of gastroenteritis with stomach cramps and severe diarrhea (not a sign of typhoid). His physicians certified that he died of "*cholera morbus*" (physician's term for really, really bad diarrhea and vomiting).

A fourth pandemic began in the Ganges Delta in 1863 and also spread to America, causing epidemics until it died out in 1869. This disease spread rapidly across the United States by way of the railroads. Thus it reached New York in May 1866, Galveston in July, San Antonio in September, Albuquerque in October, and San Francisco in December [8]. The most recent pandemic to reach the Americas originated in 1961, not in Bengal but uniquely in Sulawesi, Indonesia. It was somewhat different from the previous ones because it was caused by a new biotype called "El Tor" after the Egyptian quarantine station where it was first identified.

Cholera and Polk

During the third cholera pandemic, the disease arrived in New York City and New Orleans in December 1848 and persisted until 1854. It arrived in Nashville in January 1849. President James K. Polk stepped down from the presidency in March 1849. He had arranged for a triumphal tour of the South on his way to Nashville where he planned to retire. He and his wife went south along the coast and then west to New Orleans. He was worried about cholera because it was said to be common in that city. After a few days he sailed north on the Mississippi, but several passengers died of cholera and Polk felt so ill that he went ashore at Natchez for 4 days. However, a doctor reassured him that he did not have cholera. He therefore sailed on to Nashville, arriving there on April 2. The Polks settled into their new home but he fell ill again with what appeared to be cholera. He died on the afternoon of Friday, June 15, 1849 at the age of 53. His body was first buried in a mass grave in the Nashville City Cemetery in accordance with the city's laws. A year later his body was moved to the grounds of his home and subsequently to the grounds of the Tennessee State capitol. During 1849, there were 311 deaths from cholera when Nashville had a population of about 10,000.

Cholera and immigration

Until the mid-1820s, immigration into the United States was an informal affair managed by each state. The country had lots of available space and immigrants were generally welcomed. The arrival of thousands of Irish in the late 1840s caused some concern because they were Catholic and carried cholera. The arrival of large numbers of Chinese on the West Coast eventually resulted in Exclusion Acts in 1882 and 1892 designed to prevent Chinese immigration, largely for racial reasons—they were not repealed until 1943. In 1875 immigration laws were established to exclude criminals and other undesirables. As the century progressed and it was realized that diseases such as cholera were transmitted by infectious agents, the link between immigration and disease outbreaks was established and as a result, immigrant regulation began to focus on preventing the introduction of infectious diseases. The Immigration Act of 1891 allowed those with contagious diseases to be deported. Health inspections and quarantines became routine and relatively efficient. The Marine Hospital Service was established and quarantine stations constructed [9]. Given that New York City was the main port of entry for immigrants to the United States, most of the action occurred there. This was an era of mass immigration, and over 50,000 immigrants arrived each month, mainly from Germany, the British Isles, and Russia. Ninety percent of these immigrants passed through New York.

Many of the Russian immigrants were fleeing persecution in Czarist Russia. There had been violent pogroms and a severe famine. Jews were expelled from Moscow and as a result many sought a new life in the New World. They traveled by train across Europe. They could travel through Germany but were not permitted to leave the train. When they reached the port city of Hamburg, they were housed until ships could take them to America. An epidemic of cholera developed in Russia in 1892 with cases occurring in Moscow that July. Many of the refugees may have brought it with them, whereas others would have contracted it after arriving in Hamburg. (Hamburg's sewers emptied into the River Elbe. That is where the city also got its water supply. The ebb and flow of the tides allowed the sewage to enter the city's drinking water.) Cholera broke out in the city although, as was common, city officials kept it a secret for as long as they could. The American consul in Hamburg eventually found out about it, but by then it was too late. The refugee ships had already sailed. The immigrants not only drank polluted Elbe water while waiting for a ship but also drank it during the voyage and even when the ship was held in quarantine off New York.

On August 30, 1892, the steamer *Moravia* arrived in New York from Hamburg, Germany [10]. She had on board 385 steerage passengers, many of whom were Russian Jews. The *Moravia* anchored off Staten Island near the quarantine station. Although the Captain assured quarantine officers that everyone was healthy, a check of the ship's surgeon's records reported that there had in fact been 24 cases of "cholerine" and 32 deaths during the 13-day voyage! The ship was obliged to steam to "lower quarantine" in the outer bay.

When the news of the *Moravia's* arrival with cholera on board was announced, President Benjamin Harrison issued a proclamation declaring that all ships sailing from infected European ports must be quarantined for 20 days after their arrival. Thereafter, arriving vessels were thoroughly inspected.

On September 3, 1892, two other ships from Hamburg arrived and were inspected. Both the *Rugia* and the *Normannia* had had multiple deaths while at sea, all affecting "steerage" passengers only. These steerage passengers (in effect, third class) were sent to the quarantine station on Hoffman Island where they were bathed while the ships were disinfected. Police patrolled the waterfront to prevent anyone from swimming ashore. The quarantine facilities were seriously overextended; the island was small and crowded with hundreds of steerage passengers. Clean water was in short supply. Privies directly drained into the harbor. The sanitary committee who inspected the island reported, "the northern side was in a state so deplorable and unsanitary that it is difficult for your Committee to describe it in temperate language." The staff were overwhelmed. Sick passengers were meant to be transferred to the hospital on nearby Swinburn Island but often had to wait in hallways until this could be done. The next day two more infected ships arrived from Hamburg, the *Wyoming* and the *Scandia*. In the *Scandia* there had been 32 deaths during the crossing out of about a thousand passengers. Two more died after arrival at Swinburne Island. (Swinburne Island is a small artificial island in New York Bay, just off Staten Island. It was used as a quarantine station to house immigrants determined to have contagious diseases [Figs. 12.3 and 12.4].)

Healthy passengers were horrified at having to remain on board along with the sick, and as a result there was a debate about where to house them. There was much confusion. Some were transferred between ships. An effort was made to establish a quarantine camp on Fire Island, but this failed when a mob gathered to prevent the passengers from the *Normannia* from landing. The pilot refused to take the vessel into the dock. They were finally permitted to disembark in Hoboken on September 17.

The citizens of New York were horrified when the threat of cholera suddenly materialized as well. By this time, its cause and mechanism of spread were generally accepted [11]. It was understood to be carried by the immigrants. As a result, massive resistance to Jewish immigration developed and anti-Semitic sentiment exploded. In the end, 130 people died of cholera in New York, 66 of them on quarantined ships in the harbor and 44 in the quarantine station (Box 12.2).

Although a crisis had been averted in New York, there was a national outcry and the government responded. In January 1893 the Harris-Rayner

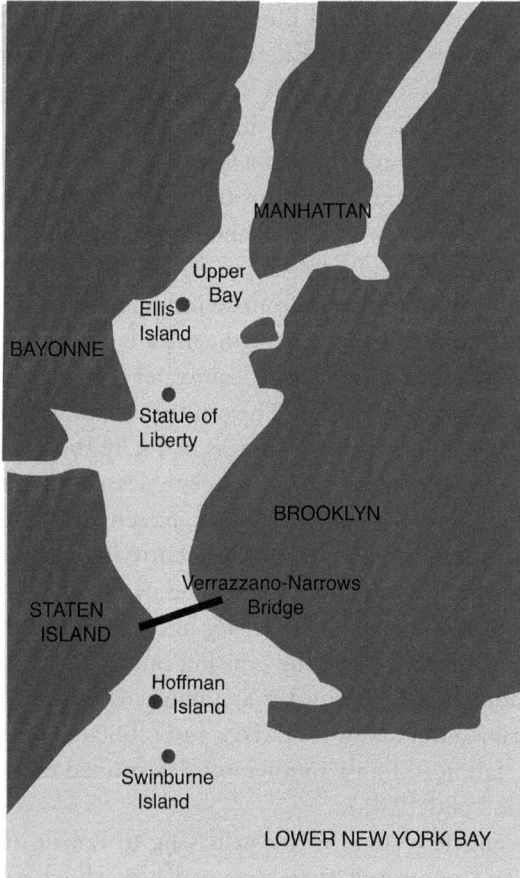

Fig. 12.3 *The locations of the New York quarantine stations.*

National Quarantine Act was passed. It provided regulations for ship inspections as well as for disinfection of both ships and immigrants. It also required specific health documentation for ships before they sailed for the United States. The Marine Hospital Service was given the power to intervene and prevent disease entry. The Canadians followed suit when Quebec and Ontario imposed their own quarantine rules. The quarantine facilities at Grosse Île installed steam disinfectors that could sterilize the luggage of 600 immigrants daily as well as a "sulfur (di)oxide blaster" to fumigate ships. The Canadians and the Americans also began to coordinate their quarantine efforts to prevent immigrants from carrying the disease across the land border [12].

Fig. 12.4 *The quarantine hospital on Swinburne Island.* Harper's Weekly, *September 1879. Courtesy National Library of Medicine.*

The last cholera epidemic in the United States began in October 1910 when a ship carried the disease from Naples in Italy to New York City. The passengers were quarantined and treated on Swinburne Island where 10 passengers and one hospital attendant died.

Cholera in the environment

Although the cholera pandemics historically originated in Asia, it now appears that *V. cholerae* is endemic in certain aquatic environments, including US coastal waters. For example, Dr. Rita Colwell, a microbiologist at the University of Maryland, and her colleagues demonstrated that *V. cholerae* was present in water in Chesapeake Bay despite the fact that cholera had not been recorded in the United States since 1911. Sixty-five pathogenic isolates of *V. cholerae* non-O1 were isolated from the Bay between 1976 and 1978 [13].

The first reported case of cholera in the United States since 1911 was isolated from a patient living in Port Lavaca, Texas in 1973; it was identified as *V. cholerae* O1 biotype El Tor [14]. In 1978 another isolate was found in a patient in Louisiana associated with eating locally harvested crabs [15]. In 1986, 13 cases of severe diarrhea occurred in the same region of the Mississippi Delta. This outbreak was traced to consumption of inadequately cooked crabs and shrimp from the local bayou. *V. cholerae* was detected in

BOX 12.2 Quarantine stations in New York.

From the first cholera outbreaks to the Spanish flu, New York City was the primary port of entry for aspiring immigrants. As a result, much of the Marine Hospital Service's efforts at quarantine centered on the area of New York Harbor. The first quarantine station was established in 1799 on the northeastern shore of Staten Island close to where the Staten Island Ferry docks today. Ships with sick passengers had to wait offshore at anchor for 30 days. The station handled smallpox, typhus, and cholera cases. However, the passengers' friends were allowed to visit them (remember there was no concept of germs at the time). As a result, the quarantine was not very effective. Eventually, however, the growing population of Staten Island expressed unhappiness with the presence of the quarantine facility by burning it to the ground in September 1858!

The federal government next built a quarantine station on the two man-made islands located in the upper bay—Swinburne Island in 1860 and Hoffman Island in 1873 (Fig. 12.3). A third station was established on the hospital ship *Illinois* anchored near Sandy Hook in the lower bay.

In theory, incoming ships from tropical ports in Africa or South America first anchored near the *Illinois* where they were boarded and inspected. Any passengers with yellow fever or cholera were then transferred to the hospital on Swinburne Island. Ships from other ports could move to the upper bay for inspection. The workload on the quarantine stations was enormous because this was the peak of European migration to the New World. Inspections were often hasty and superficial. Rich passengers could expect to pass through the inspection process very rapidly indeed.

As described in the cases of the 1892 cholera scare and the 1910 Naples affair, the sheer numbers involved overwhelmed the Swinburne and Hoffman facilities. A new quarantine station was urgently needed.

In 1890, the Federal Government transferred Ellis Island to the immigration service and built a large, two-story, purpose-built processing station and hospital. It opened in 1892 but burned down in 1897. A second, fireproof building was opened in 1900. Huge numbers of immigrants were processed. On its first day alone, 2251 went through the inspection process! The Ellis Island station underwent almost continuous expansion and improvements in its facilities. Whereas immigration slowed during World War I, by 1921 business had improved and Ellis Island processed 560,000 immigrants that year.

Congress took steps to restrict European immigration in 1921 and 1924 and as a result, the number of arriving immigrants dropped significantly. Ellis Island changed its role to that of immigrant detention center. Inspections were performed aboard ships. The hospital closed in 1930. The facility finally closed in November 1954.

the water as well as in crabs and shrimp from the Bayou. All the isolates were *V. cholerae* O1 biotype El Tor. However, these aquatic isolates have some biochemical differences from strains isolated elsewhere. Since then more cases have occurred in Texas, involving both O1 and non-O1 strains.

An outbreak of gastroenteritis in Apalachicola Bay, Florida in 2011 resulted from eating raw oysters containing *V. cholerae* O75. The disease was relatively mild because this strain did not have the ability to produce cholera toxin.

It is clear that *V. cholerae* exists naturally in brackish water, estuaries, and salt marshes and may contaminate Gulf seafood. Non-O1 strains may be commonly found in some tropical freshwater sources. These marine cholera bacilli also have a close relationship with certain microscopic plankton called copepods [16]. Vibrios attach and grow on the chitin exoskeleton of the copepods and grow within their gut. They also attach to the copepod eggs. As they circulate in currents, cholera travels with them. The copepods do well in warmer water so it is possible that the prevalence of cholera will increase as climate changes. When events such as El Niño cause plankton blooms, then the cholera load may also increase. Vibrios may also play a role in decomposing dead crustaceans.

Thus, although *V. cholerae* is endemic in the United States, it rarely causes significant disease [17]. It should be pointed out, however, that other vibrios present in salt and brackish water are a major cause of seafood-borne bacterial food poisoning in the United States. These mainly result from eating undercooked seafood, especially raw oysters. *Vibrio parahaemolyticus* can cause mild gastroenteritis and causes the majority of such cases.

Outbreaks of cholera occurred in Peru and Mexico in 1991 and appear to have originated in Asia. Recent African outbreaks have also originated in Asia. On October 21, 2010, the presence of cholera was confirmed in Haiti and the Dominican Republic [18]. By December, the Florida Department of Health confirmed five cases of cholera in residents arriving from Haiti. It was O1 biotype El Tor, serotype Ogawa. The cholera almost certainly originated from water supplies contaminated with fecal material originating from United Nations peacekeepers from Nepal. They had been sent to Haiti to help maintain order following a major 7.0 earthquake. It has been reported that 820,461 people have sickened and 9792 have died of cholera in Haiti since 2010, whereas in the neighboring Dominican Republic, there were only 21,400 cases and 363 deaths. This reflects how a quality water supply, good sanitation infrastructure, and appropriate responsiveness can reduce the effects of a cholera outbreak. Recent analysis suggests that at least two serotypes of *V. cholerae* are circulating in Haiti, one Asian, from Nepal, and one local. With improved access to potable water and sanitation, early detection and treatment, vaccination programs, and education about the disease, the number of cases of cholera was eventually reduced to zero in 2019. But the environment remained contaminated.

On September 30, 2022, after more than 3 years with no confirmed cholera cases, the Haitian Ministry of Public Health was notified of two patients with acute, watery diarrhea in Port-au-Prince. Within 2 days, this was confirmed to be cholera. As of January 3, 2023, more than 20,000 suspected cholera cases had been reported throughout the country (Box 12.3).

Treatments

A victim of cholera can lose as much as 2 to 3 gallons of fluid daily, along with huge losses of sodium, potassium, and bicarbonate. This extreme dehydration and loss of essential salts lead to an electrolyte imbalance, affecting muscles (leg cramps) and other organs (cardiac irregularities) that can lead to shock and death within hours. In less severe cases, victims with continued loss of fluids and electrolytes can succumb in a couple of days (Box 12.4).

The primary treatment of cholera is now vigorous but careful oral rehydration therapy using fluids containing glucose, sodium chloride, potassium chloride, and sodium bicarbonate. In emergencies, these fluids can be given intravenously. This simple but highly effective treatment for cholera was developed by Captain Robert Phillips, USN, in the late 1960s [19]. The antibiotic tetracycline shortens the duration of the disease and the duration of fecal shedding. However, multidrug resistance is emerging in Asia. Several safe and effective vaccines are available. They should be used by travelers visiting areas where cholera is endemic.

BOX 12.3 Cholera and climate change.

Warming temperatures are associated with the spread of some pathogens. For example, between 1992 and 2022 there was a progressive increase in the frequency of Vibrio reported from the Gulf of Mexico in Florida. In addition, a spike in Vibrio infections was reported following Hurricane Ian in September 2022. Most of these isolates were *V. parahaemolyticus* and *V. vulnificus*. These increases appear to have resulted from a rise in sea surface temperature and chlorophyll concentration, providing an optimal environment for the growth of Vibrios and related pathogens [20].

BOX 12.4 Flesh-eating bacteria.
The bacterium *Vibrio vulnificus* lives in estuaries and thrives in warm water. It is especially prevalent during heat waves when sea surface temperatures are elevated. A person with an open wound can become infected by contact with contaminated salt water or raw seafood. *V. vulnificus* can cause life-threatening wound infections. The bacterium produces a potent toxin. This toxin kills the tissues around the infected wound. The media call these organisms "flesh-eating bacteria." As a result, many people with *V. vulnificus* infection require intensive care or limb amputations, and about one in five people with this infection die, sometimes within a day or two of becoming ill.

References

[1] Daly WJ. The black cholera comes to the central valley of America in the 19th century — 1832, 1849, and later. Trans Amer Clin Climatol Assn 2008;119:143—53.

[2] Tuite AR, Chan C, Fisman DN. Cholera, canals, and contagion: rediscovering Dr Beck's report. J Publ Hlth Policy 2011;32(3):320—33.

[3] Maizlish SE. The cholera panic in Washington and the compromise of 1850. Wash Hist 2017;29(1):55—64.

[4] Haggard JV. Epidemic cholera in Texas, 1833-1834. Southwest Hist Q 1937;40(3): 216—30.

[5] Morens DM. Commentary: cholera conundrums and proto-epidemiologic puzzles. the confusing epidemic world of John Lea and John Snow. Int J Epidemiol 2013; 42:43—52.

[6] Pyle GF. The diffusion of cholera in the United States in the nineteenth century. Geogr Anal 1969.

[7] Roth M. Cholera, Community and Public Health in Gold Rush Sacramento and San Francisco. Pacific Historical Review 1997;66(4):527—51.

[8] Humphreys M. How four once common diseases were eliminated from the American South. Health Affairs 2009. https://doi.org/10.1377/hlthaff.28.6.1734.

[9] Tognotti E. Lessons from the history of quarantine, from plague to influenza A. Emerg Inf Dis 2013;19(2):234—59.

[10] Blank M. *Cholera in New York Harbor 1892*. http://www.maggieblanck.com/cholers1892.html.

[11] Jackson PS. Fearing future epidemics: the cholera crisis of 1892. Cult Geog 2012;20(1): 43—65. https://doi.org/10.1177/1474474012455017.cgj.sagepub.com.

[12] Hamlin C. "Cholera forcing": the myth of the good epidemic and the coming of good water. Am J Publ Health 2009;99:1946—54.

[13] Colwell RR, Kaper J, Joseph SW. *Vibrio cholerae, Vibrio parahaemolyticus,* and other vibrios; occurrence and distribution in Chesapeake Bay. Science 1977;198:394—6.

[14] Islam MS, Drasar BS, Sack RB. The aquatic environment as a reservoir of *vibrio cholerae;* a review. J Diarrhoel Dis Res 1993;11(4):197—206.

[15] Loharikar A, Newton AE, Striuka S, et al. Cholera in the United States, 2001-2011: a reflection of patterns of global epidemiology and travel. Epidemiol Infect 2015;143(4): 695—703.

[16] Kaper JB, Bradford HB, Roberts NC, Falkow S. Molecular epidemiology of *vibrio cholerae* in the US Gulf coast. J Clin Microbiol 1982;16(1):129—34.

[17] Lowry PW, Pavia AT, McFarland LM, et al. Cholera in Louisiana: widening spectrum of seafood vehicles. Arch Intern Med 1989;149:2079—84.

[18] CDC Update on Cholera — Haiti. Dominican Republic and Florida. MMWR 2010; 59(50):1637—41.

[19] Phillips RA. Water and Electrolyte Losses in Cholera. FASEB Proc 1964;23(3): 705—12.

[20] Brumfield KD, Usmani M, Santiago S, et al. Genomic diversity of *Vibrio* spp. and metagenomic analysis of pathogens in Florida Gulf coastal waters following Hurricane Ian. M mBio 2023. https://doi.org/10.1128/mbio.01476-23.

Poliomyelitis—1841

Contents

The term poliomyelitis refers to inflammation of the gray matter (polio-) of the spinal cord (myelitis). The gray matter contains the nerve cells—the neurons. (The other component of the spinal cord, the white matter, consists of myelinated nerve axons.) Although poliomyelitis is a disease known since ancient times, it does not appear to have caused major epidemics until the 20th century. Prior to that time, under unhygienic conditions in rural communities with primitive sanitation and poor water quality, the virus circulated freely. As a result, most children were infected with the virus when very young and so developed lifelong immunity. Indeed, many babies would have become infected while they were still protected from the virus by antibodies transferred from their mothers in the womb.

Poliomyelitis is caused by human enterovirus C (poliovirus), a positive-sense RNA virus belonging to the genus *Enterovirus* of the family *Picornaviridae*. Its genome consists of a single strand of RNA containing about 7500 nucleotides. Three types of poliovirus, types 1, 2, and 3, each with a different capsid protein, have been identified. (These different types can be distinguished by their reaction with specific serum antibodies and are thus called serotypes.) All three poliovirus types were strictly human pathogens.

Great American Diseases
ISBN: 978-0-443-31404-9
https://doi.org/10.1016/B978-0-443-31404-9.00013-6

Poliovirus type 1 was historically the most common but is currently confined to remote rural parts of Pakistan and Afghanistan. The last isolation of poliovirus type 2 occurred in India in 1999. It has not been detected since and was declared eradicated in 2015. Poliovirus type 3 was last detected in 2012 in Nigeria and Pakistan and declared eradicated in 2019.

The disease

Poliovirus first infects the cells of the intestinal tract. An individual usually ingests the virus in fecally contaminated food or water (the fecal-oral route) and the virus then grows in the lymphoid tissues of the intestine. As a result, infected humans shed large amounts of virus in their feces. In societies without sewage treatment facilities and thus contaminated drinking water, poliovirus spreads rapidly. Public swimming pools were also long recognized as a significant source of infection in the United States and Canada. In about 95% of infections, the virus causes only a transient viremia (when the virus can be found in the bloodstream) and these individuals show no evidence of disease. In the remaining 5% of infected humans, however, there is an initial incubation period of 3 to 30 days while the virus invades other tissues such as muscles and lymphoid tissue. The virus may then cause a secondary viremia, resulting in a fever, sore throat, stiff muscles, and headache. This also generally resolves rapidly with no residual problems.

In fewer than 1% of infected individuals, however, the poliovirus travels through the bloodstream to the brain stem and spinal cord. There it selectively destroys a specific subset of motor neurons (called anterior horn cells) located in the patient's spinal cord. As a result of this nerve cell destruction, the patient's arm and leg muscles can no longer be controlled, and the patient becomes paralyzed. The paralyzed muscles in the limbs relax and become flaccid. This paralysis may be temporary or permanent but normally only persists for several days or weeks. Bulbar poliomyelitis is a severe variant of the disease that results from the destruction of the motor neurons located in the brain stem. These neurons innervate the muscles in the pharynx and soft palate, and their destruction prevents both breathing and swallowing. The most severe disease affects the respiratory centers in the brain medulla. Neuronal destruction in that part of the brain results in paralysis of the respiratory muscles and consequently suffocation and death. Sensory nerves are unaffected by the virus so touch and pain sensations in the limbs remain intact.

Poliomyelitis rarely caused disease in the very young, so it was not considered a major issue in babies despite being called infantile paralysis. However, in "clean" societies where water supplies were normally virus free, children tended to become infected at a later age and it was they who developed paralysis and death. Death usually resulted from paralysis of the respiratory muscles. These patients therefore required assistance in breathing and had to be placed in an "iron lung" if they were to survive. Most but not all paralyzed patients recovered some functionality. Nevertheless, the disease affected many victims for the rest of their lives; many needed wheelchairs, crutches, or leg braces. Some survived only to spend the rest of their days in an iron lung.

History

The earliest evidence of poliomyelitis is probably seen in wall paintings and carvings from ancient Egypt that show people with flaccid foot paralysis or withered legs. A picture of a crippled beggar painted in 1559 by Pieter Bruegel shows a similar lesion. However, although suggestive, these cases may have been due to other causes. In 1773, the future Sir Walter Scott, the noted Scottish novelist, developed an infection that left him permanently lame in one leg. His detailed description strongly suggests that this was caused by poliovirus. A few years later, in 1789, a British physician, Michael Underwood, provided the first clinical description of "a debility of the lower extremities, usually attacks children previously reduced by a fever." Because the disease primarily infected children it was commonly known as infantile paralysis.

Improved sanitation, sewage disposal, and water quality changed this early infection pattern, and children began to encounter the poliovirus for the first time at a later age. Thus, as living conditions and social hygiene improved, poliomyelitis (hereafter polio) began to show itself with increasing frequency [1].

The United States

The first recorded US outbreak of what was probably polio occurred in 1841 in West Feliciana Parish, Louisiana, when 8 to 10 cases of infantile paralysis were reported [2]. The next reported outbreak occurred in Boston in 1893 when there were 26 cases. The third major polio outbreak was reported in Rutland County, Vermont, in 1894. Eighteen fatal cases and 50 cases of permanent paralysis resulted out of 123 children being affected.

The local doctor, Charles Caverly, published his observations in the *Yale Medical Journal*. He described the pattern of a fever and headache followed, in some cases, by paralysis of variable severity. He noted that most victims were male and that infantile paralysis was a misnomer. Most of his cases were in children aged between 1 and 6. He recognized that the disease did not always cause paralysis and some children recovered rapidly. Caverly was persuaded that the disease was not contagious; in only one instance did he find more than a single case in a family. He argued therefore that since many households had multiple children the disease could not have been contagious.

Outbreaks occurred almost annually thereafter, predominantly in the Northeast, but also in California in 1896 and 1899. The outbreaks occurred with increasing frequency and involved many more individuals—it was an emerging disease. For example, in 1907 there was a major outbreak in New York City with 2500 cases. Other outbreaks occurred across the northeastern states between 1910 and 1914. As these outbreaks increased in severity, it was obvious that this was a disease of summer. It occurred predominantly in hot weather and peaked in August and September. These summer epidemics resulted in the closure of swimming pools, amusement parks, and other children's activities, and children were confined indoors by their parents.

It was only in 1905 that Ivar Wickman, a Swedish physician, first suggested that infantile paralysis was contagious. He too recognized that the disease could appear in a mild form in some individuals. In 1908, two Austrian microbiologists, Karl Landsteiner and Erwin Popper, filtered the spinal fluid from a lethal case of polio through a porcelain filter impermeable to bacteria. They injected the filtered fluid into two rhesus monkeys, who then developed the same disease symptoms as the donor. The agent was too small to be a bacterium—it had to be a virus!

In 1916, the greatest polio outbreak of all occurred in Brooklyn, New York [3]. The outbreak began in a densely populated area called Pigtown, where the first two affected children were diagnosed in May of that year. By the end of the month there were 17 new cases, and by the end of June there were 646 cases. When the epidemic faded in December there had been 8900 cases of paralysis and 2448 deaths in New York City alone. The paralysis rate in 2- to 3-year-old children reached 2%, which was unusually high.

As with so many infectious diseases at the time, it was natural to attribute the outbreak to "filth." Recent Italian immigrants were blamed for introducing the disease. This was, we now know, the opposite of the truth. Analysis of the New York data shows that the paralysis rate was higher in more affluent districts. Locations with the highest infant mortality and prevalence of infantile diarrhea had the lowest polio mortality, whereas the cleaner districts had more polio, and it developed in older children. Nevertheless, the city authorities undertook the standard response—clean up the cities and quarantine the afflicted.

The 1916 polio outbreak resulted in widespread panic, and thousands left the city and moved to the Catskills and Adirondacks to get away from the disease. Public meetings almost ceased; beaches, swimming pools, amusement parks, and theaters closed. Newspapers reported the daily disease toll. By August, polio cases had also been reported in New Jersey, Pennsylvania, upstate New York, and Connecticut. It is believed that it eventually caused 23,000 cases and 5000 deaths in New England and the Mid-Atlantic States. Thereafter, polio outbreaks were almost an annual summer event somewhere in the United States. Summer was the "polio season."

During these epidemics, hospitals were overwhelmed, and because nothing was known about the agent, treatments were generally symptomatic and mainly involved good nursing care [4]. Any new or innovative treatments tried were universally ineffective. Some, not having heard of Landsteiner and Popper, still thought that polio was caused by a bacterium. Unfortunately, by the time children arrived at a hospital their motor neurons were already destroyed. Some children were injected with the disinfectant hexamethylamine by intraspinal injection. Hydrogen peroxide was given as an intranasal spray. Adrenaline, quinine, or urea was given intraspinally. Some were given salvarsan (the syphilis treatment!). Others received strychnine. Some were given tetanus or diphtheria antitoxin. Lumbar puncture was used in an effort to remove cerebrospinal fluid and reduce intraspinal pressure. This was also believed to remove "toxins" [5]. As the New York polio outbreak waned, the great influenza pandemic of 1918 eventually overshadowed it in public memory.

Iron lungs

Death from polio is usually a result of suffocation due to paralysis of the intercostal muscles. These muscles lift up the ribs and the diaphragm and so expand the chest cavity, causing inhalation. Thus, in order to keep victims alive, they had to be physically helped by an artificial respirator. The first iron lung was developed in 1928 by Philip Drinker and Louis Shaw at Boston Children's Hospital. Patients were encased in an airtight metal drum with their heads sticking outside through a rubber seal (Fig. 13.1). The chamber was attached to motor-powered bellows. When the air was pumped out of the chamber into the bellows, the negative pressure caused the chest and lungs to expand to fill the vacuum, and so the patient inhaled. When air was readmitted to the chamber, the chest would collapse, and the patient exhaled.

Improvements in technology gradually followed. Patients with bulbar polio were placed in iron lungs until they recovered the ability to breathe on their own. It was hoped that this would be a temporary expedient. Unfortunately, this might be a lifetime for some individuals. For example, in 1952, 6-year-old Paul Alexander from Dallas, Texas, contracted polio, became paralyzed from the neck down, and was placed in an iron lung. He died, 72 years later, in March 2024, still in an iron lung. The machines were large because they had to contain the patient, and they were also very expensive to purchase and run. Huge numbers were needed as the epidemics grew, and hospital wards were often lined wall to wall with iron lungs. Iron lungs are no longer manufactured. They were eventually replaced by much smaller positive-pressure ventilators.

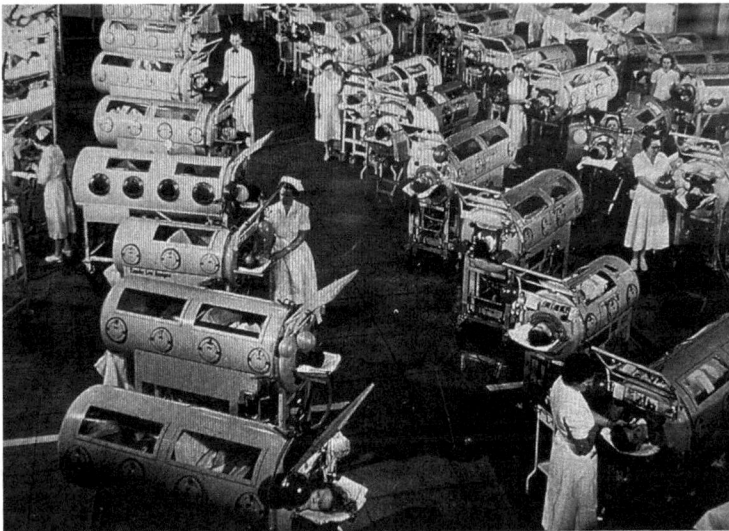

Fig. 13.1 *Polio patients suffering from bulbar poliomyelitis in iron lungs during the 1940s and 1950s epidemics. Courtesy WHO Global Polio Eradication Initiative. Tulchinsky TH. John Enders, Jonas Salk, Albert Sabin and eradication of poliomyelitis. Case Studies Public Health 2018:383—406.*

There was much debate about the best way to manage paralyzed limbs. Some physicians promoted complete lack of movement. This enforced rest, sometimes achieved by enclosing patients in plaster casts for months, led to severe muscle atrophy. Others, especially Sister Elizabeth Kenny, promoted the use of strips of blanket soaked in very hot water to relax and thus relieve muscle spasms as well as regular exercise to prevent muscle atrophy. (She believed that polio was a muscle spasm disease, not a nerve disease.) Kenny was a nurse who first established her physical therapy clinics in Australia but was severely criticized by the medical establishment there [6]. On the other hand, patients flocked to her. She moved from Australia to the United States in 1940 in the hopes of receiving a better reception. Studies in Minneapolis confirmed that her procedures involving warm wraps and exercise worked much better than muscle rest and gradually came to be accepted. She was the most famous physiotherapist of her era although considered to be controversial, outspoken, and difficult to work with. Hollywood made a movie about her in 1946. Sister Kenny Rehabilitation Institutes were established across the United States, where her innovative techniques were studied and improved.

Various other treatments were applied to polio patients, but all were ineffective. The first encouraging results were obtained by passive immunization with hyperimmune serum. Using serum from polio survivors that contained virus-neutralizing antibodies, a trial was conducted in 1951—1952. It appeared to reduce the incidence of paralytic cases by about 80%. Unfortunately, this immunity was short lived, and the antiserum was not easy to produce. Attention soon turned to active immunization using polio vaccines.

Franklin Roosevelt

Franklin Roosevelt had led a sheltered life as the son of a wealthy family and had probably avoided many childhood infections as well as exposure to poliovirus. He was building a successful political career when, in July 1921, he visited a Boy Scout camp in the Hudson Valley near his family home in Hyde Park. Then he went to Campobello Island in the Bay of Fundy off Maine, where his family owned a summer home. He went swimming on August 10 but began to feel tired and chilly that evening. The next day, he developed a fever and the beginning of a gradual paralysis, bladder and bowel dysfunction, severe pain, and loss of feeling in both his legs. He eventually recovered a little, but the disease left him paralyzed in the

lower legs. All of these features are consistent with a diagnosis of poliomyelitis. On September 16, 1921, *The New York Times* reported that 39-year-
old Franklin D. Roosevelt had been transferred to the Presbyterian Hospital
in New York "suffering from poliomyelitis." (It has since been suggested
that Roosevelt was actually suffering from Guillain-Barré Syndrome, an
autoimmune neurologic disease. This is highly improbable because
Guillain-Barré patients have an 80% recovery rate. They also have persistent
sensory loss, whereas Roosevelt's were transient [7].)

The March of Dimes

As a result of his disabilities, Franklin Roosevelt gave up his law practice but
persisted in his political career. Thus he was elected Governor of New York
in 1928 and President in 1932. He tried various treatments for his paralyzed
legs, but they did not help much. However, in 1924, he began treatment at
Warm Springs, Georgia, soaking in the warm 88°F mineral-rich water [8].
The high mineral content increased its buoyancy, and he felt that this helped
him move his legs. As a result, in 1926, Roosevelt purchased the property
using most of his savings and established the nonprofit Warm Springs Foundation for the care of polio victims. Polio patients flocked there from across
the country. The Roosevelt Warm Springs Institute for Rehabilitation still
provides services for people with a range of disabling conditions, such as
stroke, amputation, orthopedic disorders, and numerous neurological
disabilities.

On January 3, 1938, the worsening polio epidemic, as well as criticism of
his focus on polio, prompted Roosevelt to establish the National Foundation for Infantile Paralysis. He appointed his former law partner, Basil
O'Connor, as its first president. Uniquely, rather than appealing to a few
rich donors, the foundation sought to raise funds by seeking small donations
from large numbers of the public. O'Connor recruited celebrities such as
Judy Garland, Jimmy Stewart, and Eddie Cantor to the cause. The comedian
Eddie Cantor suggested that they use the phrase "March of Dimes" when
encouraging radio listeners to send their dimes to Roosevelt in the White
House. It was an overwhelming success. Dimes arrived at the White House
"by the truckload" (Fig. 13.2). In the first year it raised over two and a half
million dimes—a quarter of a million dollars (equivalent to 4.4 million dollars in 2020).

Over the next years the March of Dimes provided support for over 80%
of polio patients as well as funding research. Shriner's hospitals were also

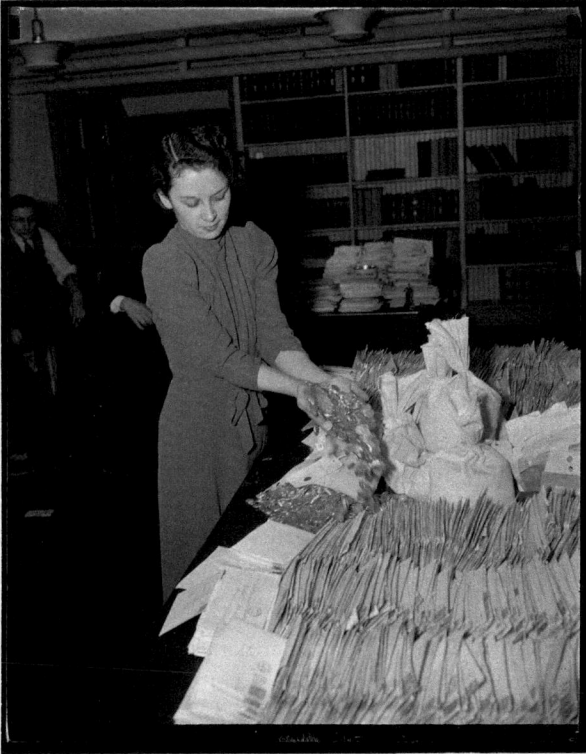

Fig. 13.2 *The White House was overwhelmed as enormous quantities of dimes arrived to support the Foundation for Infantile Paralysis. Courtesy Library of Congress.*

dedicated to care for children with polio. The March of Dimes pioneered several innovative fundraising activities, such as the annual Poster Child and the enormously successful "Mothers March," in which neighborhood mothers went from house to house one evening collecting donations. Although close to financial collapse at times, the National Foundation was a remarkably effective organization. With the eradication of polio in the United States, the organization's focus has changed. It officially changed its name to the March of Dimes Birth Defects Foundation in 1976.

During the World War II, unlike other infectious diseases, the number of polio cases continued to rise. There were 9000 cases in 1941, and this rose to 19,000 in 1944. There were several major outbreaks in 1944, the most notable being in Hickory, North Carolina. In 1946, cases numbered 25,000.

In 1945 there were under 1000 cases in Texas. In 1948 this rose to 1765, with major outbreaks in Houston, San Antonio, Galveston, and the Rio Grande Valley. In 1949 it rose again to 2355 and stayed high until Jonas Salk's vaccine was introduced. In the summer of 1949 polio hit the town of San Angelo in West Texas. It was the most severely affected town in America with 420 cases, 84 children paralyzed, and 28 dead. This was 1 case for every 124 residents. The city closed all schools, churches, movie theaters, local swimming pools, and swimming holes in the Concho River and prohibited indoor gatherings where children under 15 might be present. During the epidemic it was said that motorists were afraid to open their car windows when they drove through the town. The town hospital had 160 hospital beds, and half were filled with polio patients. The town was sprayed with the insecticide DDT, just in case.

As hygiene increased and sewage systems improved, the age at which children first encountered the poliovirus also increased. Thus, by 1950, the peak age for polio development was between 6 and 9 years of age, and a third of the cases occurred in individuals over 15. At the height of the epidemic in 1950–1954, polio paralyzed about 22,000 American children each year. Thousands were left permanently disabled and many died. The worst year was 1952, when there were 57,628 cases, 3145 deaths, and 21,269 cases of paralysis.

Research progress

Although the March of Dimes was established in 1938 to support polio research, progress remained depressingly slow. Over time, attention was increasingly focused on what seemed the only possible solution—a vaccine. Basil O'Connor, as the Director of the National Foundation for Infantile Paralysis, decided where the research support should be directed. O'Connor considered two possible vaccines: a killed vaccine being developed by Jonas Salk and a live, attenuated vaccine being developed by Albert Sabin. O'Connor decided that Salk's vaccine was likely to be available sooner than Sabin's, so the Foundation allowed itself to go into debt in order to support Salk.

Jonas Salk

Jonas Salk was born in New York in 1914. His mother and father were Russian immigrants who had migrated to East Harlem. He was a gifted student and obtained his medical degree from the New York University

School of Medicine in 1939. In one of his electives he was mentored by Thomas Francis, a professor of bacteriology at NYU and the discoverer of type B influenza virus. After completing his internship at Mt. Sinai Hospital, in 1941, Salk rejoined Thomas Francis at the University of Michigan in Ann Arbor, where he became deeply involved in the development of a killed influenza virus vaccine. (Until this time the major vaccines, smallpox, rabies, and yellow fever, all contained live attenuated viruses.) Memories of the influenza pandemic at the end of WW1 made this a matter of high priority. This killed flu vaccine was developed in 1943. It contained formalin-inactivated influenza virus that stimulated a strong antibody response. This experience led Salk to try a similar strategy in developing the polio vaccine. In 1947, he moved to the University of Pittsburgh School of Medicine and established a Virus Research Laboratory that was funded by seed money from the National Foundation for Infantile Paralysis (Fig. 13.3).

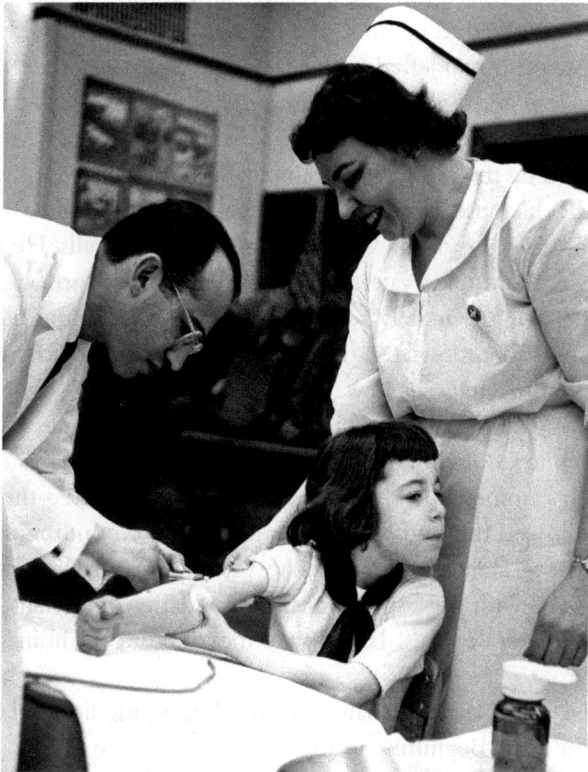

Fig. 13.3 *Jonas Salk administering polio vaccine to a young girl.* Compare the child's reaction to that in Fig. 13.4. *Courtesy U.S. National Library of Medicine.*

The keys to making a new polio vaccine were threefold. The first was to determine if there was more than one poliovirus. In 1948, Australians MacFarlane Burnet and Jean McNamara identified the three different types of poliovirus and showed that immunity to one would not protect monkeys against the other two. Jonas Salk's first task was to collect poliovirus from multiple sources and determine if they were identical. His group confirmed Burnet and McNamara's findings. Any effective vaccine would have to contain viruses of all three types.

The second need was the ability to produce a constant supply of virus. In 1949, Thomas Weller, Frederick Robbins, and John Enders, working at Boston Children's Hospital, became the first to grow poliovirus in non-neuronal cell cultures. They used several different cell types including skin and muscle cells as well as monkey kidney cells. In 1954 they received the Nobel Prize for this discovery. A reliable supply of vaccine virus was thus assured.

The third step was to develop a reliable and consistent way of inactivating the virus. This had to be done at a defined temperature with a carefully calibrated dose of formaldehyde that inactivates every virion in the solution. Salk's experience with inactivating influenza virus played a key role in successfully developing the inactivation process.

Encouraged by the results of the Enders' group, in 1948, using the same techniques he had used for influenza virus, Salk developed methods of growing large amounts of the virus in monkey kidney cells. He then determined that he could inactivate this virus by the use of formaldehyde. The vaccine containing the inactivated poliovirus was then administered by injection. Salk first demonstrated that his vaccine was safe and could protect experimentally infected monkeys.

In the spring of 1952, Salk quietly began small-scale trials of his vaccine in immune children (children who had already recovered from polio) at a local children's home and showed that the vaccine boosted their antibody levels. He followed this up by vaccinating a group of nonimmune children. Salk injected the children himself. The vaccine appeared to be very safe. Two years later, with the support of the Foundation for Infantile Paralysis, Salk began large-scale trials. Basil O'Connor recruited Thomas Francis to manage them.

The 1954 polio vaccine trials were the largest medical experiments in history to that time. Beginning in April 1954, more than 1,349,135 children aged between 6 and 9 participated as "Polio Pioneers" (they received a Polio Pioneer card and a badge). The trials occurred in 211 counties in 44 states.

The counties were selected by each state's health officers based on population, disease prevalence, and available resources. Two hundred and ten thousand children received the placebo, and 440,000 received the vaccine. An additional 1,180,000 children served as unvaccinated controls. Children received two doses, 2 to 4 weeks apart, followed by a third 7 months later.

The results of the trials were announced to the press by Thomas Francis on April 12, 1955—the tenth anniversary of Franklin Roosevelt's death. Francis declared that the vaccine was safe and very effective. The trials had indicated that the vaccine was about 70% effective against type 1 poliovirus and 90% effective against types 2 and 3. It was 94% effective against bulbar poliomyelitis, the most severe form of the disease. The vaccine was licensed within hours. Overnight Salk became famous and an instant hero— a scientific superstar. His name was posted across the front pages of newspapers in banner headlines. Church bells pealed across the country. In the face of huge demand there was a rush to vaccinate. Although the vaccine had been rapidly licensed, the government had made no arrangements to handle the enormous demand. The high price and short supply made the government look foolish. Large-scale vaccination eventually began in 1957, and as a result, the prevalence of polio dropped precipitously from nearly 58,000 cases in 1952 to 3200 cases in 1960.

Albert Sabin

Meanwhile, a second polio vaccine was also being developed. Albert Sabin was born in Poland in 1906 and emigrated with his parents to Paterson, New Jersey, at the end of WWI. He received his medical degree from NYU in 1931. After a short period working in England, he obtained a position at the Rockefeller Institute in New York. When working at the Rockefeller Institute he was impressed by Max Theiler, the developer of the very effective live-attenuated yellow fever vaccine (Chapter 11). Sabin moved to the University of Cincinnati in 1939. Thereafter he spent most of his career at the University of Cincinnati as Professor of Research Pediatrics. He worked on tropical diseases in the Army during World War II. After the war, Sabin returned to Cincinnati and continued the studies on polio that he had initiated in 1936. He demonstrated that poliovirus multiplied in the intestinal tract before spreading to the central nervous system. Unlike Jonas Salk, Albert Sabin believed that nonpathogenic viruses that grew in the intestine would be much more likely to provide prolonged strong immunity than injected, killed viruses. As a result, beginning in 1951, he worked to develop an oral polio vaccine (OPV) using a virus that had lost its ability to cause disease (Fig. 13.4). This loss of virulence is called attenuation. Sabin searched for mutant members of each of the three types of poliovirus that had lost the ability to cause disease but still induced high antibody levels. He grew these viruses in monkey cell tissue cultures and passed them many times until they had lost the ability to cause disease in humans. He first tested his vaccine on himself and his family members, members of his research team, and 30 volunteers from a nearby penitentiary. They all developed antibodies and nobody got sick. The enthusiasm for the newly available Salk vaccine made it very difficult for Sabin to get his vaccine tested in the United States. Sabin therefore had to look for a large population of children that had not received the Salk vaccine.

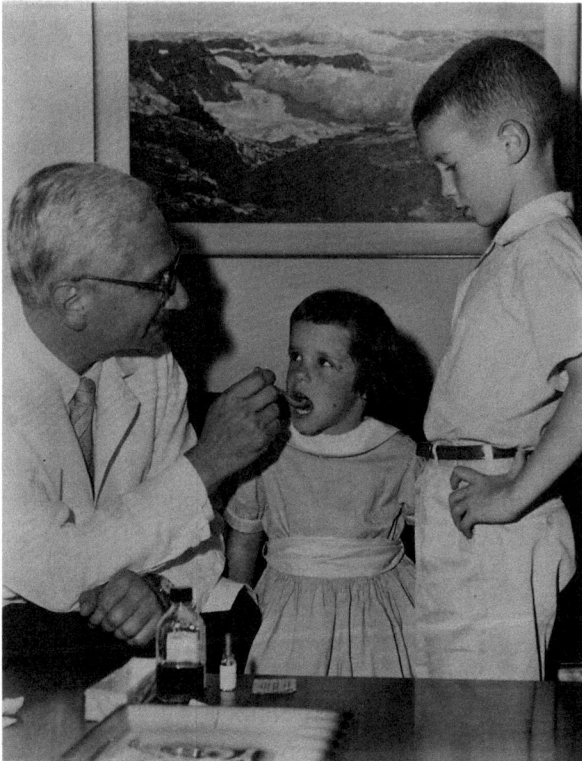

Fig. 13.4 *Albert Sabin administering his oral vaccine to two young children.* Compare this with Fig. 13.3! *Courtesy The Hauck Center for the Albert B. Sabin Archives, Henry R. Winkler Center for the History of the Health Professions, University of Cincinnati Libraries.*

By the 1950s, polio was a growing problem in the Soviet Union, and consequently, the Russians were looking for a vaccine. They considered Salk's vaccine, but it was expensive and difficult for them to produce and administer. Sabin, on the other hand, was keen to help and have his vaccine tested. He convinced the Health Ministry of the Soviet Union to conduct these studies. By 1957, clinical trials began involving 2 million Russian children. The OPV doses were either wrapped in candy or simply dropped directly into the children's mouths. At the end of the year the results were so good that the Russians decided to vaccinate 77 million individuals under the age of 20. Other trials took place around the same time in Eastern Europe, Singapore, the Netherlands, and Mexico. In 1959, Dr. Dorothy Horstmann was sent by the World Health Organization to assess the results of the Russian trials, and she reported that Sabin's OPV was safe and effective. (Dr. Horstmann was the scientist who first showed that poliovirus underwent a transient viremic stage before entering the nervous system. This meant that it would be vulnerable at that time to destruction by antibodies in the blood. She was the first woman Professor of Medicine at Yale.)

In 1956, the Foundation for Infantile Paralysis adopted a new slogan, "Polio isn't licked yet." Salk's vaccine required three initial injections and an annual booster. Although this vaccine had reduced the prevalence of paralytic polio in the United States by 90%, there was some evidence that it was less effective against type 3 poliovirus. Millions of children had never been vaccinated.

Sabin's OPV was cheaper, generated rapid intestinal immunity (it was shed in the stools and, as a result, had the potential to immunize contact children), and could be administered without a needle. It was commonly administered in drops or dropped onto a sugar cube (a novelty for many rural children) that was then eaten. Because it induced rapid immunity it could be used effectively at the beginning of an outbreak. In 1960 permission was granted for Sabin to run a vaccine trial on 180,000 children in his own city of Cincinnati. The community joined in enthusiastically, and as a result, polio was eradicated from the city! Sabin's vaccine was licensed in the United States in 1961 despite competition from other potential vaccines. In 1962, Congress passed the Vaccination Assistance Act that supports the purchase and administration of a range of childhood vaccines.

OPV effectively replaced the Salk vaccine in the United States. By 1964, only about 60 polio cases were recorded in the United States (Fig. 13.5).

The last natural cases of paralytic poliomyelitis in the United States occurred in 1979 in an Amish community in Minnesota when a type 1 virus

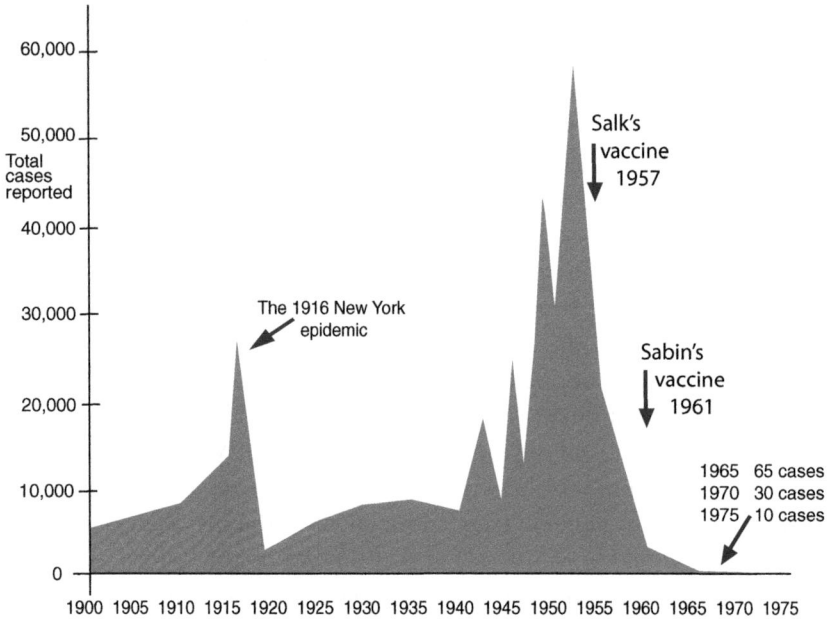

Fig. 13.5 *The number of polio cases reported in the United States and the effect of vaccination between 1900 and 1975.* It was clearly an emerging disease in the middle of the 20th century. *Data from published reports and the CDC.*

carried by a visitor from the Netherlands infected 13 unvaccinated individuals. The Western Hemisphere was declared free of polio in 1994.

Sabin's live vaccine was given orally and was thus more acceptable to young children and their parents than Salk's injection. The attenuated viruses invaded the recipient's intestinal tract and triggered a local immune response in the intestine that prevented further invasion by poliovirus. In contrast, Salk's injectable vaccine stimulated an antibody response in the bloodstream and provided shorter-lasting immunity. As a result, Sabin's vaccine was the vaccine generally employed in the United States, whereas Salk's vaccine was favored in Canada. Two problems were associated with Sabin's vaccine. One was the fact that the vaccine virus grows in the intestine and is shed in the patient's feces. It may be transmitted to other individuals and can also be detected in the environment. Analysis of wastewater samples is now widely employed as a surveillance method to detect circulating virus (Box 19.2).

A greater problem with Sabin's vaccine was that it can, on occasion, actually cause polio. This was not common, occurring in 1 in 2.6 to 8 million recipients. Although infrequent, polio vaccine is given to millions of recipients, resulting in six to eight cases annually, and paralytic polio is a devastating disease. Sabin would never admit that his vaccine could cause disease. In order to encourage vaccine use, Congress established a fund to compensate anyone injured by the virus. Nevertheless, by 1999, wild poliovirus was eradicated from the United States but vaccine-associated disease persisted. As a result, in 1999 the US Federal Advisory Committee for Immunization Practices changed its vaccine recommendations so that children would receive two doses of the inactivated Salk vaccine before receiving two doses of oral Sabin vaccine. This was reconfirmed in 2009.

Jonas Salk received many international awards and was globally recognized as a hero by the public, though not so with many in the scientific community. In Salk's own words, he was "looked upon as a hero by the public, but to be looked upon by certain of your scientific colleagues as if you perpetrated the crime of having become a hero, as if it was something you sought." Sabin and Salk were generally hostile and never cooperated [9]. Indeed, in 1948 Sabin had even suggested that Salk's formaldehyde inactivation procedure could not kill poliovirus. The feeling between the two men was so bitter that some believed that this was why neither received the Nobel Prize. Jonas Salk died in 1995 in La Jolla, California; Albert Sabin died in 1993 in Washington, DC. Neither Salk nor Sabin ever sought to patent their vaccines. When asked about the patent, Salk replied, "There is no patent. Could you patent the sun?"

The feud between the two men persisted even after their deaths with a dispute as to which vaccine was better. Nevertheless, they both got their pictures on US postage stamps in 2006. (Salk was depicted on the 63-cent stamp, the rate for 2-ounce First-Class mail, whereas Sabin was on the 87-cent stamp, for 3-ounce First-Class mail!)

The Cutter incident

Because of the huge demand for Salk's polio vaccine and the need for large-scale production, several different vaccine companies were recruited to manufacture it. Some were large and experienced, whereas others were less so. Cutter Laboratories in Berkeley, California, was one of the latter. In April 1955 they produced a batch of 120,000 doses where they had deviated from the Salk protocol and the virus had not been completely inactivated by formaldehyde. The manufacturing and inspection processes had failed. In April 1955, children, primarily in western states, received the vaccine produced by Cutter. The first case of paralytic polio developed within days. The number of affected children began to mount alarmingly. Investigations revealed that Cutter Laboratories was the source of the vaccine. The mass vaccination campaign was temporarily halted. Of the children who had received the faulty batch, 40,000 developed the self-limiting infection and remained healthy, about 56 individuals developed paralytic polio, and 5 died. Secondary spread to families and unvaccinated children resulted in a further 113 cases of paralysis and 5 more deaths. A congressional inquiry concluded that, in the rush to produce vaccine, the FDA had failed to test all the vaccine batches. It is now considered likely that Cutter allowed some virus batches to clump so that when the formalin was added it could not reach and inactivate every viral particle. Production soon resumed but it also raised the public perception that the Salk vaccine was unsafe and paved the way for Sabin's vaccine. (In an effort to restore public confidence, Elvis Presley was vaccinated on the set of "The Ed Sullivan Show" on national television in October 1956!) The March of Dimes established "Teens against Polio" with the catchy slogan, "Don't Balk at Salk." The Cutter incident led to greater federal regulation of vaccine production and safety so that today, human vaccines are quite remarkably safe.

The Cutter incident also opened the whole vaccine area to litigation. Cutter was liable to pay damages to those affected by this vaccine. The wave of litigation almost put vaccine manufacturers out of business. The National Vaccine Injury Compensation Program was introduced in 1986 to protect vaccine manufacturers from litigation. Unfortunately, vaccines are not very profitable and are complex to make. As a result, risk aversion and litigation have discouraged many companies, and only a small number now manufacture vaccines in the United States.

Postpolio syndrome

Patients who had recovered from clinical polio in the 1950s began to develop a characteristic set of symptoms in the 1970s and 1980s. These symptoms include pain, breathing or swallowing difficulties, sleep disorders, muscle twitching, gastrointestinal problems, and especially muscle weakness and fatigue. The symptoms usually affected muscles that had been previously affected by polio. Other problems may include nerve pain, arthritis, and sometimes additional muscle atrophy. This is known as postpolio syndrome. It is believed that it results from the "overuse" of their surviving nerve cells.

Global eradication

In May 1988 the World Health Assembly passed a resolution seeking to eradicate poliomyelitis from the globe by 2000. Their strategy was based on routine vaccination of entire populations followed by surveillance programs and "mop-up" campaigns as necessary. The Sabin oral vaccine was the basis of this program because of its ease of application and its long-lasting protection. It was also much cheaper to produce than Salk's. There was no need to use expensive needles and syringes. Initially, the eradication program made rapid progress. By 1994 the Americas were polio free. By 2000 the Western Pacific and in 2002 Europe were declared free. India, which had had 200,000 cases annually through the 1990s, was declared polio free in 2014. Globally, the number of polio cases has dropped by 99%, and in 2020 the disease persisted in only three countries, Pakistan, Afghanistan, and Nigeria.

This is where the campaign stalled. In each of these countries, warfare, political instability, and religious opposition have prevented completion of the task. In Nigeria, Muslim clerics suggested that the vaccines contain contraceptives, contain HIV, or are part of a dastardly Western plot to eliminate Muslims. Nevertheless, in August 2020 Africa was declared free of "wild" polio. Unfortunately, recent isolates of poliovirus in water supplies in Lagos, the capital of Nigeria, have proved to be mutants of Sabin's oral vaccine strain of type 2 poliovirus. The mutant type 2 vaccine has regained some virulence and has begun to circulate across West and Central Africa. Work is urgently underway to develop an improved oral type 2 vaccine that cannot revert to virulence so easily. This novel oral polio vaccine (nOPV2) is currently completing "fast-tracked" clinical trials, and preclinical results are encouraging. In the meantime, the vaccine-related outbreak has spilled over from Nigeria into neighboring countries such as Niger, Ghana, and the Democratic Republic of the Congo. An outbreak of wild poliovirus, imported from Pakistan, occurred in Mozambique in 2021 and 2022.

In Pakistan and Afghanistan, outbreaks of polio due to the wild-type virus are still continuing in remote tribal areas where government authority is minimal and local opposition to vaccination remains strong. Thus there were 146 polio cases in Pakistan in 2019 caused by the wild-type strain. The real number is probably higher. Because of potential violence, vaccinators cannot enter large parts of the region. Conspiracy theories have increased parental resistance by suggesting that the vaccine is harmful. Resentment against vaccinators and their police escorts is high. Deadly attacks on vaccinators and their escorts are continuing. With many parents steadfastly refusing to vaccinate, herd immunity is insufficient to result in viral elimination. The disease has also spilled across the porous western border of Pakistan into Iran. The COVID-19 pandemic closed hospitals and stopped polio vaccination in Afghanistan and Pakistan between February and August 2020, resulting in an increase in polio cases.

The collapse of the Afghan government in mid-2021 and the rise of the Taliban also delayed the final eradication of this terrible disease. Afghanistan reported two wild polio cases in 2022 and five in 2023 through June. All were located along the border with Pakistan. Many of these children had been vaccinated, raising serious concerns about its efficacy in this region. Nevertheless, wild polio cases dropped significantly in 2023 and 2024. Indeed, the global polio eradication program eradicated wild virus from the city of Karachi, a known polio hotspot! There were no cases in 2 years! Unfortunately, wastewater surveys have now shown that wild polio has returned. It was probably reintroduced by people from the Afghanistan border region. It has also spread to Gaza in the Middle East, probably introduced by health-care workers!

References

[1] Nathanson N, Kew OM. From emergence to eradication: the epidemiology of polio-myelitis deconstructed. Am J Epidemiol 2010;172(11):1213—29. https://doi.org/10.1093/aje/kwq320.

[2] Oshinsky DM. Polio: an American story. Oxford University Press, Oxford, England; 2006.

[3] Wyatt HV. The 1916 New York City epidemic of poliomyelitis: where did the virus come from? Open Vaccine J 2011;4(1):13—7. https://doi.org/10.2174/18750354011 04010013.

[4] Trevelyan Barry, Smallman-Raynor Matthew, Cliff Andrew D. The spatial dynamics of poliomyelitis in the United States: from epidemic emergence to vaccine-induced retreat, 1910—1971. Ann Assoc Am Geogr 2005;95(2):269—93. https://doi.org/10.1111/j.1467-8306.2005.00460.x.

[5] Wyatt HV. Before the vaccines: medical treatments of acute paralysis in the 1916 New York epidemic of poliomyelitis. Open Microbiol J 2014;8(1):144—7. https://doi.org/10.2174/1874285801408010144.

[6] Becker BE. Sister Elizabeth Kenny and Polio in America: doyenne or demagogue in her role in rehabilitation medicine? PM R 2018;10(2):208—17. https://doi.org/10.1016/j.pmrj.2017.11.012. Available from: http://www.pmrjournal.org/.

[7] Goldman AS, Schmalstieg EJ, Freeman DH, Goldman DA, Schmalstieg FC. What was the cause of Franklin Delano Roosevelt's paralytic illness? J Med Biogr 2003;11(4):232—40. https://doi.org/10.1177/096777200301100412.

[8] Verville RE, Ditunno JF. Franklin Delano Roosevelt, Polio, and the Warm Springs experiment: its impact on physical medicine and rehabilitation. PM R 2013;5(1):3—8. https://doi.org/10.1016/j.pmrj.2012.11.007.

[9] Katz SL. From culture to vaccine—Salk and Sabin. N Engl J Med 2004;351(15):1485—7. https://doi.org/10.1056/NEJMp048217.

The Civil War—1861—1864

Contents

The American Civil War was no different from most previous wars in which deaths from infectious diseases greatly exceeded those in battle. For example, it has been estimated that during the Napoleonic wars between 1803 and 1815, about eight times as many soldiers died of disease than in battle. During the Mexican-American War of 1846—1848, a record of sorts was made when out of 12,535 US casualties, no less than 10,986 (88%) died of disease—a ratio of seven to one (Box 14.1). During the Civil War, it is estimated that the Union armies suffered a total of 646,392 casualties, both dead and wounded. Of these, 67,058 were killed in battle but 224,586 (63%) died as a result of disease [1]. Additionally, the Confederate armies probably suffered about 454,000 total casualties, of whom 94,000 died in battle and 164,000 died from disease. The common diseases that killed soldiers on both sides included dysentery/diarrhea, typhoid, measles, malaria, and pneumonia, as well as smallpox and tuberculosis. It is estimated that in the Union army, diarrhea caused at least 44,558 deaths (3% mortality); typhoid caused 34,833 deaths (23% mortality); pneumonia caused 19,971 deaths (26% mortality); and measles caused 5177 deaths with 7% mortality.

Great American Diseases
ISBN: 978-0-443-31404-9
https://doi.org/10.1016/B978-0-443-31404-9.00014-8

BOX 14.1 The Mexican-American War 1846–1848.

The Mexican-American War began in December 1845 when the United States annexed Texas. The US troops consisted of about 27,000 regulars and 59,000 volunteers. In this short war, no less than 88% of US deaths were a result of disease, predominantly diarrhea and dysentery (the terms were interchangeable at the time), both bacterial and protozoal (amoebic dysentery)—an extreme example of *tourista*! Other diseases such as typhoid, typhus, malaria, yellow fever, smallpox, and measles also occurred but were much less significant. Not only did diarrheal diseases kill, but survivors were often debilitated and unfit for further military duty. They had to be discharged [4].

These disease outbreaks were primarily due to a total lack of hygiene in the army camps. The army, as noted above, was a mixture of regular troops and volunteers. The latter were famously resistant to discipline, but basically everyone in the army was indifferent to sanitation. They never bathed or changed their clothes. This was compounded by the heat of summer. Camps were infested by hordes of flies that also served to spread the infection. Most volunteers were farm boys who had never been exposed to these diseases. And why worry about sanitation? Everybody knew that diseases were caused by inhaling miasmas! With so many troops having diarrhea it was inevitable that camp water supplies became contaminated. Thus, with hindsight, these outbreaks were a clearly a result of incompetence, mismanagement, overcrowding, lack of proper sanitation, and callous indifference to the suffering of the soldiers.

Unfortunately, when the Civil War began 15 years later, the US Army was just as unprepared and ignorant.

(From Cirillo VJ. "More fatal than powder and shot": Dysentery in the US Army during the Mexican War, 1846-48. Persp Biol Med 2009;52(3):400–13.)

At the beginning of the war, the Union army had 98 doctors whereas the Confederates had 24. By the end of the war, the Union army had 13,000 and the Confederates about 4000. These increased numbers of "doctors" did not appear to result in an increase in the effectiveness of medical care. Medical schools at this time, although numerous, were not science based and produced graduates with no knowledge of infectious diseases [2]. The miasma theory dominated medical thinking. The Civil War was the last major conflict fought before the development of the germ theory. As a result, miasmas and "noxious effluvia" were still considered the causes of disease, and hygiene was rarely considered a priority.

Surgeons, although technically skilled, also lacked any knowledge of hygiene. Handwashing was a rarity, so postsurgical infections were almost inevitable. The chronic shortage of doctors in both armies also meant that they were overwhelmed by the workload. They usually had so many sick soldiers to look after that they were quite unable to monitor or regulate basic camp hygiene. Military objectives were greater priorities than sanitation.

The armies themselves consisted almost exclusively of males, aged between 20 and 30, most of whom had grown up in relative isolation in rural communities. Many would never have experienced the common childhood diseases until they joined the army. Diseases such as measles that are relatively mild in children are often much more severe in adults and would have had a disproportionate effect on military effectiveness.

Enteric diseases

By far the most significant affliction affecting soldiers on both sides was diarrhea.

"War is about nine hundred and ninety-nine parts diarrhea to one-part glory."
—Walt Whitman

In a society that had yet to make the connection between fecal contamination, water quality, and diarrheal diseases, it is easy to see just how diarrhea became a chronic problem in army camps. Huge training camps were set up whose population densities would have been comparable to large cities—without a sewage system. This was especially so at the beginning of the war when soldiers and officers alike took a very casual attitude towards sanitation and water quality. None of them had any training in sanitation or hygiene. Likewise, there would have been limited opportunities to maintain personal cleanliness. As a result, the men were often unwashed, were unable to wash their clothes, and were infested by fleas and lice. This was especially problematic during the winter months when the soldiers struggled to keep warm.

Because armies moved frequently, and digging latrines was tiresome work, officers often failed to ensure that the trenches were deep enough or to enforce the requirement about covering the stools with a dirt layer. Latrines were often located close to sleeping and eating areas [3].

Despite orders, exhausted soldiers would often avoid the latrines and go behind the nearest bush. As a result, anyone wandering around the camp had to be very careful as to where they stepped. The campgrounds became "saturated with noxious elements." Nobody appeared to care that they drained towards the camp water supply, and minimal precautions were taken to protect drinking water quality.

That is not to say that lack of hygiene was not regarded as a significant factor in the development of diarrheal disease. Many European authorities recommended strict hygienic precautions in army camps. As early as the 1750s, rules regarding the use of adequate latrines, proper drainage, and avoidance of marshes were published in the earliest text on military hygiene by the English authority Sir John Pringle (*Observations on the Diseases of the Army, in Camp and in Garrison*, 1752). Benjamin Rush (the yellow fever guy) published a book on the same subject, *Directions for Preserving the Health of Soldiers*, in 1778. Rush's text focused on five items: dress, diet, cleanliness, encampments, and exercise.

On July 4, 1775, on assuming command of the newly formed Continental army in Cambridge, George Washington issued these general orders:

> *All officers are required and expected to pay diligent Attention to keep their Men neat and clean, to visit them often in their quarters, and inculcate upon them the necessity of cleanliness as essential to their health and service. They are particularly to see, that they have Straw to lay on, if to be had, and to make it known if they are destitute of this article. They are also to take care that Necessarys (latrines) are to be provided in the Camps and frequently to be filled up to prevent their being offensive and unhealthy.*
> *No person is to be allowed to go to Fresh-water Pond a fishing or on any other occasion as there may be a danger of introducing the small pox into the army.*

The greatest killer of soldiers was enteric disease with its associated diarrhea and dysentery [4]. (Dysentery is best described as bloody diarrhea.) During the Civil War, it is estimated that it killed about 45,000 Union soldiers and about 50,000 Confederates. Both diarrhea and dysentery were called the "flux," "the trots," or the "runs." They were also called the "Tennessee quickstep," reflecting the need to move quickly when the symptoms struck. One surgeon commented, "No matter what else a patient had, he had diarrhea." Once a man entered the army, it was unusual for him to have normal stools.

The causal agents of these diarrheal diseases were mainly enteric bacteria, especially strains of *Escherichia coli*, *Salmonella*, and *Shigella*. A protozoan parasite, *Entamoeba histolytica*, causes amebic dysentery. There are also some enteroviruses that may also have caused acute disease. *Salmonella typhi* is of course the cause of typhoid and would have contributed to the problem, although it tends to cause constipation rather than diarrhea. However, it is acquired in exactly the same manner as the others as a result of fecal contamination of water. Flies would have been attracted to these contaminated areas and are known to spread these bacteria. They regurgitate or defecate after feeding on human stools. Their hairy legs can also get contaminated and spread the infection.

The usual treatment for diarrhea was opium. This slowed intestinal movements (opium addicts get constipated). If the patients were too constipated, they received a toxic mixture of chalk and mercury (the "blue mass") as a laxative.

A good example of the importance of diarrheal disease occurred prior to the battle of Shiloh in April 1862. The Federal troops had sailed up the Tennessee River into Confederate territory. While their leaders pondered how to proceed, the overcrowded troopships anchored at Savannah, Tennessee. Over 100 vessels, with thousands of troops on board, were tied up at the riverbank. The men simply used the river as a toilet. However, it was also the source of their drinking water. As a result, an outbreak of acute diarrhea soon broke out and incapacitated many. After the men were disembarked at Pittsburgh landing, further upstream, and encamped in the woods around Shiloh church, only perfunctory steps were taken to provide toilet facilities. Shallow ditches were soon filled with diarrhea, and the massive losses from this sickness continued unabated. One physician wrote "The pestilential atmosphere of the country about Shiloh was producing an amount of sickness almost without parallel in the history of war."

When the battle broke out on April 6, 1862, many Union officers and their troops were incapacitated by disease. Even after the battle, when the Union forces advanced to besiege the town of Corinth, the problems remained. Huge numbers of soldiers camped in low-lying swampy areas with insufficient clean water. As a result, General Sherman could muster only half of his troops. The other half were sick.

Another feature of camp life was the presence of huge numbers of horses. Just because army regulations demanded the rapid burial of dead horses and removal of their dung does not mean that it was done. In practice, such regulations appear to have been widely ignored.

Prisoners of war

If conditions for soldiers camping in the field were bad, they could, if necessary, march away and leave the "miasma" behind. Such was not the case for prisoners of war. The most notorious example of this was the Andersonville

Fig. 14.1 A plan of Andersonville Prison in Georgia where about 40,000 federal prisoners were housed. The river that ran through the center of the camp served as a source of water for both drinking and washing as well as a sanitary sewer. It is unsurprising that about 13,000 prisoners died, mainly from diarrheal diseases. *Courtesy US National Library of Medicine.*

Stockade in Sumpter County, Georgia, where thousands of Union prisoners were housed (Fig. 14.1). One stream flowed through the camp. It was

filled with the filth and excrement of 20,000 men, the stench was disgusting and overpowering; and if it was surpassed in unpleasantness by anything, it was only the disgusting appearance of the filthy, almost stagnant, waters moving between the stumps and roots and fallen trunks of trees and thick branches of reeds, with innumerable long-tailed, white maggots, swollen peas, and fermenting excrement and fragments of bread and meat.

It is unsurprising that about 60% of the deaths in the camp were due to diarrheal diseases.

Measles

Measles remained one of the major disease problems throughout the Civil War. The United States was an agrarian, somewhat underpopulated, country at this time. As a result, most of the soldiers who fought in the Civil War had grown up in relative isolation in rural areas. As in the First World War (Chapter 9), they were brought together in huge numbers for training. Given their lack of immunity these "country troops" were very susceptible to diseases such as measles. Although measles can cause lethal complications in small children, it is much more likely to do so in adults. This was especially the case with poorly fed and exhausted soldiers. The most significant complications included pneumonia and hepatitis. In the Union army there were about 67,000 cases of measles and 4000 deaths as a result. Robert E. Lee commented that "The soldiers everywhere are sick. The measles is prevalent throughout the whole army...."

During the Civil War, measles infected both Union and Confederate forces. So pervasive was the disease, it obtained the distinction of being called "Camp Measles" in official documents.

> The first troublesome illness my regiment experienced was what almost everyone in the army alike suffered from, viz., measles. It is surprising to find as we did, how many adults there are in the community who have never had this disease. One hundred and twenty-five cases in my own regiment; and such measles! The like was never known! ... Many of the men had congestion of the lungs and pneumonia following measles, as often happens.
> **(War Papers Read Before the Commandery of the State of Maine, Military Order of the Loyal Legion of the United States, 1898)**

New recruits were especially prone to measles, a disease that killed and disabled many men before they even set foot on a battlefield. This was vividly illustrated in Margaret Mitchell's 1936 novel *Gone with the Wind* when Scarlett O'Hara receives word that her husband, Charles Hamilton, has died of measles just 2 months after joining the Confederate army and ingloriously, prior to any combat.

From the beginning, measles was a scourge to both the Union and Confederate troops and never really subsided. As the war continued, however, measles became a greater problem to the Confederacy. Measles is generally a childhood disease more prevalent in cities and towns. Infection with measles confers lifelong immunity to the disease; you usually only get it once. As a consequence of the Confederacy's reliance on recruits from rural versus urban (communities with populations ≥ 2500) areas, it had a greater percentage of susceptible soldiers than the Union. Generals were well aware of the issues of disease: of the greater susceptibility of new recruits; outbreaks in areas of initial congregation, such as training camps; and the potential for disease outbreaks in combat regiments due to the introduction of new recruits, thus reducing both the unit's and the army's combat readiness. In fact, General Robert E. Lee and others suggested that new recruits be assembled in camps for at least 2 months prior to being permitted to join the armies on the front lines. This would allow for the recruits to contract camp diseases, such as measles, go through the effects, and recover. They would thus be less of a burden and less a potential introducer, that is, carrier, of the disease when they joined troops in the field.

In *Outlines of the Chief Camp Diseases of the United States Armies as Observed during the Present War* by Joseph J. Woodward, US Army medical officer of the Army of the Potomac, Dr. Woodward stated about measles,

> Epidemic measles which has prevailed so extensively among recently-formed regiments … has left the patient in an exhausted and prostrate condition from which he enjoys at best a tardy convalescence, interrupted by attacks of acute inflammations of a typhoid type (diarrhea), bronchitis, pleuro-pneumonia, and pneumonia.

In summary, not only the acute symptoms of measles, but the subsequent chronic diarrhea, respiratory problems, and other complications, greatly diminished the combat readiness of the soldier and the army. Measles would also kill. About 7% of those infected with measles died as a result.

When measles was introduced into a regiment, it quickly infected all its susceptible members, the size and functions of a regiment being very similar to a city or a school. Persons who previously had measles do not become infected because infection produced a lifelong immunity and protection. When there were no longer be any susceptible people, the disease would disappear, only to reappear again when new susceptible recruits were mustered into the regiment.

Fevers

Although it was long known that the development of a fever was a sign of disease development, its extent was simply judged by placing a hand on the forehead. The clinical thermometer was invented in 1867 by the English physician Thomas Allbutt. It was about 6 inches long and took at least 5 minutes to measure a patient's temperature. They did not come into widespread use until much later in the 19th century. The two major causes of fever were typhoid caused by the bacterium *S. typhi* and malaria caused by protozoan parasites belonging to the genus *Plasmodium*. The clinical signs of these two infections were sufficiently similar as to cause some clinical confusion. As a result, many of these infections were simply classified as typho–malaria.

Malaria

Four million men served on both sides of the conflict between 1861 and 1865 and during that period, there were approximately 1.3 million cases of malaria resulting in 10,000 deaths [5]. In addition, malaria's lingering debilitating effects on the troops greatly reduce their combat effectiveness. After combined diarrhea and dysentery, malarial fevers were the most common diseases in Union hospitals.

The cause of malaria and its vector was not determined until 30 years after the Civil War was over. As a result, both armies were essentially helpless in the face of the disease. Nevertheless, physicians were well aware of the geographical and seasonal correlations with disease occurrence.

Dr. Woodward, Assistant Surgeon, USA, wrote in 1863,

> In temperate regions, moreover, these fevers prevail most abundantly toward the close of the summer and throughout the autumn, when the warmth of the season has had time to operate upon the decaying vegetation, and diminish in frequency and intensity, often almost disappear, after the winter frosts have set in.

Thus malaria, or marsh miasm as it was often called, was a disease caused by miasma—those poisonous vapors. It was suggested to the Adjutant-General that because the "...(river) flats were more insalubrious than the high woodlands..." and "...malarial fevers do prevail on the slope towards the river...", the army should "...remove the camps beyond the first crest, so as to afford the protection of the hills against infected currents of air." In addition, to help reduce the soldier's exposure to the night miasmas, it was suggested that reveille not occur until after sunrise.

During the Peninsular Campaign in Virginia, 1862, it was suggested that to reduce marsh miasm,

> ... that camps be formed not in the woods but at a short distance from them, where a free circulation of pure air can be procured, and where the ground has been exposed to the sun and air to such an extent as to vitiate (impair the quality) the noxious exhalations from damp ground saturated with emanations from the human body and from the decaying vegetation.

Even though the relationship between the mosquito *Plasmodium* and malaria were yet to be revealed, Civil War physicians, like the ancient Greeks and Romans, recognized the association between the environment and the occurrence of intermittent fevers, or malaria, and took appropriate measures.

At the time of the conflict, the *Scientific American* summarized what was known of its cause.

What malaria is nobody knows. It may consist of organisms, either animal or vegetable, too minute for even the microscope to detect or it may be some condition of the atmosphere in relation to electricity, or temperature, or moisture: or it may be a gas evolved in the decay of vegetable matter. The last is the most common hypothesis, but it is by no means proved, and it has some stubborn facts against it. There is no doubt, however, that malaria is some mysterious poison in the atmosphere, and that it is confined strictly to certain localities.
                                                     —Scientific American, July 20, 1861

Both armies were seriously irritated by the presence of large numbers of mosquitoes. For example, they were considered a major nuisance after the Battle of Gettysburg, and as a result, the Union army went to the trouble of purchasing large quantities of mosquito netting. Nevertheless, many soldiers were obliged to sleep out in the open with only a blanket covering them. Soldiers were exposed to mosquitos anytime they camped in damp or marshy areas. However, there was never any direct link made at the time between mosquitos and malaria. The identification of *Plasmodium* as the cause of malaria occurred more than 20 years after the war was over.

Army physicians classified malaria according to its severity. The mildest was "dumb ague," a febrile illness not associated with chills or profuse sweating. However, it developed every 1 to 3 days about the same time and lasted for 1 to 2 hours. A more severe form was "ordinary fever and ague." This common form of malaria was associated with chills and shaking. This was followed by a fever lasting 3 to 4 hours and was followed by massive sweating. The most severe form was bilious fever where the fever persisted possibly for days with occasional temporary remissions.

The use of quinine to treat the intermittent fevers of malaria, as well as being used for dozens of other fevers and complaints, was familiar to physicians during this period. It was used for prophylaxis, mixed with whiskey, as a means of preventing the disease (Fig. 14.2). But according to the Medical Director of the Army of the Potomac, "much prejudice and aversion, however, had to be overcome in inducing the men to take it, and I scarcely think it would have been practicable to have forced it upon the whole army." The efficacy of quinine use preventing malaria was not uniformly agreed upon, which may have been due to lack of compliance by the soldiers and giving too low a dose. The Union had access to quinine, the Confederacy did not. The Union blockade prevented its importation into the southern states. Confederate soldiers were forced to rely on untried herbal remedies. Thus one of the reasons that the Union Siege of Vicksburg was successful was

BEFORE PETERSBURG—ISSUING RATIONS OF WHISKY AND QUININE.—[Sketched by A. W. Warren.]

Fig. 14.2 Federal soldiers at the siege of Petersburg receiving their rations of whiskey and quinine. *Courtesy US National Library of Medicine.*

Confederate debility as a result of malaria [6]. (A similar outcome had happened at the siege of Yorktown 80 years previously.)

As would be expected, troops from the northeastern states, previously unexposed to malaria, were highly susceptible and developed severe disease. As a result, federal troops had little or no immunity, whereas many of the confederate forces had had prior experience with the ague and some degree of protection. There were many in the confederacy that hoped that malaria would do their job for them and kill Union soldiers as they moved south. It certainly appeared that way during much of the war. It was confidently expected that the hot climate and the disease would force the Union troops to withdraw.

On the other hand, many troops from the northwestern states—Illinois, Indiana, Missouri, Iowa—and from the southern states had previous exposure and brought the parasite with them, really in them, to the encampments. Thus there was a spectrum in the virulence of the symptoms, depending on previous exposure, health status, nutrition, and the species of *Plasmodium* involved.

The armies of both sides had to suffer. The sickly season was the summer and fall along the coastal wetlands of the South. As a result, the North lost significantly more men than the confederacy to the ague, especially in the campaigns along the Atlantic and Gulf coasts. Even before the conflict started it had had an effect. For example, Ulysses Grant, the future Union commander, had tried farming. In his memoirs he records,

I managed to keep along very well until 1858 when I was attacked by fever and ague. I had suffered very severely and for a long time from this disease, while a boy in Ohio. It lasted now over a year, and, while it did not keep me in the house, it did interfere greatly with the amount of work I was able to perform. In the fall of 1858, I sold out my stock, crops, and farming utensils at auction, and gave up farming.
—US Grant Memoirs

Grant's experience with the ague (a term used for a disease that produces chills, fever, and profuse sweating that recur at regular intervals—e.g., malaria) perhaps served him well when he launched the Vicksburg campaign down the Mississippi in the sickly season with a generous supply of quinine.

The distribution of malaria corresponded approximately to the boundaries of the confederacy. The location of most fighting within the southeastern states ensured that malaria was a constant issue for both sides. Malaria was present along the coast and profoundly affected the coastal campaigns. Some of the areas most severely affected by malaria were along the major western rivers including not only the Mississippi but also the Yazoo, Big Black, Red, and Arkansas rivers. Arkansas was the state where the most severe outbreaks occurred. The Union forces that occupied Little Rock, Arkansas had moved from the Mississippi region and brought the parasite with them and, as a result, suffered especially severely. Between July 1863 and June of 1865, over 72,000 Union troops posted in Arkansas were diagnosed with malaria. As a result, in October 1864 only 17,618 out of 44,506 soldiers in the Union forces in Arkansas were available for battle!

Typhoid and typho-malaria

Given the total absence of diagnostic tests to identify the causes of disease, especially the highly prevalent intermittent fevers and agues, military doctors were often unable to decide whether a patient had typhoid or malaria or both [7]. They therefore classified many of these cases as typho-malaria. This was especially true in cases where the patient had typhoid accompanied by persistent chills and rigors. In 1897, Sir William Osler, a professor at Johns Hopkins School of Medicine, described the clinical signs of malaria in considerable detail. He was also highly knowledgeable about typhoid, so he was able to point out the diagnostic features that enabled the diseases to be distinguished. He not only defined the clinical stages of malaria but, in a paper published in 1897, he identified the critical differential diagnostic features of both. He showed that the seasonality was different, with typhoid being a fall disease and malaria occurring over the summer months. There is no doubt that mixed infections did occur. Once it became possible to confirm malaria by examining a blood smear for the *Plasmodium* parasite and look for antibodies to *S. typhi* by the Widal test, it became clear that most cases of typho-malaria were pure typhoid with occasional mixed infections. However, these techniques did not come into practice until the late 1890s. It is estimated that there were about 150,000 cases and 35,000 deaths as a result of typhoid alone during the Civil War.

Epidemic typhus

Although Civil War soldiers were commonly lousy, epidemic (louse-borne) typhus was almost unknown. This was a surprising observation, especially in the prisoner of war camps where the overcrowded conditions, malnourishment, and filth would have provided optimal conditions for its spread. The official medical history of the war asserts that there was not a single episode of typhus recorded. "Typhus was fortunately a stranger to our camps." The official number of typhus cases in the Union army was 2624 with 958 deaths. As described in Chapter 10, it is not known why this was the case. In general, typhus in North America was almost exclusively a disease associated with poor recent immigrants living in crowded urban settings.

Yellow fever

When Union troops occupied New Orleans in May 1862, they were, of course, highly resented. The activities of General (Spoons) Butler added to the unpleasantness. (The nickname related to his alleged propensity for stealing the silver spoons from the houses he occupied.) Butler initiated martial law, required citizens to take an oath of allegiance to the Union, and tolerated no disrespectful actions toward the Union or Union soldiers and no displays of loyalty to the confederacy. As a result, many of the inhabitants hoped for an outbreak of yellow fever that would destroy the invaders. They were disappointed. The regular outbreaks that had occurred in antebellum New Orleans depended upon its importation by ships trading with the West Indies where it was endemic. The Union blockade effectively stopped this trade and, with it, the importation of yellow fever.

This was not the case everywhere. For example, a successful blockade runner had brought yellow fever to Sabine City, Texas in July 1861. By that September, a major epidemic had broken out and many of its inhabitants had fled to Beaumont and Houston. Most of the Confederate garrison was withdrawn. Then the Union navy in the form of a gunboat squadron under the command of Frederick Crocker decided to seize the city. The

Union takeover was rapid and bloodless. Once they learned that yellow fever had killed half the town's population, however, they decided that they could not afford to stay and departed within a month. The yellow fever outbreak spread along the Texas coast to Matagorda, Indianola, and Brownsville and effectively prevented aggressive Union activities in this region. Interestingly, the Union navy also sought to capture Galveston. They bombarded the city and demanded its unconditional surrender. The confederate negotiators reported falsely that there was yellow fever in the city. As a result, the Union commander agreed to a short truce while civilians were evacuated. The Confederates took advantage of this truce to save several guns (under the terms, this was not explicitly forbidden). Union forces never occupied the city in force, and it was recaptured by the confederates in January 1863.

In August 1864, however, yellow fever did indeed break out in Galveston and as a result its garrison suffered severely. Some tried to desert but were unsuccessful. Law and order broke down. Help was sought from Houston, but the disease was present in that city also. The Union navy could likely have taken the city at that time but prudently declined. By the time the epidemic ended in November, 111 soldiers and 158 civilians had died.

Parasites

The almost total lack of personal hygiene in many soldiers ensured that lice, flea, and tick infestations were common. Upon enlistment, new recruits got a perfunctory medical examination. The doctors were instructed not to approve recruits carrying lice (these were also called "graybacks" and "bluebellies"). Lice were, however, inevitable and abundant. The heavy wool uniforms provided an optimal environment for lice. Flea, mite, and tick infestations were also an inevitable consequence of an unwashed outdoor life.

Hookworms, *Necator americana*, were endemic in the southern states. As a result, most rural southerners were infected with hookworms. Hookworm eggs are passed in the feces of an infected person, and a single female hookworm can produce up to 10,000 eggs a day. As noted previously, fecal management and sanitation were not of the highest caliber in either army. Thus the soil contained abundant hookworm larvae that could penetrate bare feet; the army of Northern Virginia was perpetually in need of shoes! Infected individuals could develop stomach pain, diarrhea, loss of appetite, excessive fatigue, and anemia. Hookworm infections were an additional burden on troops that were fatigued, hungry, and had other diseases [8]. Because most battles of the Civil War occurred in the South, troops of the Union would also have been exposed to hookworms, though they generally had appropriate footwear.

Military hospitals

During the war, large military hospitals were built close to major cities. These were single storied wood buildings that were well ventilated and heated in winter. The largest was at Chimborazo in Richmond, Virginia. It grew to have 150 wards, housed 4500 patients, and eventually treated about 76,000 soldiers. Because of the persistent belief in miasmas, as well as the terrible smells of the wounded and dying, these newer hospitals were often designed to be well ventilated, which may have improved patient survival. Ice was not usually available, so hospitals generally fed their patients food that was stable on storage such as salted beef, that was often very fat, and dried beans. Fresh fruit and vegetables were rarely available. Scurvy was therefore not unknown.

Surgical fevers

When one thinks about medicine during the Civil War, pictures of soldiers with amputated arms and legs come quickly and vividly to mind (Fig. 14.3). No wonder, since limb amputation was the most common operation performed; approximately 45,000 of the 60,000 surgeries (over 75%) were limb amputations! Why so many?

One reason was that over 70% of combat wounds were to the extremities and approximately 90% were inflicted by rifle fire. Though the Union and the Confederacy used a variety of rifles and muskets throughout the war, the basic weapon of the soldier was the Enfield .577 and Springfield .58 rifles. Both had a low muzzle velocity and used the hollow-based Minié ball (or Minnie ball) as projectiles. When these hit a person, the tip of the soft-lead Minnie ball would flatten out, sometimes breaking into pieces. Due to the low velocity and pliability of the Minnie balls, they tended to tear through tissue and organs, causing massive soft tissue damage and splintering bones; Minnie ball wounds were more often than not contaminated with shards of clothing and dirt. Arm and leg fractures resulting from these wounds resulted in limb amputation in about 50% of cases.

Standard procedure for wounds was to explore the wound and remove any bone fragments, the bullet, and any other foreign material. Wound exploration was most frequently done with the bare hand—stated to have

Fig. 14.3 Photograph of Alfred A. Stratton of Co. G, 147th New York Infantry Regiment, whose arms were amputated following wounds sustained at Petersburg, Virginia in 1864. *Courtesy US Library of Congress.*

a much better sensitivity and ability to remove objects, although sometimes a metal or flexible probe was used. Hands and surgical instruments were not washed and disinfected between patients; at most they might be rinsed off or dunked in a bucket of water that was reused often. At that time, the medical profession in the United States had very limited understanding of the importance of antisepsis and disinfection and their correlation with wound infection. The principles of aseptic technique in medicine and surgery were still in their infancy. In the 1840s and 1850s, Oliver Wendell Holmes Sr.,

American physician, poet, and novelist, and Ignaz Philipp Semmelweis, Hungarian physician, independently advocated antiseptic practices and handwashing to reduce puerperal fever. The principles of aseptic surgical practices were introduced in 1867 by Joseph Lister, a British surgeon, but were not widely adopted by American surgeons until the late 1880s. Thus secondary infections following surgery or due to open wounds were common.

Although seemingly drastic, amputation aimed at preventing wound infections. Surgeons found that amputations performed quickly after the injury, within the first 48 hours, resulted in a better survival and a reduced chance of deadly infections. Along with the many regular soldiers, Union General Daniel Sickles and Confederate General John Bell Hood (commander of the Texas Brigade) had limbs amputated. Sickles sustained the injury to his arm due to cannon fire at Gettysburg, and Hood was shot twice in the leg during the Battle of Chickamauga. Both lived productive lives afterwards (although Hood eventually died of yellow fever). Confederate General Thomas "Stonewall" Jackson was not as fortunate. Having his left arm shattered by Confederate rifle fire, it was amputated the following day. He never recovered and died of pneumonia 7 days after the surgery.

Complications due to and following amputation included death (36% died due to amputation associated causes), neuroma, bone protrusion, nonhealing of the residual limb or stump, and secondary osteomyelitis, pyemia, erysipelas, or hospital gangrene. Hospital gangrene, erysipelas, and pyemia were classified as "surgical fevers." These were frequent in overcrowded hospital wards housing patients with open sores, wounds, surgical incisions, and amputations.

Gangrene is the death of muscle cells and other tissues due to lack of blood flow or bacterial infection. Hospital gangrene, typically due to a Streptococcal bacterial infection, was a contagious, often fatal infection that occurred in contaminated wounds and surgical incisions. The specter of amputation and gangrene inspired terror and fear of hospitals among the wounded and healthy alike. When hospital gangrene occurred, the skin and muscle tissue became inflamed and died. The growing bacteria generated a sickening fetid odor. If the infection was not eliminated, the infection spread through the body. Chills and sweats, a weak rapid pulse, and a sickly-sweet breath heralded the patient's imminent death.

It is important to note that hospital gangrene is entirely different from gas gangrene that affects deep muscle tissue and is caused by anaerobic gas-producing bacteria, most commonly *Clostridium perfringens*.

In an era of nonaseptic techniques, practically all battle wounds became infected and pus was therefore expected. Surgeons optimistically looked forward to the production of "laudable pus," a thick, creamy, free-flowing purulent discharge typically caused by *Staphylococcus aureus*. Laudable pus was considered a sign of proper wound healing and an indicator of a good prognosis—not what we who live in the era of antibiotics, aseptic techniques, and disinfection and sterilization are conditioned to expect! If the pus stopped draining or became dark and blood tinged with a putrid, fetid smell (a stinky, oozing wound), this was an indicator of poor wound healing, poor prognosis, and forerunner of pyemia and death.

Pyemia means, literally, pus in the blood. It is a form of blood poisoning. Affected patients became systemically ill with a high fever and rapid pulse. Death was almost inevitable (mortality rate was 97%) and occurred rapidly, perhaps within hours.

Erysipelas is a severe bacterial skin infection caused by streptococcal bacteria. It affected many wounded soldiers but was so contagious that nurses, doctors, and orderlies could also be infected if they had any breaks, even a small scratch, in their skin. Erysipelas, as the term was used in the Civil War, is usually a local inflammation at the site of infection. If such an infection was not controlled, it could rapidly spread to other parts of the body through the blood system, resulting in pyemia. Approximately 41% of the cases of erysipelas ended in death. Due to the extremely contagious nature of this disease, affected soldiers had to be segregated from other patients and, if possible, provided with their own sheets and towels.

The accepted theory for the cause of the surgical fevers during the Civil War blamed miasmas. These poisonous vapors could contaminate wounds, producing gangrene and erysipelas, and then spread to others because it was believed that the miasmas were emitted from the contaminated wound or the breath of an infected person. Surgeons observed that good ventilation decreased the prevalence of erysipelas and other infections; it was less rampant in tent hospitals as compared to indoor facilities. In an effort to combat the miasmas, bromine was vaporized and sprayed in the air of the hospital wards, a practice that did reduce the prevalence of surgical fevers, as we now know that it has antibacterial activity. The fight against miasmas also influenced hospital design and management. Greater, more effective ventilation became a guiding principle; new hospitals were designed so that that patients could be segregated with improved ventilation within and between the wards (Fig. 14.4).

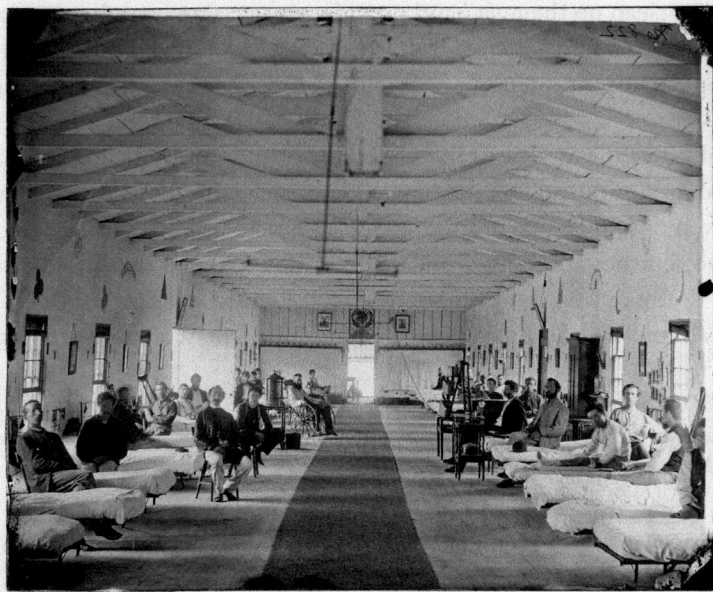

Fig. 14.4 Inside a federal military hospital. The design of union hospitals, long narrow wards with many opposing pairs of windows, improved ventilation to combat the miasmas. The improved ventilation and segregating patients by disease condition did much to improve patient outcomes. *Courtesy US Library of Congress.*

The soldiers in the Civil War had the misfortune to fight their battles at a time when medicine was just beginning to emerge from its dark ages. The new discoveries, especially in microbiology, had yet to have an impact. The development of antiseptic surgery by Joseph Lister came too late for those who were wounded in battle. Fortunately, it was one of the last wars fought under these primitive conditions.

Lessons learned

The Civil War experience made it obvious to observers that poor sanitation was a major cause of soldier's deaths. As a result, the Army established the United States Sanitary Commission, whose role was to persuade officers and soldiers alike to clean up the camps. They were quite successful and, towards the end of the war, death rates dropped sharply in camps that enforced good sanitation. When the Army medical workers returned home after the war, they brought these hard-won lessons with them. Public Health Standards were progressively established and enforced in civilian life. Cities began to "clean up their act" and provide sewage systems and a clean water supply. The lessons of the Civil War were not lost but laid the foundations for the great improvements in the health of American cities that followed the Civil War.

Only after the advent of the germ theory and knowledge about contagions, transmission, and sanitation were embraced did the disease to combat ratio even out; in WWI, death of US service men had a 1:1 disease to combat ratio. Further advances in medicine, antibiotics, and sanitation lowered disease deaths during war even more, to a 1:20 disease to combat death ratio in WWII and to 1:50 in the Korean War, Vietnam War, and Iraq War.

References

[1]   Gilchrist MR. Disease and Infection in the American Civil War. Amer Biol Teacher 1998;60(4):258—62.
[2]   Councell CE. War and Infectious disease. Public Health Rep (1896—1970) 1941; 56(12):547—73.
[3]   Cook GC. Influence of diarrhoeal disease on military and naval campaigns. J Roy Soc Med 2001;94:95—7.
[4]   Cirillo VJ. "More fatal than powder and shot" dysentery in the US Army during the Mexican War, 1846-48. Persp Biol Med 2009;52(3):400—13.
[5]   Bell AM. Trans-Mississippi miasmas: malaria and yellow fever shaped the course of the civil war on the confederacy's western theater. East Tex Hist J 2009;47(Issue 2). Article 6.
[6]   Freemon FR. Medical care at the siege of Vicksburg, 1863. Bull NY Acad Med 1991; 67(5):429—38.
[7]   Cunha BA. Osler on typhoid fever: differentiating typhoid from typhus and malaria. Inf Dis Clin N Am 2004;18:111—25.
[8]   Anderson AL, Allen T. Mapping historic hookworm disease prevalence in the southern US., comparing percent prevalence with percent soil drainage type using GPS. Inf Dis Res & Treatment 2011;4:1—9. https://doi.org/10.4137/IDRT.S6762.

Plague and San Francisco—1900

Contents

Plague is an example of a zoonosis—a disease transmitted from animals to humans. Although we think of the plague as a great killer of mankind, it is primarily a disease of wildlife. Endemic plague foci are found in many different countries where the plague bacteria circulate in wild rodents. These rodents dig burrows and form large underground colonies where the bacteria can persist indefinitely. These bacteria are transmitted between the rodents by biting insects, especially fleas, so that plague is maintained in these colonies by the continuous spread of infection to healthy individuals. When conditions are good and hosts are plentiful, the organism may readily infect other rodents. On the other hand, if rodent numbers drop, the fleas will crowd onto the survivors. The flea density may be so high that they may be forced to feed on less desirable hosts such as nearby domestic animals or even humans. Under some conditions the humans may then transmit the infection to others by the respiratory route and in this way trigger huge outbreaks—pandemics.

The plague bacillus

Plague is a bacterial disease caused by a gram–negative, nonmotile coccobacillus called *Yersinia pestis*. It was named after its discoverer, Alexandre

Great American Diseases
ISBN: 978-0-443-31404-9
https://doi.org/10.1016/B978-0-443-31404-9.00015-X

Yersin, a Swiss scientist working for Louis Pasteur. The organism was first identified during an outbreak of plague in Hong Kong in 1894. Two eminent bacteriologists, Shibasaburo Kitasato from Japan and Alexandre Yersin from France, arrived in Hong Kong that June to determine its cause. Yersin arrived 3 days after Kitasato. He found a gram-negative bacillus was present in huge numbers in the fluid withdrawn from the plague lesions (buboes). The bacteria stained most intensely at the tips of the bacillus, so-called bipolar staining (like a safety pin)—this is now known to be characteristic of the plague bacillus. Kitasato also discovered a bacillus but he declared it to be gram positive! Both investigators announced their discoveries within a few days of each other. Kitasato published first but his descriptions were vague, and his identification of a gram-positive organism suggested that he may have grown a mixed culture containing the very similar pneumococcus [1]. Yersin's description was therefore considered more accurate, and he received credit for the discovery. Yersin originally called the organism *Bacterium pestis*. In 1900 it was renamed *Bacillus pestis*, in 1923 it was renamed *Pasteurella pestis*, and finally in 1970 it was determined to be sufficiently unique to form a new genus, *Yersinia*, and so is called *Y. pestis*.

Yersin noticed the association of the plague with dead rats but did not know how it was transmitted to humans. Four years later a colleague, Paul-Louis Simond, investigated the rat problem while studying the plague in Karachi [2]. He noticed that the very first lesions on human skin looked like insect bites and that they commonly occurred on the patients' ankles and legs. In a simple transmission study in June 1898, Simond showed that brown rats infested with the oriental rat flea (*Xenopsylla cheopis*) served as the source of the infection. Plague-infected rats without fleas would not transmit the disease. This finding was met with scorn and disbelief! However, subsequent investigations fully vindicated Simond and his findings. They also formed the scientific basis for the eventual eradication of the plague in US cities.

The plague

There are three distinct clinical forms of the plague depending upon the body systems affected. They are all caused by the same agent and so can be considered different presentations of the same disease. Thus bubonic plague affects the lymph nodes, pneumonic plague affects the lungs, and septicemic plague invades the bloodstream.

The infection is transmitted from rats to humans through the bite of fleas. On ingesting a blood meal from an infected individual, the flea stomach (proventriculus) becomes blocked by bacteria and clotted blood. The flea responds to this blockage by repeatedly biting its host in an attempt to dislodge the obstruction. When this happens, a bolus of bacteria and blood is regurgitated into the host's skin tissues and blood vessels.

Thus the bubonic plague is acquired by the bite of a flea that has previously fed on an infected rodent. When the flea subsequently feeds on a human it injects the contaminated blood into its victim. *Y. pestis* is carried by the flow of lymph fluid to a draining lymph node. The function of lymph nodes is to filter out any foreign invaders in the lymph fluid. The bacterium is therefore trapped in the lymph node where it is exposed to the defensive cells of the body. These cells normally eat and kill any bacterial invaders. Unfortunately, *Y. pestis* fights back. It produces potent toxins that kill the white blood cells. These toxins also cause blood vessels to rupture so blood seeps into the lymph nodes that then turn purple and black. The first signs of the disease are therefore dark swollen lymph nodes (called buboes). Which nodes are first affected depends upon the location of the flea bite. Thus they commonly occur in the neck, armpit, or groin. Buboes are very painful and can swell to the size of an egg or an orange. If they rupture, they release foul-smelling blood and pus. The bacteria can spread progressively along a chain of lymph nodes invading each in turn. It can also escape into the tissues where it destroys cells and causes black or purple spots to appear. The presence of the infection triggers a fever, leading to delirium. As the organism continues to grow and spread through the body, it will eventually result in multiple organ failure and death. The swollen blackened buboes are characteristic of bubonic plague. The gross lesions of the disease caused historians to call it the Black Death.

As the plague bacteria spread through the bodies of their victims, they will invade many different organs including the lungs. The bacteria destroy the lung tissue so that when the victim coughs, blood or sputum droplets loaded with bacteria are expelled and can infect others nearby. The individuals who inhale these droplets will develop pneumonic plague. The first symptoms include coughing up blood followed by fever and chest pains. The massive lung damage results in heart failure and death. The progression of pneumonic plague is much faster than bubonic plague, often killing victims within 3 days. In natural outbreaks such as the Black Death, the first plague cases in a community were usually bubonic; however, once established in a dense human population, respiratory spread occurred and the pneumonic form became increasingly important. Pneumonic plague can also arise as a secondary complication of bubonic plague. The R_0 for bubonic and pneumonic plague appears to range from 5 to 7.

On occasion, the plague bacillus gets into the bloodstream and causes septicemic plague. The bacteria grow rapidly within hours, trigger blood clotting, and as a result, block blood vessels. This results in dark patches developing in unoxygenated skin and eventual organ failure. Gangrene of the extremities such as fingers, toes, and the nose develops as a result of this lack of oxygenation. The victim may die as soon as 4 hours after the onset of infection.

Until recently, it was believed that the rat flea was the only vector of the plague. If this is the case, it is difficult to explain how it spread so fast during epidemics. Analysis now suggests, however, that other human ectoparasites, most notably the human flea (*Pulex irritans*) and body lice (*Pediculus humanus*), were largely responsible for this rapid spread especially during the medieval Black Death. Thus rats may have been less important in that pandemic, which may account for why nobody at the time associated rats with the disease. This may also explain why the Black Death was much more lethal than current plague outbreaks.

Y. pestis is a natural pathogen of wild rodents. There are several regions in the world where rodent plague is endemic, but the most significant one is in Central Asia. From there the plague spread westward toward the Middle East and Europe where it has caused two disastrous pandemics, the Justinian Plague of 541 and the Black Death of 1347. In both of these great pandemics, ships played a major role in their spread. Thus sailing ships spread the Justinian plague along the shores of the Mediterranean Sea. At that time, merchant ships did not venture far from land and often put into a port every night for shelter. With short voyages such as this, a seaman with the plague or even an infected rat could survive the daily voyages and hence spread the infection. Likewise, during the Black Death, voyages were short. The plague reached both Sicily and England when a ship with a dead and dying crew drifted ashore. Crossing the Atlantic was, however, a different matter. It took Columbus more than 2 months to cross the ocean. The *Mayflower* took 66 days in 1620. During the 18th century the average crossing time was 6 weeks. The first paddle steamer, the *Great Britain*, changed all that by making the crossing in 14 days. The *Titanic*, had she arrived, would have taken about 7 days. Thus the plague was unable to spread from the Old World to the New until sailing ships were replaced by steamers. Plague killed too quickly for its victims or the rats to survive long voyages. Although there were occasional reports of plague deaths in the New World during the sailing ship era these are not well substantiated. That had all changed by the beginning of the third pandemic in the mid-19th century.

History

Despite being predominantly a disease of rodents, the most ancient traces of the plague have been found in skeletons of humans who died 4900 years ago in Sweden during the Neolithic Age [3]. Between 5000 and 6000 years ago Neolithic societies declined in Western Europe. It is believed therefore that the plague may have contributed to this decline in the development of large human settlements and delayed their growth [4]. The host species in the Black Death was the black rat, *Rattus rattus,* that probably acquired the infection from other Asiatic rodents. The brown (sewer) rat, *Rattus norvegicus,* replaced the black rat as the main host of the plague in the 18th and 19th centuries.

On at least three major occasions, the plague bacillus escaped from its enzootic rodent populations in Central Asia and jumped into humans. The result each time was the development of a disastrous pandemic that wiped out 30% to 40% of the affected populations [5].

The first pandemic

The first major plague pandemic is known as the Justinian plague after the Roman emperor who ruled the Eastern Roman Empire at that time, around 541–543. This plague likely originated among the burrowing rodents in Central Asia and eventually reached India. From there it spread westward by the trading ships that carried silks and spices from China to Egypt. The European records reported the first cases in Egypt, especially the great city of Alexandria at the mouth of the Nile. At that time Egypt, with its fertile delta, produced most of the grain for the Roman Empire. Ships carried the grain around the Mediterranean as well as to the eastern capital of the Roman Empire, Constantinople. Black rats were, of course, regular passengers on these grain ships as well, so the plague went with them. It reached Constantinople in late 541 where it killed about 40% of the population. It was claimed that deaths reached 5000 to 10,000 every day in the city at the pandemic peak. The plague also attacked Justinian's armies, so they were also reduced by 40%. Roman attempts to recover lost territory were abandoned,

The disease spread all around the shores of the Mediterranean. North Africa was especially hard hit. It reached Spain and Portugal and eventually

Britain, Ireland, and Denmark. The Eastern Roman Empire entered a spiral of state failure and dashed hopes, leading eventually to its doom.

The second pandemic

The second plague pandemic, dubbed the Black Death by later historians, also originated in rodent plague foci in Central Asia. It spread westward along the Silk Route until in the late summer of 1347 it reached Kaffa, a trading port in the Crimea, and from there it was carried by ship to Constantinople, Messina, Sicily, and Marseilles, France. It then began an inexorable spread northward. It affected all of Italy, France, England, and Ireland and eventually reached Scandinavia in 1349 and Eastern Europe a year later. Like the Justinian plague, up to 40% of the affected population were killed. (This varied by location. Some regions were very severely affected whereas others escaped unscathed—at least for a while.) Of those infected, perhaps 70% died. Unlike the Justinian plague, once established in Europe, the second pandemic persisted for hundreds of years. The final outbreak of this pandemic occurred in Marseilles, France in 1720. Disease at this time was seen as a deserved punishment from God for our sins, but the Black Death was especially severe. The social transformation caused by this massive population decline led to a loss of confidence in the claims of the Catholic church and eventually to the Protestant reformation.

The plague spread rapidly across medieval Europe. Medieval cities were extraordinarily filthy by our standards and would have housed very large concentrations of black rats. People not unnaturally ran away from affected towns and cities in an effort to escape the miasma. Some resorted to the countryside and others simply went to the next town. No doubt many would have carried some rat stowaways with them so that once the plague killed the rats, the starving fleas would have sought human blood. It is clear, however, that this is not the whole story. The humans themselves would have carried their own fleas and lice. Recent analysis of plague outbreaks suggests that within a community the first cases in an epidemic would have been transmitted by fleas and lice and thus would likely be of the bubonic form. However, once established in a community the pneumonic form would have been the predominant route of transmission [6].

The third pandemic

The third plague pandemic started in South Central China around 1855. It gradually spread eastward. In 1883 it killed huge numbers of people in

Canton (Guangzhou) and reached Hong Kong in 1894. (It was here in Hong Kong where Yersin first identified the bacterium, B. pestis.) Hong Kong was then, as now, a major trading port with cargo ships sailing around the globe. It was therefore perhaps inevitable that the disease broke out in Honolulu in 1899.

San Francisco

Traveling by steamship, the plague eventually reached San Francisco. It is suspected that the cargo ship *Australia* that arrived on January 2, 1900 from Hawaii might have been the source. The vessel was thoroughly examined by quarantine inspectors. They checked out passengers for the presence of buboes but didn't know to look for it in the ship's rats. Under pressure from impatient passengers, the ship was allowed to dock. This dock was located close to the site where the sewers from Chinatown emptied into the bay. Some of the rats managed to get ashore and settle in Chinatown. It was later noted that there had been a mass die-off of rats in San Francisco soon after. On March 6, 1900, the first human plague victim, 41-year-old Wong Chut King, a Chinese immigrant, was found unconscious in his room in the Globe Hotel on Dupont Street. After several days with a high fever, he went into a coma, was carried to a nearby coffin shop, and died there.

The police surgeon who examined Wong's body noted swollen lymph nodes and an insect bite on one of his legs. He notified Dr. W. H. Kellogg, the bacteriologist at the San Francisco Health Department. Dr. Kellogg aspirated fluid from the lymph nodes by syringe and examined the extracted fluid under the microscope. It was full of short bacilli that to Kellogg looked very much like B. pestis. Dr. Joseph Kinyoun was employed by the Marine Hospital Service and managed the local quarantine station on Angel Island across the bay from the city. Hardworking and competent, Kinyoun was a trained bacteriologist having studied under both Louis Pasteur and Robert Koch. (Kinyoun had made the first bacteriologic confirmation of cholera in New York City in 1887.) He worked with Kellogg on Wong's case and confirmed that the patient had died from plague. He reported this to his superiors in Washington. Wong's body was promptly cremated (Box 15.1).

At that time, the results of Paul Simond's rat-flea transmission studies in India were not widely known or, if known, were disbelieved. The authorities were therefore unaware of the important role of rats and their fleas in transmitting the plague. San Francisco was famously filthy and overcrowded

BOX 15.1 The Marine Hospital Service.

In 1798, while still in Philadelphia, President John Adams signed "An Act for the Relief of Sick and Disabled Seamen." This act established a Marine Hospital Fund to provide hospitals to care for sick sailors. It was funded by a tax on seamen of 20 cents a month. A year later Congress extended the Act to cover officers and men in the US Navy. Originally restricted to the East Coast, it expanded as the country grew. Eventually a network of Marine hospitals was established on both coasts as well as on some rivers and the Great Lakes. In 1870, the fund was completely reorganized and renamed the Marine Hospital Service (MHS). It was managed from Washington and staffed by career physicians who could be posted where needed. In 1878 it obtained quarantine authority and managed quarantine facilities for the federal government. For example, the MHS was responsible for medically inspecting immigrants at Ellis Island in New York Harbor and those on Angel Island in San Francisco Bay. For many years, it was a bureau of the Treasury Department under the supervision of the Surgeon General. In 1902 it was renamed the Public Health and Marine Hospital Service, but as its responsibilities shifted away from the marine hospitals it was renamed once more, becoming the Public Health Service in 1912. It was initially given the authority to investigate diseases such as tuberculosis, leprosy, hookworm, and malaria. In 1946 the Communicable Disease Center was established within the Public Health Service. It is now the Centers for Disease Control and Prevention. The United States Public Health Service is currently an agency within the Department of Health and Human Services.

with the streets overflowing with garbage. The city's initial growth had been caused by an influx of migrants during the 1848–1855 gold rush. Huge numbers of Chinese immigrants were brought over to help build the transcontinental railroads. However, anti-Chinese prejudice was such that in 1882, further immigration was blocked by the Chinese Exclusion Act. Likewise, they were segregated into "Chinatowns." Chinatown was the most impoverished squalid and congested area in the city of San Francisco. Poorly built, small wooden buildings were crowded together into a small area.

When Wong Chut King was found dead, the city acted fast. A quarantine was imposed. The police stretched ropes around a dozen square blocks nearby. White citizens were told to leave, but Chinese were forbidden to do so. Chinatown was quarantined because it was contaminated by a "miasma." The city authorities blocked movement in and out of Chinatown. There was extreme prejudice at the time and a belief that Chinese were more susceptible to plague than Whites. Some even suggested that Chinatown should be burned down to stop the disease.

This first effort at quarantine simply didn't work. The population of Chinatown refused to acknowledge that Wong had died of the plague. It was poorly enforced, mocked by the press and the business community, while the Chinese consul and businesses threatened to sue the city. There had been no more cases. The local Board of Health, under intense public pressure, lifted the quarantine after only 2 days.

But then three more individuals died, then six. By mid-May 1900 there had been nine such deaths—all of them in Chinatown. The Chinese began to hide their sick relatives from the health inspectors! The Board of Health recognized that there was plague in the city, but local leaders and the Governor of California denied that the plague was present. The Federal government imposed a travel ban on Chinese residents. They could not leave the city. The Chinese sued; the government lost. The judge ruled that the government had failed to prove that Chinese were more likely than Whites to contract the disease. The California State Health Board ordered the city to reimpose the quarantine. A second quarantine was therefore imposed on May 28. This was rigorously enforced; people could not go in or out, food ran short, and the Chinese rioted. They argued that the quarantine was due to racial bias, not disease. They took the city to court, and on June 15 the judge ruled against the quarantine again—it had to be lifted. Joseph Kinyoun continued to do his job; he enforced the travel ban and was charged with contempt of court.

A battle then developed between Kellogg and Kinyoun and the city. The city refused to follow their suggestions for inspection and selective quarantine. City officials were determined to hide the presence of plague in their city. It was bad for business and would wreck tourism. Politicians, the leading newspapers, the Chinese in Chinatown, as well as the business community refused to acknowledge that the plague was in the city and did what they could to keep it a secret. A Chinese benevolent society, "The Six Companies," brought a lawsuit against the Marine Hospital Service and claimed that Kinyoun was not a real doctor because he only used a microscope! The local medical community also resented this federal intrusion. Local taxpayers resented paying the costs of quarantining Chinatown. Kinyoun was, however, aloof and brusque and no diplomat. It is fair to point out that the Mayor and the San Francisco City Board of Health supported Kellogg and Kinyoun despite the abuse. As word of Kinyoun's findings got out, however, other states such as Texas and Louisiana threatened to close their borders to Californians. The Governor of California, Henry Gage, wrote to the Secretary of Health, John Hay, about the "plague fake." He even threatened to sue anyone who reported otherwise.

By August 1900 Kinyoun had confirmed yet more plague deaths in Chinatown. As a result, the US Surgeon General, Walter Wyman, felt compelled to appoint an independent federal commission to make an inquiry into the diagnosis of the plague. He appointed Simon Flexner from the University of Pennsylvania as chairman. The other members were Lewellys Barker from The University of Chicago and Fredrick Novy from the University of Michigan. The commission's bacteriologist, Frederick Novy, confirmed that the plague was indeed present in San Francisco but that it did not have the devastating epidemiology of the medieval Black Death. Not everybody agreed. Physicians who knew of the Black Death disputed his report. The cases were not typical of the disease. A couple of the cases even had no buboes!! Governor Gage refused to accept the report, called it unfair, and ordered a news blackout! Gage denigrated the university background of the commissioners rather than their laboratory experience. Nevertheless, Gage initiated sanitation measures in Chinatown following the federal commission's report. He had made a deal with Wyman to do so providing the report was kept secret! Eventually the report leaked, and the local press gave Gage credit. Gage bitterly resented the costs of these sanitary measures, so he set up a "State Special Health Commission" to refute the presence of plague in Chinatown [7]. They examined 145 deaths between April and October 1901 and declared that none were caused by the plague. As a result, the sanitary measures in Chinatown were rescinded. However, the health department continued to monitor plague deaths while both the state and city pledged funds to fight the disease.

In February 1901, Kinyoun was sent off to obscurity in Detroit where he denounced California for prioritizing business interests over public health. He was replaced by the US Public Health and Marine Hospital Service with an equally competent but much more diplomatic federal inspector, Rupert Blue (Fig. 15.1). Initially Blue acted as a mediator between those who did and did not believe that plague was present in San Francisco. He organized a conference between the disputants to discuss the sanitary situation in Chinatown. Blue also found out just how filthy much of Chinatown was and consequently prioritized cleanup. He managed to secure funding to implement vermin eradication and maintenance to make buildings less rat friendly. However, families still persisted in hiding sick relatives, and ongoing disputes damaged relations between the Marine Hospital Service

Fig. 15.1 Dr. Rupert Blue dressed in the uniform of the United States Public Health and Marine Hospital Service, at his desk, during the San Francisco plague campaign. *Courtesy US National Library of Medicine.*

and both the state and city. There had been no cases for several months, so California declared the city plague free in June 1902. A cluster of cases, however, surfaced again in July 1902 in a Chinatown brothel and more occurred during the summer and fall of that year. As a result, in January 1903, threatened sanctions forced the state to reintroduce strong preventative measures.

It is satisfactory to note that his failure to manage the plague resulted in Governor Gage not being nominated for reelection in 1902. His successor, George Pardee, promised to reintroduce control methods and take vigorous action. The threat of sanctions on the state was lifted. The problem appeared to be solved so Rupert Blue asked for a transfer back to Milwaukee. His wish was granted but he was recalled in January 1903 when new cases occurred. He immediately stepped up the rat eradication campaign. He was given greater authority to enforce the law. Governor Pardee together with the City and State Board directed Blue to work on improving sanitation and eliminating rats. Blue established an office in Chinatown and was careful to cooperate with the Chinese organizations and societies. Tens of thousands of rats were trapped, killed, and necropsied. Basements were emptied and disinfected, streets were paved, garbage removed, cesspools filled, and decrepit buildings torn down.

The city was cleaned up rapidly, but plague cases continued to occur until February 1904. There were none for the next 11 months and Blue, believing his job done, closed his laboratory and left the city for a second time in April 1905. A total of 121 cases of the plague had been reported with 113 deaths. Almost all of the victims of this first outbreak were Chinese, and most cases were concentrated in a restricted area in Chinatown. The unusually poor survival rate suggests that many cases may have been concealed or misdiagnosed.

The plague had not, however, gone away. On May 27, 1907, Oscar Tomei, a sailor on a harbor tugboat, died of the plague. More cases occurred that August. Cases also broke out in the City Hospital isolation ward. This second wave of plague, unlike the first, was not centered on Chinatown but spread all across the city, resulting in 159 cases and 77 deaths, almost all in Whites. The Mayor asked Washington for help and the Public Health Service once again sent Rupert Blue, who arrived on September 12. By this time, it was well accepted that rat fleas spread the infection, so Blue focused on rat and flea control. He established his headquarters and a plague laboratory in the Lower Haight—"The Rattery" (at 401 Fillmore Street). He hired teams of rat catchers. The rat catchers operated across the city and trapped 13,000 each week. In the Rattery, the staff looked for the plague bacillus by dissecting and culturing thousands of rats. They filled as many as 10 garbage cans daily with the carcasses (Fig. 15.2). Blue gave public lectures on how rat control would stop the plague. Kill the rats by poison or starvation. No food scraps, sealed garbage cans; never touch a rat and avoid their fleas.

New cases of the plague progressively declined over the fall and winter of 1907 and the last San Francisco case occurred on January 30, 1908. The second outbreak had lasted for less than a year. Two million rats had been killed, and the city was finally declared plague free at Thanksgiving in November 1908. Blue's successes resulted in him being promoted to surgeon in 1909, and after Walter Wyman's death in 1912, Rupert Blue was appointed by President Taft to be the fourth US Surgeon General. He served two 4-year terms. Blue died in 1948.

But the plague bacillus had escaped into the California countryside!!

Fig. 15.2 *Blue's team dissecting rats at the "rattery" in San Francisco.* Note the absence of gloves! *Courtesy US National Library of Medicine.*

The spread through California

In August 1903, a blacksmith from the town of Pacheco in Contra Costa County on the north shore of San Francisco Bay was admitted to the German Hospital in San Francisco suffering from the plague [8]. He died. The victim had not traveled to the city within the previous month but had been hunting ground squirrels 3 to 4 days prior to his illness! It was suspected that he had contracted it from a ground squirrel. The next month, a railroad construction worker from nearby San Ramon also died of the plague. He had been living in a railroad camp where ground squirrels were used for food. Following another case in the same area in 1908, infected squirrels were detected close to the river wharves. Clearly, a rat arriving directly from China or indirectly from San Francisco had introduced the disease to rural California. An eradication campaign was launched immediately, but the infection persisted and spread. In 1907, immediately after the great earthquake, a boy developed typical bubonic plague. He had been shooting squirrels near Berkeley 3 or 4 days before he became ill. He reported that he had had to put his arm into squirrel burrows to extract the animals he had shot! Fortunately, he recovered. A plague epizootic occurred in ground squirrels in 1908 and resulted in thousands of squirrel deaths and a massive but temporary drop in their numbers. In August 1908, another human plague case occurred in the outskirts of Los Angeles and an infected ground squirrel was captured near a railroad yard where grain cars were unloaded. By 1909, ground squirrel eradication programs, shooting, poisoning, and trapping were in full swing. About 0.3% were positive for plague. After initial successes, funding was reduced in 1915, complacency set in, and funds were redirected to New Orleans (see below). In August 1919 a small plague outbreak occurred in Oakland and infected rats were found at the Oakland City Dump. A widespread survey suggested that this was a singular event and no more cases developed. This state of affairs persisted until the 1924 Los Angeles outbreak. In that outbreak, infected ground squirrels were found within city limits, and it appeared that it had arrived not by sea, but from inland areas. The spread of plague was relatively slow but inexorable. It moved over the Sierras and into the Rockies. In 1935, a survey by the US Public Health Service indicated that the plague had spread across all the western states reaching as far east as Montana, Wyoming, and New Mexico.

New Orleans

In May 1914 a 49-year-old Swedish sailor, Charles Lundene, arrived in New Orleans (it is not known where he had come from). He lived in a homeless shelter (The Volunteers of America Home on St. Joseph Street) where he developed a high fever. He was transferred to the Charity Hospital. His fever worsened and his lymph nodes in the groin began to swell. He died on June 28, 1914. An autopsy confirmed that he had died of the plague. By 1914 it was well recognized that plague was spread by rat fleas. The city and state began an aggressive rat-trapping program and in a few days caught their first infected rat, so it was likely that Lundene had caught the disease after his arrival. (No additional positive rats were ever detected.) The Mayor, Martin Behrman, recognized that this needed to be handled aggressively. The lessons of San Francisco had been learned well. He acknowledged the existence of the disease and brought in the US Public Health Service, then under the charge of Rupert Blue. The mayor and his political allies cooperated to get rid of the disease as fast as possible.

Following Lundene's death, 30 more people fell ill with the plague and 10 died over the next months. The last case occurred in September 1914. This was managed by an aggressive rat-killing campaign. The federal government hired rat catchers. They repeated Blue's strategy that had worked well in San Francisco. He set up a headquarters, mapped out the vulnerable areas, and sent out the rat-killing squads in a massive campaign. They caught and necropsied hundreds of rats daily. In 8 months, they processed 375,000 rats. In addition to catching rats, they also worked on eliminating their food supply and nesting locations. A city ordinance required that all new buildings had to be rat proofed. Garbage had to be placed in closed metal cans. Wooden floors were to be raised off the ground and the space underneath kept empty. Buildings were sealed by wire mesh and concrete. In 18 months, 75,000 buildings were rat proofed and thousands of derelict shacks and other structures were torn down. Ships were fumigated. Buildings were inspected, 3,000,000 poison baits were laid down, and 17 infected rats were identified. Despite the inconvenience, these measures were effective and generally supported by the citizens of New Orleans.

It was found that the center of infection was around Stuyvesant Docks, where the first infected rat had been detected in 1914. It is perhaps no coincidence that this was where grain from the Midwest was transshipped from grain elevators to railroad cars and ships. It was suspected that the plague may have arrived in a vessel from the Canary Islands. Meanwhile, the residents of the Volunteers of America Home were quarantined in a house outside the city. Their furniture was burned in a large bonfire, and armed guards were placed outside. Infected individuals were housed in an isolation ward and treated with an antiplague serum. (Passive immunization was well accepted by that time.) Another outbreak of plague occurred in New Orleans between 1919 and 1921. It caused 25 cases and 11 deaths. It was effectively controlled by the same strategy but was targeted specifically at black rats.

It is interesting to note that black rats are also known as "roof rats" because they tend to live at higher levels than the Norway rats that live on the ground. Roof rats are excellent climbers and can readily enter homes and encounter humans. Norway or brown rats are primarily ground dwellers and prefer to live in burrows. They often live in sewers and generally avoid humans. They tend to occupy the lower floors of buildings. The 1919 campaign also targeted ships that were then recognized as the likely source of the infected rats.

Galveston

Outbreaks of the plague occurred in 1920 in the Texas Gulf ports of Beaumont and Galveston [9]. Although their origin was unknown, it was believed that they originated from a single source, likely a ship. The first case occurred on June 8, 1920 in a 17-year-old boy, Emil Horridge, working in a feed store in Galveston. Over the following months, 18 additional plague cases were admitted to the John Sealy Hospital, with the peak occurring in August (6 cases). In two cases, affected individuals had left Galveston before falling sick and as a result were diagnosed in Port Arthur and Houston. Seven of those affected survived. Patients were passively immunized with an antiserum, "Mulford's Antiplague Serum." When supplies ran out, additional antiserum was obtained from the Pasteur Institute in Paris. Of 12 patients who were treated with the antiserum, 5 survived. Only one of seven untreated cases recovered. One physician who conducted an autopsy accidentally pricked herself with a needle and developed plague. She was treated with antiserum and recovered. In the meanwhile, an aggressive "war on rats" was launched. More than 500 rats were killed daily in Galveston for several months. Eventually more than 46,000 rats were tested and 67 were found to be positive. Plague-positive rats were detected until May 1922, but the last human case was diagnosed on November 13, 1920. As in New Orleans, derelict buildings were torn down whereas others were rat proofed by constructing raised concrete barriers around their foundations. Homeowners were required to pull up the floors of their homes once a week to ensure that they had no nesting rats. Buildings, railcars, ships, and even rocks along the sea wall were fumigated with cyanide. Seventeen of the cases were classic bubonic plague, but one was pneumonic!

Los Angeles

The last urban plague outbreak in the United States occurred between October 1924 and January 1925 in Los Angeles [10]. According to the state board report,

> on October 19, 1924, Lucena Samarano, a Mexican woman living at 742 Clara Street, died after an illness of four days. An autopsy was not held, as no unusual interest was attached to the death. This woman lived with her husband and family at the above address, where she maintained a rooming and boarding-house.

On October 22, 1924, Lucena's husband, Guadeloupe Samarano, and Jessie Flores, a practical nurse who had nursed Lucena, were taken ill, the latter dying at her home on Clara Street, the former dying in the Los Angeles General Hospital. An autopsy was performed on Guadelope Samarano, and the cause of death was given as "double lobar pneumonia."

Plague was not suspected. However, a few days later on October 29 a physician requested an ambulance to transport two critically ill patients from Macy Street in central Los Angeles to Los Angeles General Hospital. He did not know its cause, but he recognized that it was highly contagious because several other individuals were suffering from similar symptoms. The next day 13 more cases, all friends of Lucena, were admitted to hospital with high fever, severe pneumonia, bloody sputum, and cyanosis (turned blue to a lack of oxygen). Three of the patients died that day. On October 31 an autopsy on one of the victims revealed the characteristic gram-negative, bipolar bacilli of plague. That evening the hospital superintendent enquired from state and federal officials about the availability of plague serum. On November 1 the Public Health Surgeon in Los Angeles, learning of the enquiry, called to confirm that plague was indeed present and then informed the Surgeon General officially. The Surgeon General sent one of his senior surgeons to Los Angeles to investigate. On the same day, the secretary of the State Board of Health read about the strange disease that had broken out in Los Angeles in the newspaper. He wired the Los Angeles Health Department asking about the cause of death of Lucena Samarano. He got a polite reply, "Death L.S. caused by *Bacillus pestis!*" By November 2, eight blocks in the Macy Street location were quarantined. Food was supplied; a priest and public health nurses made house-to-house inspections. Only on November 6 was the true cause of disease released to the press. In total, there were 32 cases of pneumonic plague, 7 cases of bubonic plague, and 30 deaths in this outbreak. The first patient who developed the disease, Lucena Samarano, was visited by friends and relatives, of whom 16 developed pneumonic plague. The nurse, the family priest who had administered the last rites, and the ambulance attendant who took her to hospital were also infected. Because Lucena was near death when admitted to the hospital, it was not possible to determine just where she contracted the infection. Although it was initially believed that the disease had originated from rats in the Los Angeles port of San Pedro, the rats in the harbor area were uniformly negative, and San Pedro was 22 miles from Clara Street. On the contrary, evidence suggested that the outbreak was preceded by an epizootic among rats in the district several weeks previously. The first rat to be shown to be infected was found dying in a grocery store one block from Lucena's home. Dead rats were found underneath the floor of a nearby house. Unlike New Orleans, Los Angeles city council was reluctant to pass a rat-proofing building ordinance. However, this was the last outbreak of human-to-human transmitted plague in the United States—so far.

Across the west

But the plague had spread to wild rodents. It rapidly spread eastward to reach Texas by 1920. By 1950, it was enzootic west of the 100th meridian (from California to Montana and West Texas) [11]. There have been no known cases associated with urban rats since 1924. However, *Y. pestis* is maintained in the western United States by a diverse rodent population such as fox squirrels, prairie dogs, ground squirrels, woodrats, chipmunks, voles, and mice—and their fleas [12]. Predators such as cougars and domestic cats often die as a result of eating infected rodents. The primary measures used to control the disease include the use of insecticidal dust at burrow entrances. Insecticidal bait stations have also proved effective. Nevertheless, sporadic cases occur in humans, primarily in hikers and ranchers [13]. Since 2000, there has been an average of 7 cases each year reported in the United States (1—17 cases per year). Of these cases, over 80% were bubonic, implying flea transmission. Most of these cases were plausibly linked to animal exposure. About half the cases occur in persons 12 to 45 years of age, reflecting their outdoor activities. Though the plague wildlife reservoir is restricted to the western United States (Fig. 15.3), infected people can travel a great distance prior to clinical disease appearing. In 2002, for example, a married couple became ill in New York City and were diagnosed with bubonic plague. Subsequent investigations found that the couple had acquired

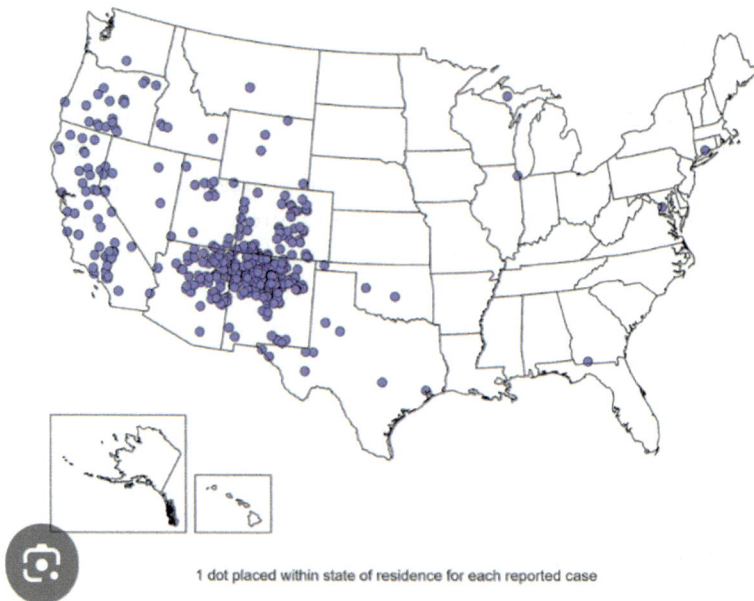

1 dot placed within state of residence for each reported case

Fig. 15.3 *Plague map. The location of plague victims in the United States.* Courtesy CDC.

the infection from *Y. pestis*—infected fleas at their home in New Mexico. The disease is readily treated with antibiotics if diagnosed sufficiently early, but 65 people (14%) have died as a result of the infection since the 1950s and 14 since 2000 (Fig. 15.3).

References

[1] Bibel DJ, Chen TH. Diagnosis of plaque: an analysis of the yersin-kitasato controversy. Bacteriol Rev 1976;40(3):633—51. https://doi.org/10.1128/br.40.3.633-651.1976.

[2] Frith J. The history of plague - part 2. the discoveries of the plague bacillus and its vector. JMVH 2012;20(3):4—8.

[3] Keller M, Spyrou MA, Scheib CL, et al. Ancient yersinia pestis genomes from across Western Europe reveal early diversification during the first pandemic (541—750). Proc Natl Acad Sci U S A. 2019;116(25):12363—72. https://doi.org/10.1073/pnas.1820447116. Available from: https://www.pnas.org/content/pnas/116/25/12363.full.pdf.

[4] Rascovan Nicolás, Sjögren Karl-Göran, Kristiansen Kristian, et al. Emergence and spread of basal lineages of yersinia pestis during the neolithic decline. Cell 2019; 176(1-2):295. https://doi.org/10.1016/j.cell.2018.11.005.

[5] Bollet AJ. Plagues and poxes; the impact of human history on epidemic diseases. Demos Medical Publishing; 2004. p. 2004.

[6] Dean KR, Krauer F, Walløe L, Lingjærde OC, et al. Human ectoparasites and the spread of plague in Europe during the second pandemic. Proc Natl Acad Sci U S A 2018;115(6):1304—9. https://doi.org/10.1073/pnas.1715640115. Available from: http://www.pnas.org/content/pnas/115/6/1304.full.pdf.

[7] Kazanjian P. Frederick Novy and the 1901 san francisco plague commission investigation. Clin Infect Dis 2012;55(10):1373—8. https://doi.org/10.1093/cid/cis693.

[8] Rucker WC. Plague among ground squirrels in Contra Costa County, California. Public Health Reports (1896-1970) 1909;24(35):1225. https://doi.org/10.2307/4563611.

[9] Boyd Mark F, Kemmerer TW. Experience with bubonic plague (Human and Rodent) in Galveston, 1920. Public Health Reports (1896-1970) 1921;36(30):1754. https://doi.org/10.2307/4576070.

[10] Viseltear AJ. The pneumonic plague epidemic of 1924 in Los Angeles. Yale J Biol Med 1974;47(1):40—54.

[11] Creel RH. Plague situation in the Western United States. Am J Public Health Nations Health 1941;31(11):1155—62. https://doi.org/10.2105/AJPH.31.11.1155.

[12] Poland JD Barnes AM. 1970 1970 1970 1970 Proceedings of the 4th vertebrate pest conference current status of plague and plague control in the United States 10.

[13] Campbell SB, Nelson CA, Hinckley AF, et al. Animal exposure and human plague, United States, 1970—2017. Emerg Infect Dis 2019;25(12):2270—3. https://doi.org/10.3201/eid2512.191081. Available from: https://wwwnc.cdc.gov/eid/article/25/12/19-1081_article.

Influenza—1918

Contents

By 1918, life expectancy in the United States was rising. Bacteriology was enjoying its early triumphs. Water supplies were improving, reducing the fear of enteric diseases. Vaccines and antitoxins were making an impact on childhood infectious diseases. Public health measures such as insect and rodent control had eliminated yellow fever and plague. The goal of eliminating the major infectious diseases appeared achievable—and then influenza struck! Beginning in 1916—1917, in the middle of the World War I, influenza began to affect troops as they fought in the trenches of Europe. From the battlefields of France, it spread worldwide. It returned to the United States and initially affected troops training in Massachusetts. It then spread westward to the Pacific Coast. Unlike conventional influenza, this strain of influenza virus was deadly and caused up to 10% mortality. Eventually it killed more Americans (675,000) than were killed by the fighting in the trenches. It lowered life expectancy by 12 years. The

Great American Diseases
ISBN: 978-0-443-31404-9
https://doi.org/10.1016/B978-0-443-31404-9.00016-1

pandemic peaked in November 1918 and then rapidly disappeared. Its severity was compounded by the war, especially by the military draft and the consequent massive overcrowding of susceptible recruits in barracks, troopships, and labor compounds, to say nothing of civilians gathering for mass meetings and processions. Influenza is a crowd disease, and the events of 1918 provided it with lots of crowds [1,2].

Influenza viruses

Influenza viruses are negative-sense, single-stranded RNA orthomyxoviruses. What makes them unique is that they have a genome made up of eight segments of RNA that collectively encode 10 to 14 viral proteins. There are three major types of influenza virus, A, B, and C. Influenza A is by far the most important and is the species described in this chapter. Its importance lies in the fact that it has animal reservoirs, especially in birds, and when transmitted to humans it can cause lethal pandemics.

Influenza A viruses have two important proteins on their surface. These are called hemagglutinins (HA) and neuraminidases (NA). The HA is required for the virus to enter a cell, whereas the NA is required for it to leave a cell. There are at least 18 different hemagglutinins and 11 different neuraminidases found among the type A influenza viruses, and they can be paired in many different ways. The viruses are therefore identified according to a simple nomenclature system. The hemagglutinin of the swine influenza virus is called H1, and its neuraminidase is called N1, so it is classified as an H1N1 virus. Only H1N1, H2N2, and H3N2 influenza A viruses have been transmitted between humans. H1N1 influenza A caused the 1918 pandemic. H2 and H3 viruses caused pandemics of much less severity in 1957 and 1968.

The global reservoirs of influenza A viruses are wild waterfowl and shorebirds. Of the 198 possible HA and NA combinations, most are found in wild waterfowl. Influenza outbreaks in humans, other mammals such as pigs, and commercial poultry result from the virus switching hosts as a result of spontaneous changes in the viral genome.

As influenza viruses spread through a population, their HA and NA genes mutate so that the structure of these surface proteins gradually changes. Influenza virus, being an RNA virus, cannot correct "mistakes" in its RNA sequence as it grows and spreads. The virus is genetically unstable, mutations accumulate, and hundreds or thousands of genetic variants may result. When the structure of these proteins changes, so does the antigenicity of the virus.

The immune response struggles to keep up with these variants, but the virus is always one jump ahead. This gradual structural change is called antigenic drift, and it permits the virus to persist for many years and cause outbreaks of seasonal influenza. It also permits these viruses to adapt to new hosts and environments. Other RNA viruses including enteroviruses and coronaviruses are also genetically unstable.

Much less commonly, influenza viruses can undergo a sudden, major genetic change in which a new strain develops whose hemagglutinins show no apparent relationship to the hemagglutinins of prior strains. Such a major change, called an antigenic shift, is not due to mutation but results from the reassortment of gene segments between two different virus strains. Thus, if two different influenza viruses infect the same host, their gene segments may be regrouped to create what is, in effect, a new hybrid virus. If this new virus can infect humans AND pass from human to human AND is pathogenic, then these naturally altered influenza viruses with a completely new antigenic structure to which no humans are immune may cause an influenza pandemic.

The disease

Influenza is a respiratory disease typically spread by coughs and sneezes that create virus-laden aerosols. The fluid droplets may come from nasal secretions or saliva. The virus is present in the droplets, and these are readily inhaled by others nearby. Once inhaled, the virus binds to the epithelial cells that line the upper respiratory tract. Using its hemagglutinin, the virus enters a cell and then replicates to high numbers before leaving the cell using its neuraminidase. The infected epithelial cell dies, leaving a bare spot. If sufficient cells are killed, the victim develops a sore throat. The virus continues to destroy cells and damages the respiratory system, resulting in coughing, fever, muscle aches, fatigue, and weakness. The raw areas may be invaded by bacteria, so causing more damage and resulting in secondary pneumonia. Influenza A is a highly infectious virus with an R_0 of about 2–3.

The disease of 1918 was much more severe than "regular flu." Some victims became delirious and went unconscious. Their lungs filled with fluid and the lack of oxygen caused their faces to turn blue or purple. These lung fluids in the form of blood and mucus drained from the victim's nose. Death soon followed. Some victims died quickly—within a day. Others could last for a week to 10 days. Even those who survived were ill for several weeks and took months to fully recover. Caregivers were overwhelmed, especially when several family members were sick at the same time. And what happened to those who missed work? Under these circumstances, the flu was especially disastrous for the poor and impoverished. They suffered and died without any help from outside. Cases were reported of the sick and dead being found together in the family's single bed.

History

It was long recognized that influenza outbreaks occurred seasonally. It therefore made sense for Italian victims to attribute the disease to the *influence* of the stars—hence influenza. However, this term was not used in England until an epidemic in 1743 and was probably not used in America until after the revolution. Assorted terms were used in earlier days, including deep colds, pleuritic fever, or malignant pleurisy. Even today, many people generally consider influenza to be a "bad cold." Because the clinical signs of influenza are very nonspecific, identification of influenza pandemics from historical records prior to the modern era is an unreliable business. It has proved especially difficult to dissect the various reports of lethal respiratory disease outbreaks from colonial North America, especially because winter colds and whooping cough were also regular events.

An influenza pandemic probably occurred in 1580. It probably originated in China and then spread westward to Europe and eventually North America. Influenza may have been spread by the Roanoke colonists to nearby Native Americans in 1585 and caused an epidemic that devastated the Accomac tribe living in the southern part of Virginia's Eastern Shore [3]. Influenza may also have affected the Pilgrims; there was a serious respiratory disease outbreak in Plymouth in 1620–1621. Noah Webster, writing in 1799, lists several epidemics in 17th-century New England that may have been influenza [4]. For example, in 1647 he notes, "This year appeared an epidemic catarrh (an accumulation of mucus in the airways) in America, and the first of which we have any account." "It began with a cold and in many accompanied with a light fever." It affected many New England towns at around the same time and spread from the French in Canada to the Dutch in New York. It killed the wife of John Winthrop, the Governor of the Massachusetts Bay Colony. Another outbreak spread throughout New England in 1655. "It was so epidemical that few persons escaped." Influenza affected England, Ireland, and Virginia in 1688. Another outbreak occurred over the winter of 1697–1698 in Massachusetts. "This catarrh began in November and prevailed until February. Its violence was in January, when whole families were sick at once, and whole towns were seized nearly at the same time." It appears to have returned the next year in 1699, when an outbreak attacked Massachusetts, and Cotton Mather reported that "many dyed, especially in Boston." Mather himself suffered "exquisite Miseryes" for about a month. In November 1711, South Carolina suffered from four disease outbreaks—"Small pox, Pestilential Feavers, Pleurisies, and Fluxes." Noah Webster describes an outbreak caused by "a species of Putrid pleurisy" in Waterbury, Connecticut, in 1712—1713. Similar outbreaks of serious respiratory disease in New England are recorded at approximately 10-year intervals in 1717, 1725, 1737, 1747, 1752, 1760, and 1772. Another global pandemic occurred in 1729. This one appeared to originate in Russia and spread across the globe, whereas yet another originated in China in 1781. In the 19th century, influenza pandemics occurred in 1803, 1833, 1847—1848, and 1889—1890 [5].

The last major pandemic of the 19th century began in 1889. It first appeared in St. Petersburg, Russia, in October of that year. This "Russian flu" spread rapidly via the railroads and steamships. It arrived in Boston and New York in January 1890. It eventually spread across the globe—a true pandemic. Like conventional flu, it mainly killed people over 50. It is estimated that it killed about 1 million worldwide. It spread faster than previous pandemics, probably as a result of improved transportation. However, it should be noted that recent molecular genetic evidence suggests that the 1889—1890 pandemic occurred at around the same time that a new coronavirus, strain OC43, appears to have jumped from cattle into humans. It has been suggested that Russian flu was really a coronavirus pandemic!

Influenza 1918

By the time the United States entered the World War I in 1917, the fighting had been raging for 3 years. The participating countries were exhausted. Societies and their armies were under severe stress. Disease susceptibility was exacerbated by the events of the time. Crowded camps, hospitals, trains, troopships, and trenches enabled infections to spread rapidly. Young, healthy men in good physical condition from rural communities were drafted and came into contact with viruses that they had never encountered previously. Diseases such as measles swept through the camps (Chapter 9), but the deadliest of all was the flu.

Origins

It is still not known where (or when) the 1918 influenza pandemic originated. The two favored origins are Étaples, France and Camp Funston,
Kansas [6]. One line of evidence suggests that the huge British army camp
located around Étaples in northern France may have been the original source
of the flu. In order to feed the troops in the trenches, the British Army had
established farms raising vast numbers of pigs and poultry. In addition, as a
transit camp, as many as 100,000 troops passed through Étaples on some occasions. There was also a major hospital in the camp with beds for 22,000
that would also have provided opportunities for disease spread. It is possible
that this is where a bird virus underwent the jump into pigs or humans. An
article published in *The Lancet* in July 1917 reported on an outbreak of severe
respiratory illness they called "purulent bronchitis" that had occurred in the
Étaples camp between December 1916 and March 1917. It had many features in common with the 1918 influenza, especially the presence of pus
blocking the airways and the "heliotrope" cyanosis that patients developed
as a result of fluid-filled lungs and a resulting lack of oxygen. (Cyanosis is the
name given to the purplish-blue color of the skin that results from a lack of
oxygen in the bloodstream. The British Army employed an artist to record
the exact color of the cyanosis and he named it after the blue heliotrope
flower.)

An alternative origin was in Kansas. An outbreak of severe respiratory
disease was first reported in rural Haskell County, in central Kansas in late
January 1918. It caused very severe pneumonia but had disappeared from
that county by mid-March. But by then, it had spread to crowded Camp
Funston within the Fort Riley military reservation, where as many as
55,000 troops destined for France were undergoing training. This was the
second largest Army camp in the country, very crowded and very cold.
The men had not been issued adequate warm clothing [7].

The first victim at Camp Funston, Albert Gitchell, a camp cook, reported sick at the base hospital on March 4. Within a week, 522 more
men also reported sick. Two hundred and thirty-seven of these men developed pneumonia and 38 died. Three weeks later 1200 men were hospitalized. The influenza spread across the United States, Europe, and Asia over
the next few months. Troops from Camp Funston were transferred to other
Army camps. The next cases appeared at Camp Oglethorpe in Georgia. One
after another, 24 of the largest Army camps reported influenza outbreaks that
spring. At the same time, the disease traveled to France, carried by

troopships, and thus appeared in Brest, France, in April, where most US troops disembarked. It then spread from Brest to reach Paris and Italy later that month. The first cases in the French and British Armies were reported in mid-April. By May 1918 the British Army had reported 36,473 hospital admissions and thousands of less serious cases. At this stage the influenza had relatively few complications and most patients recovered fully. It received little attention at the time. (Because of this lack of complications, it is possible that this spring outbreak was caused by a different virus than the one that appeared later in the year.) The disease eventually spread all across Europe that summer. One reason for the inattention was that bacterial pneumonia, especially that caused by *Streptococcus pneumoniae* and *Staphylococcus aureus,* was already common in the camps. Epidemics of these infections reflected the mixing of huge numbers of soldiers from diverse origins in grossly overcrowded conditions. Over the winter of 1917–1918, more than 30,000 US troops were hospitalized with pneumonia and, in the absence of antibiotics, almost 6000 died. The Army had already established a Pneumonia Commission to investigate the problem. A major measles epidemic had already occurred in the camps in 1917.

The Germans, however, also contracted the flu. They tried to keep the severity of the disease secret, but it resulted in a severe manpower crisis on the Western Front. Influenza swept through the German ranks, and they essentially ran out of reserves. Senior German officers were complaining about the flu by June. By May, influenza had also reached India and China, and by September, it had reached Australia and New Zealand. It was still relatively mild. Troops called it "the three-day fever." On August 10, 1918, the British declared the epidemic over—they were dead wrong!

The beginning

The number of reported influenza cases began to increase again in early August, and pneumonia cases increased a couple of weeks later. Outbreaks occurred in Brest, France, where the number of sick sailors began to overwhelm local hospitals. American troops probably acquired the infection in Brest and then spread it across France. Freetown, Sierra Leone, was a major coaling port for ships sailing from Europe to South Africa and the Far East. One ship, HMS *Mantua,* arrived in Freetown from Brest on August 15 with 200 crew members suffering from influenza. The disease first affected dock workers but soon spread to sailors and passengers. It is likely that a ship from Freetown carried the virus to South Africa and further East. From France it spread to the United States. On August 27, three sailors reported sick at the Navy Pier in Boston, Massachusetts. The next day there were eight. The next day there were 58, and 15 men had to be admitted to the Chelsea Naval Hospital. But late in August, the disease reached Camp Devens!

Camp Devens was a hastily built army camp about 35 miles northwest of Boston in Lowell County. In August 1918 it was grossly overcrowded, housing 45,000 men, although it had been built to house 36,000. Tents had to be erected to house the new arrivals. The recruits had come from all across the United States. The first cases of flu appeared around September 8 among the men of the 42nd infantry regiment. But it spread to other barracks, and by September 22, almost 20% of the men had reported sick, and of these, 75% had had to be hospitalized. The medical services were totally overwhelmed, and doctors and nurses began dying. When the leaders of the Army Medical Corps arrived at Camp Devens for an inspection on September 23 they found chaos. Sixty-three men died that day. The camp hospital built to house 2000 contained 8000 patients with cots crammed tightly together. Seventy out of 200 nurses were hospitalized. Corpses littered the hallways and were stacked up in the mortuary. Autopsies showed remarkably severe, blue, swollen lungs, and some victims bled profusely. Blood was everywhere. An initial impression suggested that this might be a "new kind of infection or plague." It was certainly of a very different character than the pneumonias that the pathologists were used to seeing. The Army responded to the reports from Camp Devens by ordering all army bases to be prepared for the epidemic. But troops were always on the move, and on September 13, a contingent was transferred to Camp Upton, Long Island, the debarkation point for France. Within 40 days, Camp Upton had sent more than 6000 men to the hospital suffering from influenza. It continued to spread from camp to camp. It spread across the country, arriving at Camp Grant, Illinois, on September 21 and Camp Fremont, California, by October 8.

It was not only the Army that was affected. The US Navy suffered as well; it is estimated that 121,225 (40%) of US Navy sailors got the flu and 4158 died. Warships of that era had fairly primitive ventilation systems and were incredibly crowded. They were ideal places for the spread of respiratory infections. Naval vessels sailed from Boston to other US ports and naval stations and carried the flu with them. Many sailors had to go to the hospital immediately upon arrival. It also spread from the army camps to the cities. The pattern of these outbreaks repeated itself in each community.

Philadelphia was, by far, the US city most severely affected by influenza in 1918. It was crowded, as workers flooded into the city to work at the shipyards and in the munitions factories. The overcrowded tenements and lodging houses provided ideal conditions for the spread of the flu. The pattern here was similar to the occurrences in many other US cities. First, the hospitals filled quickly. For example, the Philadelphia Director of Public Health had sought to reassure the public. The deaths of seamen in the Navy Hospital were no cause for alarm. As a result, on September 28, the city held a great Liberty Loan Parade to sell war bonds. While under pressure from colleagues to cancel the event, the director let the parade go ahead. An estimated 200,000 people crowded along the parade route, a perfect situation for catching the flu. Within 72 hours there were 635 new cases of influenza. Every bed in the city's hospitals was filled. They could accept no more patients. Even though lines formed outside, there was nothing the hospitals could do. On October 1, 117 people died. New emergency hospitals were set up—and immediately filled. On October 5, 254 people died.

Because hospitals were filled to capacity, many of the victims died at home [8]. People fell sick and there was nobody to take care of them. Communities pleaded with the Red Cross and the Public Health Service for help, but none was available. Millions never saw a doctor and relied on home remedies. There were no specific treatments available for influenza then. Nothing available in 1918 could help. Antibiotics had yet to be discovered and, anyhow, this was a viral disease. As a result, patients in hospitals or at home died in huge numbers. The best treatment was bed rest and good nursing, but soldiers and other young people may have resisted this. The medical profession was helpless.

Although public health was a well-established discipline by 1918, it was unclear, even then, whether influenza was caused by a bacterium or a virus. Bacteriology was thriving and the causal agents of diseases such as tuberculosis, cholera, and plague had long been identified. It was, however, not difficult to isolate bacteria from the lungs of influenza victims. Some of these bacteria were so consistently isolated that they were believed to be the cause of influenza. For example, Friedrich Pfeiffer, a distinguished microbiologist in Robert Koch's laboratory, had reported in 1892 that the "Russian flu" was caused by a gram-negative bacillus, now called *Haemophilus influenzae*. During the 1918 pandemic, pathologists found it in some but not all diseased lungs, and there was much debate over its significance. Some considered the bacillus to be the cause of influenza, and technical failures prevented its isolation in all cases. Others were less easily persuaded and suggested that some other agent caused influenza. Vaccines against pneumococcus and *H. influenzae* were eventually developed and widely employed but were clearly ineffective. An antiserum against *S. pneumoniae* may have helped some patients with secondary bacterial pneumonia. Human influenza A virus was not isolated until 1933 after a researcher accidentally transmitted his flu to laboratory ferrets. (And then a sneezing ferret sneezed on a colleague, thus returning the favor and giving him the flu!)

The Public Health Service and civilian authorities took some time to get mobilized. Federal and local funds were provided to hire physicians and nurses. The trouble was there were too few doctors or nurses available [9]. Most were with the Army and Navy. The remaining doctors and nurses continued their duties and died in large numbers. The police continued to work and died too. Over 9000 trained White nurses were in France working in the Army while thousands more were assigned to army training camps in the United States and thus unavailable for civilians. The nursing profession failed to fully use trained Black nurses or employ nurse's aides. It was well recognized that influenza was caused by an airborne agent. As a result, the authorities relied on their experience in controlling tuberculosis and tried to develop a hygienic program to control its spread. Thus isolation of flu victims was a priority. Isolation was, however, very difficult to achieve in the conditions of the time, especially among the poor in the inner cities. From its onset, influenza was associated with crowded conditions, from army camps and troopships to theaters and schools (Fig. 16.1). The authorities were reluctant to close schools and focused on more general public gatherings as well as public transit. Closing schools would have left parents with difficulties in providing child care. Closing saloons would have caused lost revenue and economic damage and, as usual, was strongly opposed by the business community [10]. In some states, such as Pennsylvania, Illinois, and Indiana, the State Boards of Health imposed bans on public gatherings, which the cities generally ignored. Another major cause for concern and dispute was the closure of places of worship. New York closed churches but not stores, and this elicited a major protest. Theaters were always, and sometimes the only, prime target for closure. Thus there was rarely a full-scale public gathering ban imposed in major cities.

There was also a strenuous effort made to minimize these closures in a concern for triggering panic and hysteria while maintaining morale. The authorities urged everyone to stay calm and not panic if they developed a mild sneeze or head cold.

Fig. 16.1 *Almost the only crowd sources that were shut down by the authorities were theaters.* However, this poster from Chicago also provided some very good advice. *Courtesy US National Library of Medicine.*

Public health authorities also focused on personal hygiene, again based on their experiences in dealing with tuberculosis. One such focus was a campaign to prevent coughing, spitting, or sneezing as well as injunctions to use a cloth handkerchief to cover the mouth when coughing or sneezing.

Chicago passed a law that made people who coughed or sneezed without using a handkerchief liable to arrest. Cloth hankies were not widely used at the time, especially among the poor, where nasal mucus flowed freely. After 1918 it became common for children to be taught their proper use in schools. (Paper facial tissues did not come into use in the United States until 1924, when they were first used to remove excess theatrical makeup with cold cream. Within a couple of years, however, they were being widely used for other purposes, including blowing your nose.)

Another measure adopted and recorded widely in photographs of the time was the use of face masks (Fig. 16.2). Mask use was promoted in an effort to reduce the spread of droplet infection. When bans on public gatherings were lifted, there was often a proviso that everyone wore masks. Masks were an obvious symbol of public support and an effort to contribute to preventing the spread of the disease. They were not, however, comfortable. The Mayor of San Francisco unhooked his face mask while watching the Armistice Day parade, and its Commissioner of Health was fined for not wearing a mask at a boxing match. (In fairness to the mayor, many other citizens of the city also assumed that the end of the war meant that they no longer had to wear masks—they were wrong.)

Fig. 16.2 *Red Cross ambulance crews standing by ready to carry off the sick and dying.* *Courtesy Library of Congress.*

The most severely affected individuals eventually died a miserable death while the grieving survivors were left without wage earners, mothers, or children, all made much harder to bear by the chaos that resulted in many cities. People died faster than the bodies could be taken care of. Dead bodies began to accumulate; undertakers were overwhelmed. Coffins were in such short supply that they were stolen and had to be guarded. In Philadelphia, the city morgue could take no more bodies, so they piled up on front porches and backyards and began to decompose. The dead lay around for days. Bodies could not be buried quickly enough, so many were stored at home before they could be removed. They were stacked three to four deep in the city mortuary. Some were collected and placed in cold storage. As in so many epidemics, grave digging by hand could not keep up with the demand, so mass graves had to be dug using steam shovels.

Almost all families were affected; people avoided contact with others, and churches, theaters, and eventually schools were closed. Nobody visited, nobody kissed, nobody brought food. In Prescott, Arizona, it was illegal to shake hands. The houses of the sick were posted with signs denoting the presence of influenza. In October, the San Francisco City Council mandated the wearing of gauze masks by everyone. They put out a full-page advertisement in capital letters "WEAR A MASK AND SAVE A LIFE." The penalty for not wearing one was a $5 to $10 fine or 10 days imprisonment! When they thought that the epidemic was over 4 weeks later, in late November, the city dropped the mask requirement, but 3 weeks later, on December 17, the number of cases began to rise again, and everyone had to put their masks back on, and they stayed on until February. A Public Health Service warning to avoid crowds came too late.

It is interesting to note that one response to the Mask Ordinance in San Francisco and elsewhere was political polarization. Despite the massive mortality, some individuals protested strenuously. There was even the establishment of an "Anti-Mask League" in San Francisco, whose members protested the requirement until the order was lifted in February. Cigar smokers cut a hole in their masks. Others simply wore masks that didn't cover their faces. Many were arrested and jailed. Conversely, the American Red Cross stated, "the man or woman or child who will not wear a mask is now a dangerous slacker!"

These events were happening everywhere. Cities closed down; neither federal nor local governments could do more to help. Rupert Blue (the plague guy) reassured, "There is no cause for alarm if precautions are observed." But people knew otherwise, as relatives, friends, and neighbors sickened and died. The problems boiled down to caring for the sick and maintaining order. Victor Vaughan, in the US Surgeon General's office, calculated in mid-September that if the epidemic continued to grow at the same rate, mankind would be eliminated in a few weeks.

What made the disease even more horrible was the fact that it attacked the young, strong, and healthy. This included parents with young children; huge numbers of children were orphaned. Pneumonia was considered an old man's disease, but this was different. Pregnant women were also unusually susceptible. Their death rate was 23% to 71%. Of those that survived, 26% lost their fetus. The number of stillbirths rose by 60%. Death could be very fast. Some died within hours of developing illness. Cases of people dropping dead on the street were not unknown. But still the deaths continued! Industry came to a halt as workers sickened. The poor and African Americans were doubly victimized by lack of access to care. Living from hand to mouth, they had no savings. As a result of lost wages, families went cold, hungry, and homeless.

The spread eastward

The World War I continued in France. The disease was incapacitating and killing large numbers of troops on both sides. The warring powers did not want the enemy to learn about the extent of the pandemic and exercised strict censorship over the outbreak. Spain, however, was not at war, and the news from that country was readily available. As a result of this increased press attention, the impression was conveyed that the disease was most severe in that country. This was untrue, but as a result, it was commonly called the "Spanish" flu.

The war continued with its insatiable need for manpower. The government was focused on finishing the war. Liberty Loan rallies continued. President Wilson ignored the disease. The government made no attempt to provide resources or organize support. The Surgeon General of the Army, William Gorgas, told the Army Chief of Staff, General Peyton March, to stop all troop transfers from the camps to France immediately. He was ignored. Eventually, on September 27 Provost Marshall was obliged to cancel the next draft of 142,000 troops. The situation in the camps was just too chaotic to house and train them. The call-ups did not begin again until October 23. On the other hand, troops continued to be shipped across the Atlantic on crowded troopships. Again Gorgas warned of the consequences and again he was ignored. President Wilson was faced with an agonizing dilemma as the war in Europe reached its climax. Pressure had to be maintained on the Germans by mounting an offensive that would force them to sue for peace. General Pershing needed as many troops as possible, in addition to more hospitals, doctors, and nurses. Wilson was in an unenviable situation, but he believed that he had no alternative if the Germans were to be defeated. The vessels carrying troops to Europe became death ships. Despite attempts at quarantine, within 48 hours of departure the sick bays were filled. The sick and dying filled more and more spaces. When there was no more room inside, men simply lay on the deck. As they died, the bodies were buried at sea. When they arrived in France, men were incapacitated for weeks. The soundness of continuing troop transports in the face of the epidemic is debatable. While the presence of dead and ill troops was a most visible factor, the use of manpower and resources required in caring for flu victims meant that these were not available for the battles.

The spread westward

Influenza raged across the United States, beginning in the Northeast from Massachusetts to Virginia and spreading westward, rapidly reaching San Francisco, Seattle, and Los Angeles, then moving to the states in the Midwest. The disease spread west and south along the railroad lines and rivers. Almost everyone lost a relative, friend, or colleague. Pennsylvania, with 883 deaths/100,000, and Maryland, with 803/100,000, had the highest mortality rates, but common factors are not apparent. Death rates in the Midwest were lower, with Minnesota having only 390/100,000 and Wisconsin, 405/100,000.

As can be inferred from the above statistics, locations where the infection arrived late in the pandemic had lower death rates than those where it arrived first. For example, Philadelphia, which first started to experience influenza cases in late August, had 47,000 reported cases with 12,000 deaths in the first 4 weeks of the outbreak and an overall death rate of 932/100,000— Philadelphia lost about 1% of its population! In contrast, in Dallas and San Antonio, where cases were first reported in late September, death rates were 250–500/100,000 and 550/100,000, respectively. This, of course, may simply reflect the fact that they had more time to prepare. In general, the Northeast and South were hit hardest: Pittsburgh, 807 deaths/100,000; Boston, 845 deaths/100,000; New York, 582 deaths/100,000; Washington

DC, 608 deaths/100,000; Richmond, 508 deaths/100,000; New Orleans, 768 deaths/100,000. The West Coast and Midwest were less so: San Francisco, 647 deaths/100,000; Los Angeles, 484 deaths/100,000; Seattle, 414 deaths/100,000; Chicago, 373 deaths/100,000; and Minneapolis, 267 deaths/100,000—one of the lowest for any US city [11]!

A third smaller wave of influenza occurred in February to April 1919. It principally affected the United Kingdom, Australia, and other Southern Hemisphere countries. It now appears that the viruses causing each of the waves were slightly different. This third wave was, in general, not as lethal as the second wave. But San Francisco, a city that had avoided the worst effects of the first pandemic, was hit severely.

Native peoples

In Labrador and Alaska, influenza killed about a third of the Native population. Thus in Nome, Alaska, over half the Native Americans died. In some Alaskan villages, mortality reached 90%. In others, no one was left alive as the few survivors scattered elsewhere. Once most individuals had died, the few weakened survivors would have starved to death if they were unable to leave.

The arrival of the SS *Harmony* bringing supplies and mail to the coastal villages of Labrador in November 1918 also brought influenza in the form of infected crewmen. The death rate among the native people was exceptional—71%! In some Alaskan villages, mortality reached 90%. As in other areas, the deaths in Labrador mainly affected young adults and parents. The old and very young died too, but from starvation and exposure. The dead could not be buried and as a result, both the dead and the dying were devoured by roaming packs of semi-wild sled dogs. When the epidemic was over, the survivors first had to hunt down and kill these dog packs and try to identify and bury any remains. The Inuit and Innu people of Labrador were almost wiped out. Even in the lower 48 states, the Navajo Nation suffered 12% mortality compared to a global rate of 2.5% to 5% (Box 16.1).

The death toll
Armed forces

The United States entered the World War I in April 1917. During the next year, the US Army increased in size from 378,000 men that April to

BOX 16.1 Swine flu.

In a remarkable coincidence, in October 1918 a government veterinarian, Dr. J. S. Koen of the Bureau of Animal Husbandry in Fort Dodge, Iowa, reported a major outbreak of a highly contagious upper respiratory tract disease in pigs, with fever, cough, nasal discharge, appetite loss, and prostration. In most pigs it was a relatively mild disease with a sudden onset and fast recovery. It looked to Koen remarkably like influenza. He called it "hog flu." The disease affected millions of pigs and killed thousands. Indeed, it was influenza, and its antigenic structure, H1N1, was the same as that of the 1918 human disease. (Serum from humans recovered from the flu reacted strongly with the swine flu virus.) It was assumed, therefore, not unreasonably, that this swine flu virus was closely related to the cause of the 1918 pandemic.

When pigs were challenged with 1918/rec virus, unlike the lethal results in mice, ferrets, and primates, it readily infected them, but it caused only a mild upper respiratory tract disease and a transient fever. There were some lung lesions in the pigs, but these were similar in severity to those caused by current swine flu viruses, except that the investigators noted that they persisted for longer!

These results support the hypothesis that the 1918 human flu virus was probably the same as the 1918 swine flu virus. It was likely inadvertently introduced into pigs during the pandemic. Alternatively, the 1918 virus may have jumped directly or indirectly from birds into humans. It is therefore possible that pigs were the intermediate mammalian host. All current swine influenza viruses appear to be descended from it. In 1930, the swine influenza virus was first isolated from diseased pigs. Interestingly, its severity was largely determined by the secondary bacterial infections.

more than 4.1 million by the end of 1918. These men were accommodated in hastily assembled and overcrowded barracks and camps. Training camps, supply depots, arsenals, and airfields were constructed rapidly. The Army established 24 large training camps, each housing from 25,000 to 55,000 troops. The Army Medical Department had great difficulty in keeping up with this extraordinary growth. Even before the flu arrived, this overcrowding was seen as a significant health hazard. In January 1918 William Gorgas testified to Congress about the need for more space. It was, however, difficult to prevent overcrowding in barracks and impossible to prevent it in the trenches. (In an interesting analysis conducted after the pandemic, it was noted that in one army camp where each soldier had 78 ft^2 of space, 2.5% of the men developed influenza; in another camp where the soldiers had only 45 ft^2, 26% fell ill. A 10-fold difference [12]!)

The Medical Department focused on good sanitation, good food, clean air, clean barracks, and personal hygiene. It established new hospitals with vastly increased capacity. About a third of the country's doctors were drafted for military service, leaving large areas of the country largely depleted of civilian physicians.

From September through November 1918, the War Department estimated that influenza affected 26% of the Army (more than a million men) and killed about 36,000 before they even got to France [13]. The Navy suffered 5027 deaths and more than 106,000 were admitted to the hospital out of a force of 600,000. (Given that mild cases were not recorded, and many men went home to their families when they sickened, the virus probably affected close to 40% of the Navy's personnel.)

While the army camps in the United States were grossly overcrowded, conditions in the trenches in France were no better. Conditions were such that the virus prospered and perhaps increased in virulence. In France, during October 1918, 70,000 soldiers reported being sick. In the US Army as a whole, there were more than 600,000 men admitted to hospitals. In the crowded camps, mortality was over 5% and in some cases over 10%. American combat deaths reached 50,280, but 57,460 died of disease. During 1918, 227,000 soldiers were hospitalized with battle wounds, but 340,000 were hospitalized with influenza. Throughout the war, influenza accounted for a third of US military deaths. The US Army estimated that 36% of its soldiers got the flu, and the Navy estimated that 40% of its sailors got it. Eighty percent of American war casualties were caused by influenza. The number of influenza deaths exceeded those killed in the Vietnam War.

The civilians

The 1918 pandemic also caused remarkable civilian mortality. The virus infected over one-third of the US population. Philadelphia was the hardest-hit city. On one day, October 10, 759 people died in Philadelphia. During the week of October 16, 4597 people died in that city. On October 23 there were 21,000 deaths across the United States. Overall, the total civilian death toll in the United States was around 548,000 to 675,000. The population of the United States at that time was about 105 million, so influenza killed about 0.6% of the population (Fig. 16.3). This was reflected by a reduction in average life expectancy of 11.8 years from 52 to about 40. Male deaths exceeded female deaths by 50% to 75%, but this likely reflected the events in the overcrowded

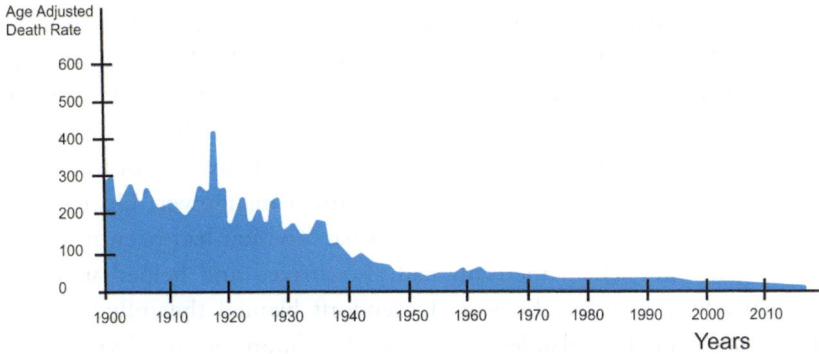

Fig. 16.3 The progressive decline in death rates from respiratory disease throughout the 20th century in the United States. The rates are expressed as deaths/100,000. The influenza death peak in 1918 is very obvious. Data courtesy CDC.

army camps. (It should be pointed out, however, that even in a "normal" year, 36,000 to 40,000 Americans are killed by the flu.)

The global death toll

The global mortality from the 1918 influenza pandemic is not known with certainty, but by one estimate about 500 million people (one-fifth of the world's population) developed clinical disease and it probably resulted in somewhere between 50 and 100 million deaths. (The vagueness of this estimate reflects the poor or nonexistent recordkeeping in many European colonies.) Influenza may have killed 17 million in India alone. In the Dutch East Indies, 1.5 million died. In Samoa, 22% of the population died in just 2 months. In Britain, 228,000 died, in France 400,000. It killed about 43,000 Canadians.

The end of the pandemic

Survivors developed immunity. Herd immunity developed. The virus simply ran out of vulnerable hosts, its R_0 dropped below 1, and the epidemic stopped abruptly. This occurred in the army camps as soon as December 1918 and in the major US cities in early 1919.

Consequences

In March 1918, the German Army, directed by General Erich von Ludendorff, began a major offensive in France that they hoped would end the war in their favor. Having signed a treaty with Russia and closed down their Eastern

Front, the Germans had a great advantage in manpower, and they wanted to decide the outcome of the war in the West before the Americans arrived. The offensive started off well and drove the Allies back almost as far as Paris. But then it lost momentum. They failed to break through the Allied lines and eventually were forced to halt their advance. One reason for this is that influenza had caused a massive loss of German manpower. Up to 2000 soldiers out of about 14,000 in each German division had been incapacitated by the flu. The Americans had also arrived and helped stem the German advance. After the war, Ludendorff blamed the influenza, the "*Blitzkatarrh*" or the "Flanders fever," for the failure of his July offensive. Some German units were depleted by 25% at a time when the Allies were mounting the major offensives that would end the war. Thus the influenza pandemic significantly accelerated the end of the war. Subsequent outbreaks of influenza also interfered with the proceedings of the peace conference in Versailles in 1919. Woodrow Wilson himself contracted the disease, was bedridden for days, and never fully recovered.

Molecular archeology

Why was the 1918 H1N1 influenza A virus so incredibly lethal? What was unique about it, and by extension, why were subsequent outbreaks of the less lethal "normal" type? In seeking an explanation for the unusually high mortality caused by this specific strain of flu, investigators have succeeded in obtaining viral RNA from the bodies and tissues of influenza victims [14].

In 1918, in the village of Brevig Mission, Alaska, during a 5-day period from November 15 to 20, 72 of the 80 adult inhabitants of the village died from influenza. They were buried in a mass grave dug in the frozen permafrost. Johan Hultin, a Swedish microbiologist, obtained permission to excavate the site in 1997. Buried about 7 feet deep, he found the body of a woman whose lungs were perfectly frozen. When analyzed later, these lungs tested positive for influenza RNA. In addition, the US Army had thoroughly investigated the 1918 outbreak. They had autopsied many victims and stored their tissues in formaldehyde at the Armed Forces Institute of Pathology in Washington DC. It proved possible to extract some viral RNA from these stored tissues as well. Similar samples have been obtained from British and German army archives. As a result of all these efforts, Drs. Ann Reid and Jefferey Taubenberger at NIH were able to sequence the complete viral HA gene of the 1918 virus. While it had some mammalian (especially pig) characteristics, the HA gene was mainly of avian origin. This viral surface protein was characteristic of an influenza A virus that had recently jumped to humans from wild waterfowl. Much of the unique lethality of the 1918 virus has been traced to this hemagglutinin. This is the protein on the viral surface that is required if the virus is to bind to host cell membranes and then enter the cell.

The viral HA gene sequences obtained from 1918 victims do not reveal any obvious causes of their great lethality [15]. No single mutation in the genes encoding this molecule can account for all its virulence. However, influenza HA are spike-like proteins that protrude, hedgehog-like, from the viral surface and cause the virus to bind to surface glycoproteins on target cells. Avian influenza A viruses grow in the intestine of birds, and their HAs only bind to cell surface glycoproteins through an $\alpha 2,3$-bond. Virus strains that infect humans, however, grow in the respiratory tract and their hemagglutinins bind to target cell glycoproteins using an $\alpha 2,6$-bond. So the only way an avian influenza virus can bind and invade a human respiratory cell is for its hemagglutinins to change as a result of mutation. The mutation results in alterations in the shape of the HA binding site. The 1918 viral hemagglutinin contains the avian-type binding sites but can still cause disease in humans! It appears that the structure of the HA binding site in the 1918 virus, while retaining its avian structure, had subtle changes that permitted it to also bind to the glycoproteins on human lung cells. Conventional influenza viruses require their HA to be processed by enzymes before they can infect cells; the 1918 virus does not. The 1918 viral NA also shared features of both mammalian and avian viruses. This suggests that an avian ancestor virus spent some time in mammals before finally infecting humans. Thus, although the 1918 virus likely originated in water birds, it probably went through an intermediate stage in pigs before jumping into humans.

Eventually, investigators succeeded in sequencing all the gene segments of the 1918 virus. Once this was accomplished, the next step was to assemble them and so generate a live 1918 virus, a very hazardous procedure!

The reconstruction of the complete 1918 virus was achieved by Dr. Terrence Tumpey and his team working under very high security conditions at the CDC. When administered to mice, this reassembled virus (1918/rec) was shown to replicate itself extraordinarily rapidly. It was highly lethal for mice, who died within 3 days with unusually severe lung lesions. (It is estimated that at that time, the amount of virus in the mouse lungs was 39,000 times greater than in the lungs of mice receiving a contemporary H1N1 virus.) 1918/rec was at least 100 times more lethal in mice than other influenza viruses tested! The mouse lesions were similar to those recorded in the 1918 victims with fluid-filled lungs and severe inflammation. 1918/rec is also highly lethal for ferrets and nonhuman primates. The 1918 virus also grows extraordinarily rapidly in human bronchial cell cultures as well as in fertilized chicken eggs.

In conclusion, although most of these unusual properties were traced to the HA molecule, there was no single mutation that made the 1918 virus so lethal. Its behavior resulted from the unique combination of all eight viral gene segments acting together.

This study caused an outcry when first published because of the risk that the virus might escape and the events of 1918 would be repeated. It was, however, conducted under stringent biosafety conditions in world-class isolation facilities, and the researchers first demonstrated that the virus was highly susceptible to modern antiviral drugs before proceeding with their investigations.

The exceedingly high mortality, as well as its lethality for young, previously healthy adults between 15 and 35, was likely due to two other unique features: an unusually high prevalence of secondary bacterial infections and an aberrant inflammatory response. For example, there was a disproportionate increase of bacterial pneumonias caused by *S. pneumoniae*, *Streptococcus pyogenes*, *S. aureus*, and *H. influenzae* in young adults. (It has been estimated that the proportion of bacterial pneumonias in 1918 influenza cases was 25-fold higher than in a "normal" flu outbreak!) Patients developed massive bacterial pneumonia spreading out from the bronchi. The virus killed the cells lining the bronchi, leaving denuded areas through which the bacteria could invade. Even areas of the lung where the bacteria had yet to reach were full of inflammatory cells. Most fatalities were therefore associated with secondary bacterial pneumonia.

In addition, patients suffered massive lung damage, and their lungs filled with fluid. The excessive immune response to the virus resulted in a "cytokine storm" leading to intravascular coagulation. A notable feature of the lung lesions was the blockage of lung blood vessels by clots (thrombi). These clots would have promoted blood vessel leakage and escape of fluid into the lungs as well as cutting off the blood supply, resulting in lack of oxygen, suffocation, and death. Some victims simply suffocated or drowned, showing the so-called "heliotrope" cyanosis. Victims sometimes suffered agonizing abdominal pain so that some doctors confused it with typhoid or cholera. Many suffered from extreme earaches. Ruptured lungs resulted in air pockets forming under the skin. Agonizing headaches also affected vision.

Because the immune response triggered intravascular coagulation, the subsequent depletion of clotting factors resulted in patients bleeding from their nose, ears, stomach, and intestine. One nurse, Josie Brown, who served in the US Navy, wrote, "They would have terrific nosebleeds with it. Sometimes the blood would just shoot across the room. You had to get out of the way, or someone's nose would bleed all over you." Hemorrhages in the eyes were common. In other words, it was a hemorrhagic fever.

The 1918 influenza pandemic mainly affected people under 65. This is not typical of influenza infections. In a "normal" flu outbreak, infants with immature immune systems and the elderly with aged immune systems die. This results in a U-shaped, age-mortality curve. In 1918, the age-mortality curve was W-shaped as a result of the deaths of young adults. (This may also reflect the large numbers of young men held in overcrowded conditions in army camps, hospitals, troopships, and trains.) Thus 49% of the deaths occurred in individuals aged between 20 and 39, 18% were under 5, and 13% were over 50. Relatively few elderly people died, and it has been suggested that many may have been immune as a consequence of an earlier pandemic. The W-shaped mortality curve was not exclusive to the United States. It was also observed in India, Australia, and South Africa. The loss of working-age employees and breadwinners adversely affected businesses as well as families. Although society recovered quickly after the pandemic, individual families were profoundly affected.

One consistent feature of epidemics throughout the ages is the fact that the "authorities" try to play down the significance of the outbreak for as long as they can. (In fairness it was also important to avoid giving information to the Germans.) Thus in 1918 both local and federal governments did their best to keep morale high. They called it the Spanish flu both for chauvinistic and disinformation morale reasons. Its severity was downplayed—"It is just ordinary influenza by another name." Newspapers and public health authorities played along. As a result, people lost any trust that they might have had in authority. People became alienated and at the same time isolated.

Present status

Almost all currently circulating human influenza viruses appear to be genetic descendants of the 1918 pandemic virus. These viruses have drifted and shifted from the 1918 virus and none have achieved its lethality—so far. Nevertheless, they still cause lethal outbreaks and kill millions of people [16].

The second "pandemic" of the 20th century occurred in 1957—1958. It was first reported in China and so was called the "Asian flu." The virus was classified as H2N2. Spread by ships, the pandemic occurred in waves, with the second wave being more severe. The major wave came in the fall of 1957 when schools reopened. This flu was relatively "mild," causing 1 to 2 million deaths worldwide and 70,000 deaths in the United States (about twice the number in an average year). It affected young people primarily, suggesting that older folks had some preexisting immunity. In fact, during 1957, as a toddler, one of the authors was hospitalized with respiratory infection, presumably influenza.

Another pandemic caused by the "Hong Kong flu" occurred in 1969—1970. It spread rapidly as a result of air travel—one of the first examples of airplane-spread pandemic disease. The virus was of the H3N2 type. It was even milder than the Asian flu. Curiously, the United States and Canada suffered most deaths during the first wave, whereas in contrast, Europe and Asia had most deaths in the second wave.

During the flu season of 1975—1976, an outbreak of "swine flu" caused a scare in the United States. It did not amount to a major outbreak, but it resulted in a loss of trust in US public health and vaccination (Box 16.2).

Cases of human infection caused by an H5N1 virus occurred in China in 1997 and clearly originated in chickens (it was therefore called "the bird flu"). Although highly lethal for affected individuals, it required a very close association with chickens, but it did not achieve significant human-to-human transmission. (It grew in the lower, not the upper, respiratory tract.) A pandemic did not ensue. Another H5N1 bird flu outbreak occurred in Asia in 2005. The virus killed millions of birds. Though only about 100 people died, it was very virulent in humans with a death rate of approximately 60%. But it was not easily transmitted to nor between humans. At that time, experts feared that the virus would mutate to become more easily transmitted between people to become a full-blown pandemic like the 1918 flu—it did not.

An outbreak of H1N1 influenza from swine originated in Mexico in March 2009 and spread to the United States but did not constitute a major outbreak.

During 2013—2022, bird flu outbreaks in animals and infections in people not only increased in numbers but have been detected over a growing geographic area as well as among a growing number of different animal categories. Bird flu outbreaks have included not only wild birds but also

BOX 16.2 The swine flu scare.

In late 1975 some soldiers came down with influenza in Fort Dix, New Jersey. The disease spread and one private died in February 1976. The Army was unconcerned but took the precaution of sending samples off to the laboratory for testing. After receiving some confusing results, the samples were sent to the CDC for further analysis. When the virus was isolated and typed, it proved to be an H1N1 strain of swine flu! But the men had been nowhere near pigs! It must have been transmitted between the troops. Serum from survivors of the 1918 pandemic reacted strongly with this 1975 strain, suggesting that they were similar if not identical! This looked bad. Was another pandemic about to emerge?

With memories of 1918, a decision was made to begin the development of a vaccine and rapidly scale up vaccine production against H1N1. Nobody could risk a recurrence of the 1918 pandemic. Was this to be done by the government or industry? Could we take the risk of doing nothing and hope for the best? Should we make the vaccine and just store it? A decision had to be made quickly. The committee studying the problem reported their recommendation to the Secretary for Health, who passed it on to President Gerald Ford (Fig. 16.4). The government allocated $135 million for vaccine production, and the administration and Congress approved the purchase of 200 million doses. Salk and Sabin, the two leading vaccine scientists in the country, also supported the decision. Despite the quick decision, vaccine production began slowly. The vaccine companies would not bottle it until it was decided who was going to pay if vaccine-related problems led to lawsuits. Congress held hearings and things were at a stalemate. Then Legionnaire's disease broke out in Philadelphia. Some thought that it might be swine flu. Congress panicked and agreed to assume liability. The law was passed in August 1976.

Vaccination began on October 1, and the first deaths in vaccinated individuals occurred 10 days later. These were likely coincidental heart attacks, but it was a bad start. In mid-November, a case of Guillain-Barré syndrome, an autoimmune nervous system disease, developed in a vaccinated person in Minnesota. By mid-December, five more cases of the syndrome in vaccinated individuals had been reported. The syndrome developed in 1:100,000 people vaccinated. In nonvaccinates, the risk was 1 in 1,000,000. By mid-December, more than 45 million doses were made available, but the acceptance rate by the public plummeted as the epidemic failed to materialize and the hazards became apparent. The swine flu vaccination campaign had to be stopped. Subsequent litigation resulted in over $9 million being paid out in damages. The unfortunate private at Fort Dix was the only known death.

Fig. 16.4 *President Ford receiving his swine flu vaccine on October 14, 1976. Courtesy Gerald R. Ford Presidential Library and Museum.*

backyard and commercial poultry and nonhuman mammals such as mink, foxes, and cats as well as a polar bear!

From January 2013 to June 2022, there were reports of 34 bird flu virus subtypes during more than 21,000 bird flu outbreaks in animals and 2000 human infections with bird flu viruses globally. In total, 16 new bird flu subtypes have been detected; 10 of these were new highly pathogenic avian influenza (HPAI) virus subtypes identified in animals, including birds. H5N8 has been the virus subtype that accounted for the most global animal outbreaks each year.

Although the recent isolates of H5N1 virus in US poultry and wild birds appear to pose a low risk to the health of the general public at this time, human infections are possible. As of August 9, 2023, more than 58.7 million poultry and more than 7100 wild birds have been affected in the United States. Only one human case of H5N1-mediated disease has been found in a person in the United States over that time.

On March 25, 2024, the presence of influenza virus H5N1 was confirmed in dairy cattle in Texas. The virus was found in mammary tissue and milk. It was also detected in cats that had died on the premises. Subsequent investigations showed that the virus had already spread to multiple other states as a result of cattle transportation. The virus has also infected four dairy workers who developed mild respiratory disease.

This outbreak has caused widespread concern because of its potential to trigger an epidemic. The virus is present in the milk of infected cows but is destroyed by pasteurization. However, many individuals prefer to drink "raw" milk. Extensive surveys are currently underway to determine the extent of the problem [17].

References

[1]   Morens DM, Taubenberger JK. Influenza cataclysm, 1918. N Engl J Med 2018; 379(24):2285—7.

[2]   Taubenberger JK, Morens DM. 1918 Influenza: the mother of all pandemics. Emerg Inf Dis 2006;12:15—22.

[3]   Mires PB. Contact and contagion: the Roanoke colony and influenza. Hist Archaeol 1994;28(3):30—8.

[4]   Webster Noah. A brief history of epidemic and pestilential diseases; with the principal phenomena of the physical world, which precede and accompany them, and the observations deduced from the facts stated. In: Two Volumes. Hartford: Hudson and Goodwin; 1799.

[5]   Saunders-Hastings PR, Krewski D. Reviewing the history of pandemic influenza: understanding patterns of emergence and transmission. Pathogens 2016;5:66—85.

[6]   Erkoreka A. Origins of the Spanish influenza pandemic (1918—1920) and its relation to the First World War. J Mol Genet Med 2009;3(2):190—4.

[7]   Caulfield E. The pursuit of a pestilence. Proc Amer Antiquarian Soc 1950;60(1):21—52.

[8]   Bristow NK. "As bad as anything can be": patients, identity and the influenza pandemic. Public Health Rep 2010;125(Suppl 3):134—44.

[9]   Keeling AW. "Alert to the necessities of the emergency": US nursing during the 1918 influenza pandemic. Public Health Rep 2010;125(Suppl 3):105—12.

[10] Tomes N. "Destroyer and teacher": managing the masses during the 1918—1919 influenza pandemic. Public Health Rep 2010;125(Suppl 3):48—62.

[11] Acuna-Soto R, Viboud C, Chowell G. Influenza and pneumonia mortality in 66 large cities in the United States in years surrounding the 1918 pandemic. PLoS One 2011; 6(8):e23467. https://doi.org/10.1371/journal.pone.0023467.

[12] Aligne CA. Overcrowding and mortality during the influenza pandemic of 1918. evidence from US Army Camp A.A. Humphreys, Virginia. Am J Public Health 2016; 106(4):642—4.

[13] Byerly CR. The US Military and the Influenza Pandemic of 1918—1919. Public Health Rep 2010;125(Supplement 3):82—91.

[14] Taubenberger JK, Kash JC, Morens DM. The 1918 influenza pandemic: 100 years of questions answered and unanswered. Sci Transl Med 2019;11:eaau5485. https://doi.org/10.1126/scitranslmed.aau5485.

[15] Tumpey TM, Basler CF, Aguilar PV, et al. Characterization of the reconstructed 1918 Spanish influenza pandemic virus. Science 2005;310:77—80.

[16] Garrett TA. Economic effects of the 1918 influenza pandemic: implications for a modern day pandemic. 2007. www.stlouisfed.org/community/other_pubs.html.

[17] Nguyen T-Q, Hutter C, Markin A, et al. Emergence and interstate spread of highly pathogenic avian influenza A(H5N1) in dairy cattle. BioRxiv 2024. https://doi.org/10.1101/2024.05.01.591751.

CHAPTER SEVENTEEN

Acquired Immune Deficiency Syndrome—1981

Contents

Beginning in July 1980, physicians in several regions of the United States began to encounter gay men suffering from multiple different infections that all appeared to be the result of profound immunosuppression. Subsequent investigations revealed that the immunosuppression was caused by a virus, eventually named human immunodeficiency virus (HIV). The virus was transmitted through body fluids. Thus it could spread through sex, it could spread through the contaminated needles used by intravenous drug abusers, and it could spread through the administration of contaminated blood and blood products. The disease was eventually named acquired immune deficiency syndrome (AIDS). AIDS has now become one of the greatest pandemics in history. Unlike influenza in 1918, it did not cause a rapid acute die-off in most of the world. In contrast, it spread much more slowly and insidiously over many years, and it is with us still.

Great American Diseases
ISBN: 978-0-443-31404-9
https://doi.org/10.1016/B978-0-443-31404-9.00017-3

In 2020 there were about 37 million people living with AIDS world-wide. Of these, 1.8 million were children under 15 years of age. Most of these victims, 25.7 million (69%), were living in sub-Saharan Africa; 5.2 million (14%) live in Asia and the Pacific region; and 2.2 million (6%) live in Western Europe and North America. In 2017 about a million people died from AIDS-related illnesses. About 1.1 million people in the United States are living with HIV, but 14% of them are unaware that they are infected. About 34,800 new cases were diagnosed in the United States in 2019. Sixty-six percent were in gay and bisexual men, 24% were in hetero-sexual men and women, and 6% were in people who inject drugs. Of known HIV-infected individuals, 63% were receiving some form of anti-HIV drug treatment.

The virus

Retroviruses

Retroviruses are members of a unique family of RNA viruses. Unlike other RNA viruses that translate their RNA directly into viral protein, ret-roviruses can do it backwards. They employ an enzyme called a reverse tran-scriptase to transcribe their RNA genome into a DNA genome. The DNA genome can then be transcribed back into RNA and then translated into viral protein (Fig. 17.1). In addition, the newly formed viral DNA can become integrated into the host's cellular DNA. This makes it almost impos-sible to make drugs that will completely eliminate the virus from its host. Another feature of a retrovirus such as HIV is that its reverse transcriptase is highly inaccurate. As a result, when the virus multiplies within its victim's

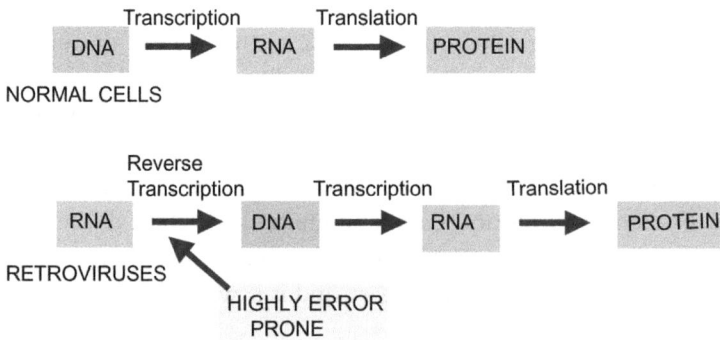

Fig. 17.1 The role of the reverse transcriptase in the replication of retroviruses.

cells, its DNA genome and hence its proteins are constantly changing. Errors (mutations) ensure that once the virus enters its victim its major antigenic proteins rapidly change their structure. The patient mounts a good immune response against these virus antigens but, because they are constantly changing, the immune system is constantly trying to catch up. The immune system can destroy many of the virus particles but not all, so the infection persists. As a result of these mutations and the resulting antigenic changes, there is not just a single AIDS virus in a victim. Instead, there are huge numbers of different variants found in their body.

The syndrome

HIV is found in and transmitted by body fluids, most notably blood, semen, vaginal fluids, and breast milk. The virus specifically invades a subset of T cells. As mentioned in Chapter 2, there are two sorts of T cells. One sort kills virus-infected cells using a cell surface molecule called CD8. The other sort "helps" the immune response using a cell surface molecule called CD4. HIV specifically uses the CD4 molecule to bind to its target cells. HIV therefore targets and selectively kills CD4-positive (CD4$^+$) T cells, especially those in the intestinal tract. As a result, HIV-infected individuals gradually lose their CD4$^+$ T cells. At first this has minimal effect because the healthy human body has plenty to spare. Eventually, however, the progressive loss of T cells results in an increasing susceptibility to infections. This progressively worsens so that the patient eventually becomes susceptible to otherwise innocuous microbes. The progressive, slow decline in CD4$^+$ T cell numbers means that AIDS has a long latent period. This latent period also provides an opportunity for intervention with antiretroviral drugs that can stop or reverse the decline in T cell numbers. Normal individuals have CD4$^+$ cell counts around 1500 cells/ml, but in the final stages of HIV infection these counts may drop to less than 200/ml.

There are three recognized clinical stages of HIV/AIDS. The first stage occurs soon after contracting infection and is called acute HIV infection. The patient may show nonspecific signs and symptoms such as a fever, headache, fatigue, sore throat, and muscle and joint pain. They may develop a rash and perhaps swollen lymph nodes as well. These signs appear 2 to 4 weeks after exposure and last for several days to weeks before disappearing.

The second stage is called chronic HIV infection. This stage is essentially asymptomatic although the virus is continuing to multiply, the patient is steadily losing T cells, and the body's ability to replace them is progressively declining. This stage may last for up to 10 years. In the absence of treatment, however, HIV infection will inevitably progress to the third, clinical AIDS stage. The signs that first develop in this stage are also nonspecific, for example, coughing, weight loss, diarrhea, fatigue, and fever. Eventually and progressively, the patient sickens as opportunistic infections take hold. The final stage of AIDS eventually develops when the CD4 cell count drops below 200/ml. Persistent high fevers, chills and night sweats, severe fatigue, massive weight loss, rashes, memory issues, and pneumonia occur. The victim is eventually overwhelmed and killed by the opportunistic infections.

Opportunistic infections

As an AIDS patient loses T cells, their ability to mount a cell-mediated immune response declines. The first signs begin to appear when CD4$^+$ cell counts drop to around 500/ml. We live in an environment where there are many microorganisms that cannot invade a healthy, normal individual but will seize any opportunity to invade a body that is not being vigorously defended. These opportunistic invaders include, most notably, the Mycobacteria. As pointed out in the chapter on tuberculosis, most healthy individuals control tuberculosis infection by mounting a T cell–mediated immune response and suffer no apparent ill effects. T cell–deficient individuals are, however, vulnerable. Thus an AIDS patient who inhales *Microbacterium tuberculosis* is increasingly likely to develop progressive disease. Likewise, there are other environmental mycobacteria that cannot invade healthy humans but will take the opportunity to invade AIDs patients. Three species, *Microbacterium avium*, *Microbacterium kansasii,* and *Microbacterium intracellulare,* are of greatest concern because they can cause progressive lung disease. In many countries, this has led to a significant increase in the prevalence of tuberculosis, especially in sub-Saharan Africa.

Airborne, microscopic fungi and yeasts are also common in the environment and readily inhaled. As with *M. tuberculosis,* cell-mediated immune responses readily destroy them. However, AIDS patients eventually become susceptible to these fungi as well. Examples include Pneumocystis, Candida, Coccidioides, Cryptococcus, and Histoplasma. Protozoan parasites are also opportunistic. These may include Toxoplasma, Isospora, and Cryptosporidium. Even other viruses can invade and cause disease. Herpesviruses especially cause chronic infections normally controlled by T cell responses. Once the T cell numbers are seriously reduced, the herpesviruses reactivate to cause disease and death. Thus Kaposi sarcoma, described below, caused by human herpesvirus 8, as well as cytomegalovirus and herpes simplex virus are important opportunistic infections in AIDS patients.

History
California

In December 1980, in Los Angeles, a 33-year-old artist called Michael was admitted to the UCLA Medical Center with significant weight loss and an extremely severe fungal infection of his mouth (called thrush) [1]. Michael eventually died in May 1981 as a result of an uncommon form of pneumonia caused by a fungus called *Pneumocystis carinii* (now called *P. jirovecii*). This fungus is common in human lungs but only causes disease if the person is unable to mount an effective T cell—mediated immune response. It classically occurred in the elderly and in organ graft recipients who had been immunosuppressed by drugs to prevent graft rejection. Blood tests indicated, however, that Michael had very few T cells in his blood, confirming that he was profoundly immunosuppressed. Although considered rare, five similar pneumonia cases occurred in the Los Angeles area that year. All were in young, gay men.

The US Centers for Disease Control and Prevention (CDC) issues a weekly publication called *Morbidity and Mortality Weekly Reports* (*MMWR*). This documents the current status of infectious disease outbreaks across the United States. The issue of June 5, 1981 reported the unusual occurrence of five *P. carinii* pneumonia (PCP) cases in young men from Los Angeles. This was just the beginning. In response to the Los Angeles report, a flood of other reports now came to the CDC reporting similar outbreaks of PCP and other opportunistic infections in gay men across the country.

Around the same time, reports also surfaced of multiple cases of a very rare skin cancer, called Kaposi sarcoma, developing in gay men in San Francisco and New York. (Moritz Kaposi was an Austrian physician who in 1876 was the first to describe this form of skin cancer.) It was recognized that there were common features linking the pneumocystis cases and the Kaposi sarcoma cases. All these cases had occurred in gay men who had also contracted other sexually transmitted diseases. Subsequent studies revealed that all these individuals were profoundly immunosuppressed. As a result, the syndrome was initially called gay-related immunodeficiency disease (GRID). The numbers of affected individuals grew rapidly. By the end of 1981 there had been 270 reported cases of GRID and 121 deaths.

By early 1982, however, it became increasingly clear that this immuno-deficiency was not restricted to gay men. It was shown that the disease could be acquired by female sexual partners of infected men. Reports also suggested that infants could become infected by their mothers around the time of birth. Infected mothers can transmit HIV to their children in several ways. The virus can cross the placenta. It can be picked up by the infant at the time of birth, and it may be transmitted to the infant through breast milk. As a result, up to half of infants may be infected when their mothers are HIV positive.

The syndrome also developed in intravenous drug users as a result of their reuse of needles contaminated with the blood of others. Heterosexual partners of these drug abusers were also infected. Cases also began to be reported in patients with hemophilia—an inherited failure of blood clotting. These individuals were not homosexual and did not use intravenous drugs. However, at that time, hemophilia could be controlled only by regular injections of clotting proteins obtained from pooled human blood plasma. Three heterosexual hemophiliacs were diagnosed with T cell immunodeficiency in early 1982. Eventually, half of the hemophiliacs in the country contracted AIDS. Another report told of the disease in 34 recent Haitian immigrants to the United States as well as an outbreak of Kaposi sarcoma in Port-au-Prince, Haiti. Most victims were heterosexual men. Cases were also reported in infants and children that had received blood transfusions at birth. The CDC began to recognize that male sexual behavior was not the only way the disease could be acquired. It was obviously spread by body fluids. It was not transmitted by casual contact, food, water, or environmental contamination. On September 24 the CDC changed the name of the disease to AIDS. It is called a syndrome because it has a common cause but victims may die in different ways as a result of many different opportunistic infections.

It was apparent that AIDS was caused by an agent that could be transmitted sexually and by exposure to blood and blood products. The appearance of AIDS in hemophiliacs also demonstrated that the blood supply had become contaminated. However, the leaders of the nation's blood banks initially refused to accept this. They resisted efforts to improve quality control of their blood products, and the government did not press them. Hemophiliacs continued to die. We now know that whole blood or fractionated components such as cells, plasma, and purified clotting factors can all contain HIV. Tests are now available to detect infected blood and blood products, and our blood supply is now safe.

The search for the cause

AIDS was not restricted to the United States. It was affecting the rest of the world as well. In 1982, Luc Montagnier and Francoise Barré-Sinoussi working at the Pasteur Institute in Paris began to look for the cause of AIDS [2]. It had been suggested that AIDS was caused by a retrovirus, and Montagnier and Barré-Sinoussi were very experienced retrovirologists. They examined lymph node samples from an AIDS patient, cultured his cells, looked for the viral reverse transcriptase, and they found it! There was a retrovirus present in his cells. The reverse transcriptase activity dropped as the cultured cells began to die. However, when they added fresh lymphocytes the reverse transcriptase activity climbed again. They also found a virus in a lymph node biopsy from an AIDS patient. They called it lymphadenopathy-associated virus. Although the virus was clearly lethal for T cells, they did not formally demonstrate that it was the causal agent of AIDS. They published their results in the journal *Science* on May 20, 1983, and the paper was almost totally ignored. In the same issue of the journal three other papers on AIDS were published, two from Robert Gallo's group in Washington DC and one from Max Essex and his colleagues at Harvard. These papers suggested that there was a link between AIDS and the previously discovered human retroviruses called human T-lymphocytotropic viruses (HTLVs). It was suggested in the journal editorial that Montagnier's group had discovered just another variant of HTLV.

Robert Gallo and his colleagues in Washington DC continued looking for the virus, and they published similar results to Montagnier's in *Science* a year later, on May 4, 1984 (Fig. 17.2). They called it HTLV type III (HTLV-III). Gallo's laboratory was the first to grow T cells successfully in the laboratory and, as a result, was able to study the viruses that affected them.

An acrimonious debate erupted between Montagnier's and Gallo's groups. This became especially fierce when both laboratories sought to patent blood tests for their virus. However, when the gene sequences of the French and American viruses were compared, they appeared to be almost identical and likely came from a common source. Subsequent investigations indicated that Gallo's virus had in fact come from Montagnier's laboratory. Montagnier had supplied Gallo's laboratory with an infected sample following a request.

It is now accepted that Montagnier's laboratory discovered the virus about a year and a half before Gallo but that Gallo's group demonstrated

Fig. 17.2 Dr. Robert Gallo. *Courtesy US National Library of Medicine*

that it was the cause of AIDS. The public dispute continued until 1987 when American President Ronald Reagan and French Premier Jacques Chirac met and agreed that credit for the discovery of HIV and any accompanying patents (for diagnostic tests, for example) should be shared between the two. Peace was restored, and in 2003 the two coauthored a paper where each acknowledged the key role of the other. Other scientists, however, had their own opinions on the matter. The 2008 Nobel Prize for Medicine was awarded to Luc Montagnier and his colleague Francoise Barré-Sinoussi for the discovery of HIV. (They shared it with Harald zur Hausen, who had shown that cervical cancer was caused by human papillomavirus.) It was a surprise to many that Gallo did not share the Nobel prize, but the committee clearly believed that Montagnier and Barré-Sinoussi had made the fundamental discoveries.

The origins of the virus

In 1983, captive monkeys in the United States were found to be suffering from an immunodeficiency syndrome that was called simian AIDS. Subsequently, in 1985, a simian immunodeficiency virus (SIV) was isolated from rhesus monkeys suffering from this disease. Analysis of this SIV showed that it was a retrovirus closely related to HIV. Subsequent surveys of other African monkey populations showed that SIVs were widespread in these animals. More importantly, each species of monkey appeared to have its own, unique SIV species. When their RNA genomes were sequenced and compared, two of the SIVs stood out as being very closely related to HIV. For example, a strain of SIV (SIV_{SMM}) was isolated from sooty mangabeys, a small monkey common in West Africa, and proved to be almost identical to HIV-2, a strain of the AIDS virus circulating in that region. Similar studies showed that a chimpanzee SIV strain (SIV_{CPZ}) was the precursor of HIV-1, the predominant strain responsible for most AIDS cases worldwide.

Chimpanzees are widely distributed across west central Africa. They occur in isolated populations, and these are often separated by wide rivers. Chimpanzees do not like to swim and as a result their populations rarely mix. Analysis of isolates of SIV_{CPZ} from different populations showed that the pandemic strain of HIV-1 called M (main) strain was most closely related to the virus found in chimpanzees living in forests in Cameroon. Other evidence, however, suggests that the site of origin is somewhat further south, across the border in the Democratic Republic of the Congo. Either way, it is clear that HIV-1 originated in a population of central African chimpanzees. It is also possible to determine approximately when this jump might have occurred. This is determined by estimating the mutation rates in the two viruses and then extrapolating back to when they had a most recent common ancestor. The consensus estimate is that the jump probably occurred early in the 20th century with a suggested range between 1902 and 1921 and a median date of 1908 [3]. The jump of SIV_{SMM} to humans likely occurred in the Ivory Coast by a similar pathway between 1932 and 1945.

Chimpanzees are forest dwellers and although said to be good to eat have historically been hard to hunt. As a result, they were not a significant source of bushmeat until firearms became readily available. The most plausible explanation for the jump of an SIV from chimpanzees to humans suggests that it was transmitted when a hunter or bushmeat handler was cut while butchering an infected animal. Infected blood flowed freely. Such wound infections are common, and it is speculated that once SIV_{CPZ} got into a human it would have adapted fairly easily to the change in host. Provided the hunter was monogamous, sporadic cases of AIDS would have passed unnoticed in remote rural societies where disease was common, medical care lacking, and human mortality high.

Once SIV_{CPZ} made the jump into humans and progressively adapted to become HIV-1, the virus may have taken several years and routes to infect more individuals and eventually gain access to the general population. This may have been achieved through the actions of the colonial powers, France and Belgium. In many areas of central Africa, the authorities recruited men from the villages to work on plantations. They were often fed with bushmeat. In the absence of their spouses, prostitution flourished. Attempts were made to control syphilis, sleeping sickness, malaria, and yaws in this workforce by injecting drugs. This probably resulted in the repeated use of improperly sterilized needles and syringes in multiple individuals, so spreading the infection through the workforce. Cities expanded as the population grew and people moved in from the countryside. The arrival of railroads and steamships on the rivers enabled infected individuals to travel long distances. This process increased dramatically when the colonial powers left and former colonies gained their independence. The cities of central Africa grew rapidly with an influx of people from the countryside. Weak governments could not enforce the laws, and the sex trade flourished. The prevalence of HIV and other STD infections in the cities grew alarmingly. Investigators found that 35% of commercial sex workers in Kinshasa, Zaire were HIV positive in 1988. It spread across sub-Saharan Africa to such an extent that two-thirds of global AIDS cases now occur in that region.

In 1982, a new disease appeared in the region of Lake Victoria. It was called "slim disease" because its progressive diarrhea and fever resulted in severe weight loss and emaciation of its victims. Cases of Kaposi sarcoma began to appear as well. It was not until 1985 that slim disease was shown to be another form of AIDS. Central Africa was in a state of constant warfare during the 1980s. As armies passed through communities, mass rape ensured that the prevalence of AIDS remained at astronomically high levels. By 1985 there were 30 million refugees across the continent.

How the disease spread to the United States from Central Africa is a matter of speculation. Following independence, the countries of Central Africa relied on assistance from the developed world. The French-speaking colonies, especially the Republic of the Congo, relied on teaching and medical assistance from Haiti. It is believed that the disease spread first to Haiti around 1965 and then to the United States via contaminated blood products and American sex tourists in the late 1960s. It may have arrived in New York City as early as 1971 and spread to the West Coast a few years later, although sporadic cases may have occurred prior to that time.

Molecular analysis of North American HIV-1 M strains has confirmed that they had an origin in the Caribbean (Haiti, Dominican Republic, Jamaica, or Trinidad) and then spread to New York City around 1970. The San Francisco epidemic began later, probably as a result of a single introduction around 1976. So-called patient "zero," who was believed by many to have introduced AIDS into the United States and was discussed at length in the book by Randy Shilts, *And the band played on*, was actually infected with a later genetic variant of HIV typical of US strains at the time and therefore could not have introduced the disease [4].

The virus eventually invaded the large male gay communities in cities such as San Francisco and New York. At that time, the risks of unprotected gay sex were not fully recognized. The cities were sexually liberal and both bathhouses and gay clubs flourished. Men had multiple sexual partners. San Francisco was especially tolerant of gay individuals, so large numbers of young men moved to the city. The long incubation period and the high transmission rates associated with their behavior allowed the epidemic to develop. At the beginning of the epidemic, the disease had a long incubation period, almost 10 years in some cases. As a result, it was not detected for many years. By the time the first deaths occurred, the virus had spread worldwide.

In addition to affecting the gay community, HIV was readily transmitted by contaminated needles and syringes to intravenous drug users. It appears that it began to kill heroin users in the late 1970s. Initially the deaths were not regarded with alarm. They were called "Junkie Flu" or "the dwindles." Patients with *Pneumocystis* pneumonia were told that they had bronchitis. These individuals had limited access to health care, and their deaths did not at first alarm the public health authorities.

Politically, however, individuals who were neither gay nor hemophiliac did not feel threatened. The perception was that this was a disease restricted to drug addicts and male homosexuals. The gay community themselves contributed to this perception by vociferous lobbying for research support. Conversely, some conservatives suggested that the appearance of the disease was a response to homosexual behavior. The television evangelist Jerry Falwell preached that "AIDS is God's punishment; the scripture is clear: we do reap it in our flesh when we violate the laws of God." It is of interest to note that similar sentiments were expressed by preachers during the 1721 smallpox outbreak in Boston and the 1918 influenza pandemic. The Reagan administration had little interest in diseases affecting homosexuals. Likewise, they had few concerns regarding those who had broken the law by using drugs. Later that year, multiple cases occurred among heterosexual Haitian immigrants in Florida, but the government again remained indifferent. In 1995, Senator Jesse Helms argued that funding for AIDS programs should be reduced. He summed up his views, unfortunately shared by others, that those who had contracted AIDS did so as a result of their "deliberate, disgusting, revolting conduct" [5]. He was presumably thinking only of homosexuals and drug users. He clearly believed that the virtuous (by his standards) did not get AIDS and as a result its victims were undeserving of sympathy and support. Fortunately, sounder minds prevailed. Funding for the Ryan White HIV/AIDS program began in 1991 started at $220.6 million but was increased to $738.5 million in 1996, to $996.3 million in 1997, and to $2.34 billion in 2018. However, issues of personal morality and funding for public health research and activities still persist in the United States and many other countries.

Societal changes

The severity of the AIDS epidemic in the United States resulted in significant social changes. In addition to needle-exchange programs, increased condom usage, and safe sex campaigns, it also led to changed attitudes within the gay community. By encouraging monogamous relationships, it subsequently created a demand for the recognition of gay marriage. In 1984 and 1985, the bath houses and private sex clubs in New York and Los Angeles that had served as centers for gay sex closed. In October 1984 San Francisco ordered the closure of the bathhouses as these were centers of hypersexual activity and the spread of AIDS. In March 1985 the first commercial blood tests became available.

In 1985, Ryan White, a 14-year-old in Kokomo, Indiana, was excluded from his middle school because he was HIV positive. He was a hemophiliac who had acquired the infection from a routine blood transfusion. The school board excluded him in response to demands by other parents who were terrified that he might spread it to their children by casual contact.

His family mounted a prolonged legal battle to protect his right to attend school. He died of AIDS-related disease in 1990 at the age of 18 (Fig. 17.3).

In 1986, the US Surgeon General Dr. C. Everett Koop issued his report on AIDS. He made it clear that AIDS could not be spread casually. He called for a nationwide campaign, increased condom use, and voluntary AIDS testing. In 1988 Koop mailed out 107 million copies of a booklet called "Understanding AIDS" to all American households. It was the largest public health mailing in history. In 1987 the WHO launched its global AIDS program in an effort to bring the disease under control in the most severely affected countries.

United States

The number of AIDS cases grew rapidly in the United States. Thus in 1983 there were 3064 cases and 1292 deaths, but by 1989 there were 100,000 cases. By 1992 AIDS was the number one cause of death in US men between the ages of 25 to 44, and in 1994 it became the number one cause of death in all Americans between 25 and 44. AIDS deaths peaked at 41,000 deaths in 1995. By 2015 in the United States, there had been 658,507 deaths since 1981, and there were 1,122,900 US residents living

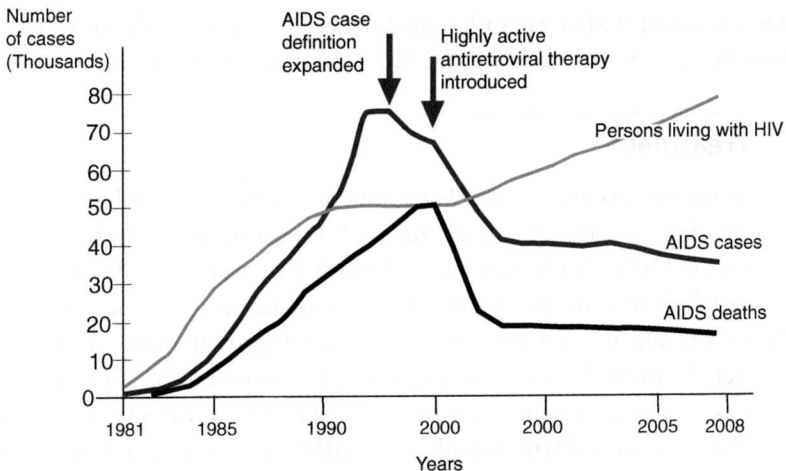

Fig. 17.3 The time-course of the early years of the AIDS pandemic. Note that cases and deaths were climbing until the advent of highly-active anti-retroviral therapy in 1997. This was remarkably effective in preventing disease and death. However the numbers of people living with the infection continued to climb. (Data courtesy of the CDC)

with AIDS, of whom about 15% were unaware of their infections. There were still 15,807 AIDS-related deaths in the US in 2016 [6].

Global spread

The number of AIDS cases worldwide also exploded rapidly from about 20,000 cases in 1985 to 2.5 million cases in 1993. By 1999, AIDS was the fourth greatest cause of death globally and the leading cause of death in sub-Saharan Africa. In 2005, AIDS deaths peaked with 1.6 million deaths that year alone. By 2017, 36.9 million people were living with AIDS worldwide and 940,000 died from AIDS-related illness. Seventy percent of these cases are in sub-Saharan Africa. It is likely that about 75 million people have been infected and 32 million have died worldwide from AIDS-related illnesses since the start of the pandemic. Tuberculosis remains the leading cause of death among people with AIDS.

By the end of 2021, it was estimated that about 1.2 million people in the United States were infected with HIV and about 13% were unaware of it! These numbers had not changed significantly by the end of 2023. In 2021, HIV caused 20,000 deaths that year. It is estimated that 32,100 new HIV infections occurred in the United States. This was 18% higher than the previous year and is attributed to an underdiagnosis of HIV in 2020 as a result of the COVID pandemic. The highest rates of HIV diagnosis occurred in the South where 52% of the cases occurred. Seventy percent of the cases occurred in gay or bisexual men. Twenty-five percent were acquired by heterosexual contact and 8% were in intravenous drug users.

Treatment

In marked contrast to the development of antibiotics for bacterial diseases, the development of specific antiviral drugs had lagged badly prior to the onset of AIDS. Antibiotics were relatively easy to develop because bacterial metabolism is distinctly different from metabolic processes in animal cells. As a result, it is possible to block bacterial growth without affecting the infected patient. Viruses have a much more intimate association with animal cells, so it is much more difficult to design safe and nontoxic antiviral drugs. All this changed with the advent of AIDS. This was an important disease with a large market. Work began immediately. The most obvious drug target was the viral reverse transcriptase. Animal cells do not have such an enzyme. As a result, drugs were designed to specifically block viral reverse transcription.

The first antiretroviral drug produced was the reverse transcriptase inhibitor azidothymidine (AZT, Zidovudine), which became available in 1987. In that year, Congress allocated $30 million for the AIDS drug assistance program in an effort to hasten the search for effective drug treatment. In 1996 the first nonnucleoside reverse transcriptase inhibitor, nevirapine, was approved. A second unique feature of HIV was that it initially synthesized all its proteins in a long continuous sequence. The virus then used an enzyme called a protease to cut this long chain into its individual components. In 1995 the first protease inhibitor was approved.

Using these three drugs, it was now possible to undertake aggressive drug treatment to destroy HIV. Given that HIV mutates very rapidly, it was highly undesirable to use just a single drug against it. In such a case, a mutation resulting in the development of drug resistance could easily develop. It is essential to use mixtures of drugs so that if resistance develops to one, then the others will still be effective. A similar strategy is used in combating multidrug-resistant tuberculosis (Chapter 6). The use of this combination of drugs was called highly active antiretroviral treatment (HAART). HAART transformed the AIDS situation. In 1997 a drug combination in a daily tablet was made available. Price remained a major issue, especially in Africa, Brazil, and India, until 2001 when the manufacturers cut prices to Third-World victims. At the same time, the World Trade Organization affirmed the right of countries to buy or manufacture generic drugs to meet a public health crisis.

In 2011 the CDC determined that a daily oral dose of antiretroviral drugs could prevent new infections in heterosexual couples. By 2012 antiretroviral treatment had reduced AIDS-related deaths worldwide by two-thirds, and for the first time people that needed treatment were receiving it. Antiretroviral treatment results in a drop in the amount of HIV within the bloodstream (viremia). In many cases the virus can drop to undetectable levels and the patient's immune system recovers. Nevertheless, that virus cannot yet be completely eliminated from the body and there have been very few confirmed cures—yet.

Susceptible populations

By 1998 it had become apparent that African Americans were disproportionally affected by AIDS. In that year African Americans suffered 49% of AIDS-related deaths. This was 10 times that in Whites and 3 times that in Hispanics. It was clear that special efforts had to be made to protect this population. In parts of the United States, the AIDS situation was as severe as in some regions in Africa. Black people continue to suffer a disproportionate share of HIV compared to other races. Thus in 2017 African Americans accounted for 43% of new cases and 52% of deaths despite constituting only 13% of the US population. The overall prevalence in this infection in male African Americans has increased by 4%. Much of this is due to poor access to health care. Latinos are also disproportionately affected. For example, in 2010 the rate for new AIDS infections in Latino males was 2.9 times that for White males. Socioeconomic inequalities, access to health care, and cultural/societal views and expectations (stigmatisms, fear, discrimination, and homophobia) are among the causes of this disparity.

The situation in Africa

AIDS in Africa has had an impact totally different from that in the United States. In 1999 the WHO announced that HIV/AIDS was the fourth greatest killer worldwide and the top killer in sub-Saharan Africa. In that year it was estimated that 33 million worldwide were living with AIDS and 14 million had died from the disease. By 2002 average life expectancy in sub-Saharan Africa dropped from 62 years to 47 years as a result of AIDS 2.4 million Africans were expected to die from AIDS in 2002. By 2006 the number of individuals receiving antiretroviral therapy reached 1 million in Africa. Those who die are often adults leaving behind children and elderly to fend for themselves. In 2020, cases in Central and South Africa accounted for 54% of all global infections.

In addition to the discouraging situation in Africa, it is also apparent that the virus is developing significant resistance to the antiretroviral drugs in some countries. More than 10% of infected adults have developed resistance to the drugs in countries such as Honduras, Cuba, South Africa, and Nicaragua. This is also a major issue in infected infants in sub-Saharan Africa.

Eradication?

The United States remains heavily infected; in 2018 there were 37,881 newly reported HIV infections. These are located primarily in the South, whereas the lowest rate is in the Midwest. There are major disparities among racial and ethnic groups [7]. In contrast to the situation in less-developed countries, however, the authorities in the United States are considering what strategies may be required to eradicate AIDS. In 2019, the US Government announced a goal of reducing AIDS cases in the United

States by 90% by 2030. Because progress in developing an HIV vaccine is slow and discouraging, the key strategy will be to suppress the virus in infected individuals by providing access to antiretrovirals. It also involves the increased use of preexposure prophylaxis in those individuals at increased risk of becoming infected. Preexposure prophylaxis involves taking two antiretrovirals in a single pill daily. It has been shown to reduce the rate of new HIV infections by at least 86%. Careful targeting of appropriate regions and susceptible adults within these areas will have a profound effect, especially once the targeted individuals can be persuaded to take the pills.

References

[1]   A timeline of HIV and AIDS. see HIV.gov.
[2]   Gallo RC, Montagnier L. The discovery of HIV as the cause of AIDS. N Engl J Med 2003;349(24):2283−5.
[3]   Sauter D, Kirchoff F. Key viral adaptations preceding the AIDS pandemic. Cell Host Microbe 2019;25(1):27−38.
[4]   Worobey M, Watts TD, McKay RA, et al. 1970s and "Patient 0" HIV-1 genomes illuminate early HIV/AIDS history in North America. Nature 2016;539:98−101. https://doi.org/10.1038/nature19827.
[5]   Murphy TF. Is AIDS a just punishment? J Med Ethics 1988;14:154−60.
[6]   Kazi DS, Katz IT, Jha AK. Preparing to end the HIV epidemic − California's route as a road map for the United States. N Engl J Med 2019;381:2489−91.
[7]   De Cock KM, Jaffe HW, Curran JW. Reflections on 40 years of AIDS. Emerg Infect Dis 2021;27:1553−60.

CHAPTER EIGHTEEN

Zoonotic diseases: rabies as a link between human and veterinary medicine

Contents

A zoonosis is a disease caused by a bacteria, virus, fungus, or parasite that is transmitted between animals, both domesticated and wild, and humans. A further definition provided by the Joint WHO/FAO Expert Group on Zoonoses is "those diseases and infections [the agents of] which are naturally transmitted between [other] vertebrate animals and man." Simply stated, zoonotic diseases infect both and are transmitted between animals and humans. The etymology of zoonosis is derived from the combination of the Greek "zóion" or "zoo," meaning animal, and "*nosos,*" meaning disease. The word was first coined in the mid-1800s by the German physician Rudolph Virchow to define these types of diseases.

Zoonotic infectious agents can be spread to humans by direct or indirect contact. These include bites; scratches; contact with skin, fur/hair, body fluids, and feces; and by fomites such as contaminated bedding, grooming equipment, feed and water bowls, or tack. Some examples of zoonotic diseases spread by contact include rabies, through exposure to the saliva of an infected animal; epidemic typhus (Chapter 10), through rubbing infected feces from a human body louse into the wound caused by the bite of the

Great American Diseases
ISBN: 978-0-443-31404-9
https://doi.org/10.1016/B978-0-443-31404-9.00018-5

louse; and the fungal disease ringworm (*Microsporum canis*), through contact with the fungus growing on the skin of infected dogs and cats.

Other zoonotic agents may be ingested in the form of contaminated food and water, or as a result of fecal contamination. These include brucellosis, tuberculosis, and avian flu as a result of drinking contaminated milk; campylobacteriosis and salmonellosis can be acquired by eating surface-contaminated food; and even raccoon roundworms (*Baylisascaris* species) can be spread by fecal contamination (the fecal-oral route).

Sometimes zoonotic organisms can be inhaled in the form of aerosolized droplets or contaminated dust. This can be the case with the influenza virus (Chapter 16), which is why face masks may be of benefit. Psittacosis (*Chlamydia psittaci*) is a respiratory disease in pet parrots, acquired by inhalation of contaminated feather dust and dander as well as aerosolized dried fecal matter. The bacterium that causes tularemia (*Francisella tularensis*), like many bacterial pathogens, has numerous routes of transmission such as ingestion, contamination of mucous membranes and broken skin, or from the bite of an arthropod vector.

A few zoonoses are vector-borne and acquired through the bites of fleas, mosquitoes, lice, or ticks that have previously fed on animals. Examples include Lyme disease (*Borrelia* bacteria), spread by the bite of the *Ixodes* tick; the infection is carried largely by deer. Rocky Mountain spotted fever (*Rickettsia rickettsia*) is spread through the bite of the *Dermacentor variabilis* tick. One major example is the plague (Chapter 15), which is transmitted by flea bites, but the disease is carried by rats and other wild rodents.

As described in Chapter 19, COVID-19 was almost certainly picked up from a captive mammal in a market in Wuhan, China, in the fall of 2019. The carrier is unknown but was probably a bat.

These are just some examples of the more than 200 known zoonotic diseases that occur in the United States, Mexico, and Canada. Many of the diseases discussed in this book are zoonotic, for example, plague, bovine tuberculosis (*Mycobacterium bovis*), typhus, yellow fever, influenza, hantavirus and other hemorrhagic fever viruses, enteric diseases, and even COVID-19. Interestingly, the measles virus, which now only infects primates, probably has its origins from the cattle disease rinderpest, thus beginning as a zoonotic disease that became primate specific (Chapter 9).

Zoonoses thus represent a major public health issue as they cross-infect domesticated pets, livestock, wildlife, and humans. Additionally, zoonotic pathogens have the potential to be used as bioterrorism agents. The United

States Centers for Disease Control and Prevention (CDC) classifies bioterrorism agents (pathogens) and diseases into three categories.

Category A—agents that pose the greatest risk and include pathogens that can be easily transmitted from person to person and can result in high mortality in infected people. These diseases (*etiological agents*) include:

Plague (*Yersinia pestis*)—zoonotic (Chapter 15)

Viral hemorrhagic fevers (Arenaviruses such as Ebola, Marburg, Lassa)—zoonotic (Chapter 4)

Anthrax (*Bacillus anthracis*)—zoonotic

Tularemia (*F. tularensis*)—zoonotic

Botulism (a toxin of *Clostridium botulinum*)—zoonotic

Smallpox (Variola virus)—humans only (Chapter 8)

Category B—agents that pose the second highest risk due to their moderate ease of transmission and moderate mortality (deaths); they require specific special diagnostics. Some of these diseases (*etiological agent*) include:

Brucellosis (*Brucella* species)—zoonotic

Glanders (*Burkholderia mallei*)—zoonotic

Psittacosis (*C. psittaci*)—zoonotic

Q fever (*Coxiella burnetii*)—zoonotic

Typhus fever (*Rickettsia prowazekii*)—zoonotic (Chapter 10)

Bacterial food-borne diseases (*Salmonella* species, *Escherichia coli* O157: H7)—zoonotic (see Chapter 10 for *Salmonella enterica* Typhi)

Viral encephalitis (Eastern equine encephalitis, Western equine encephalitis, and Venezuelan equine encephalitis)—zoonotic

Category C—emerging pathogens or pathogens that can possibly be modified or engineered to facilitate ease of dissemination and increased morbidity and mortality rates. For more information on bioterrorism agents and diseases, consult CDC emergency preparedness and response on the web—https://emergency.cdc.gov/agent/agentlist-category.asp.

Rabies

Rabies is "an acute virus disease of the nervous system of mammals." The word's derivation is from the Latin "*rabere*," to rage or rave, and the Sanskrit "*rabhas*," to do violence. Rabies is an ancient disease, with references since 6000 BCE to enraged, frenzied dogs. Ancient Greek and Roman literature contains many suggestions of rabies and rabid dogs. The Greek goddess Lytta or Lyssa is the deity of mad rage, fury, crazed frenzy, that is, rabies. Thus from the Greek "*lyssa*" and Latin "*virus*"—poison—the rabies virus is in the genus *Lyssavirus*, in the family *Rhabdovirus*, which reflects the virus's rod-like shape; the Greek term *"rhábdos"* means rod, wand, streak, or stripe.

Though not listed by CDC as a bioterrorism agent, rabies is almost 100% fatal when the clinical disease in animals and people. But it is 100% preventable with vaccination and treatment if commenced before clinical symptoms occur. Still, rabies holds a special place in society's fears because of the disease's history, its relationship to "man's best friend," and its ability to cause such a terrible death.

History

Rabies has been known to humans since ancient times. There is evidence suggesting that rabies, specifically rabies in bats, was present in the Mayan and Aztec cultures of pre-Columbian America. There were Aztec words for rabies, although canine rabies was noticeably rare in pre-Columbian America. In *Historia natural de la Nueva España* by Francisco Hernández, written between 1570 and 1577 about his travels throughout North America, there is no description of any disease that resembles rabies. An additional theory suggests that canine rabies may have been introduced into North America with the migration of man and mammals across the Bering land bridge between northeastern Siberia and Alaska some 30,000 years ago. In Inuit folklore originating long before Europeans in North America, there are references to a disease resembling rabies being transmitted from foxes to both the Inuit and their dogs. Certainly, dogs were present in the New World long before the arrival of Europeans. They appear, however, to have been quite small and nonthreatening because the Spaniards used their large mastiffs to effectively terrorize the native peoples that they conquered.

The first written reports of a disease that reported dog behavior resembling rabies in the Americas were in 1703 and 1741 from Mexico and Barbados, respectively. Dog-mediated rabies, which was highly prevalent in Europe, seems to have increased in the Americas with European exploration and colonization, whether due to increased dog importation or the introduction of rabies virus from Europe. In 1753, rabies was the first reported zoonosis in the English colonies, occurring in the Virginia colony. Soon afterward, it was reported throughout all the American colonies. The early colonists, as well as the Native Americans in the Northeast, kept dogs for hunting as well as a source of food when famine struck.

By the 1770s, rabies was enzootic in the United States, with major out-breaks reported in Philadelphia, New York, and the state of Maryland. In fact, in a diary entry by George Washington in 1769, he wrote of having to kill a mad dog that had bitten several of his dogs; there is no mention of the fate of the bitten dogs. Rabies was occurring throughout the United States. In 1772 the *Virginia Gazette* proclaimed that dogs were a public nuisance and permitted unowned dogs to be killed without any penalties. However, unlike many other diseases discussed in this book, rabies killed far fewer people—mere dozens each year. Even though rabies was an infre-quent disease and was not a major cause of death in North America, it had a high profile among people, instilling a livid fear of the disease and stray dogs. Because there was no way to know if the dog that bit a person was rabid, there was no treatment for the disease, and the disease was invariably fatal, fear and anxiety of a potentially horrible death was upon a bitten person for weeks.

During the 1800s, rabies continued to be most prevalent in dogs, with the occasional cat and livestock becoming infected. Because the main threat to humans was from dogs with rabies, prevention centered around reducing rabies in dogs and the exposure of humans to rabid dogs; thus the control of rabies focused on control of stray and roaming dogs. Muzzling ordinances were introduced in most major cities. Animal control organizations insti-tuted the capture of stray dogs. In New York City alone, tens of thousands of dogs were caught and killed annually. The destruction of these animals was neither humane nor painless; in many instances, it was accomplished by means of mass drownings (Fig. 18.1) [1].

The first rabies vaccine was developed by Louis Pasteur in France in the early 1880s, predominantly as a treatment for humans attacked by rabid an-imals. However, the vaccination of animals to prevent rabies was also inves-tigated. In the 1920s, an inactivated (killed) rabies vaccine for dogs first became available. In 1940, a live attenuated rabies virus vaccine, which had undergone 40 to 50 passages in chicken eggs to reduce the virus's viru-lence, became available. This was commercially available to the veterinary community and was used for mass vaccinations of dogs in the United States starting in 1947 and in Canada in the 1950s [2].

The practice of vaccinating dogs against rabies was instrumental in reducing the number of rabies cases in dogs and, concurrently, drastically reducing the number of human rabies cases. As can be seen in Fig. 18.5, in the 1930s and early 1940s, there were thousands of rabies cases in domes-tic animals, predominantly dogs; at the same time, human cases numbered

Fig. 18.1 *Dog control in New York City. Image from* Harper's Weekly *(July 15, 1882) depicts some of the incidents and activities of dog control in New York City. From* Harper's Weekly, *Vol. 26, Issue 1334, 1882.*

between 30 and 50 cases annually, most, if not all, attributed to dogs. However, following the introduction of a mass vaccination of dogs in 1947, cases in domestic animals precipitously declined along with a parallel reduction in human rabies cases. In fact, by 1960, human cases of rabies in the United States were virtually eliminated. Since the 1960s, only one to two cases of human rabies have occurred annually in the United States, and most of these cases have been due to rabid wildlife exposure, 70% being attributed to bats. With the introduction of vaccinating dogs for rabies and stray dog controls, the canine variant of rabies was eliminated from the United States in 2007 and from Canada in 2023. The elimination of dog-to-dog transmission of canine rabies does not mean that owners need not vaccinate their dogs and cats. Vaccination of pets has now made wildlife the major reservoir for the rabies virus in North America. Dogs can still be infected with other rabies variants that circulate among wildlife, such as the raccoons, skunks, feral cats, and foxes, so rabies vaccination of dogs and cats remains a public health priority. Even with these dramatic reductions and reservoir shifts in the United States, someone is being treated for possible rabies exposure every 10 minutes and one to two people die annually of rabies in the United States (Fig. 18.2).

This reduction in canine rabies, and simultaneously in human rabies cases of rabies, is not reflected across all the Americas. In 2023 and the first half of 2024, there were a total of 18 human deaths due to rabies throughout the Americas. Nine deaths (two in Haiti, one in the Dominican Republic, two in Venezuela, one in Peru, and three in Bolivia) were caused by rabid dogs, whereas nine cases (two in Mexico, two in Peru, three in French Guyana, and two in Brazil) were due to exposure to cats, bats, a monkey, or a cow. In the countries of Central and South America, as well as globally, dogs remain the predominant cause of rabies in humans (Fig. 18.3).

Even with the control of dog rabies and with human rabies deaths being very low in North America, there is still a societal phobia of rabies with fear

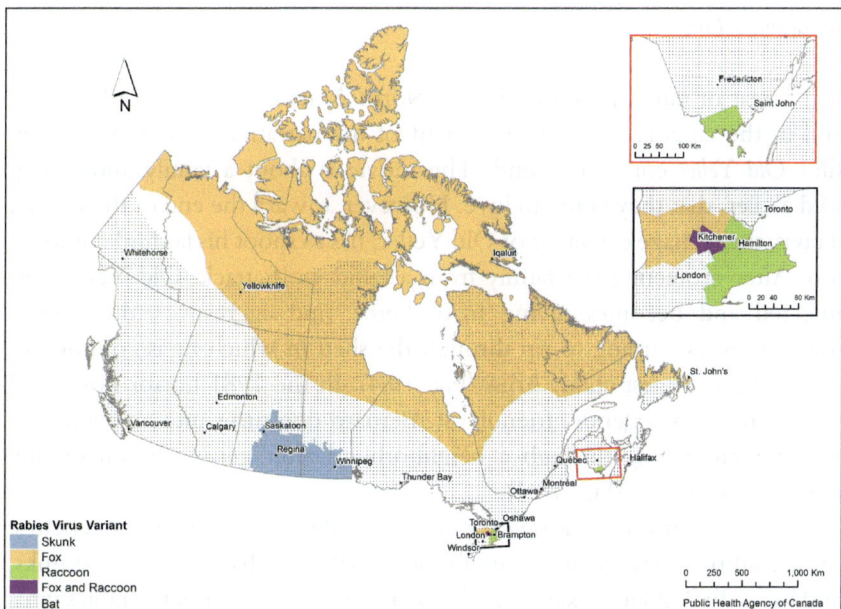

Fig. 18.2 *Geographical distribution of rabies virus variants and species reservoirs in Canada, 2016–2020. From Government of Canada. Rabies: For health professionals. 2024. https://www.canada.ca/en/public-health/services/diseases/rabies/for-health-professionals.html#a7.*

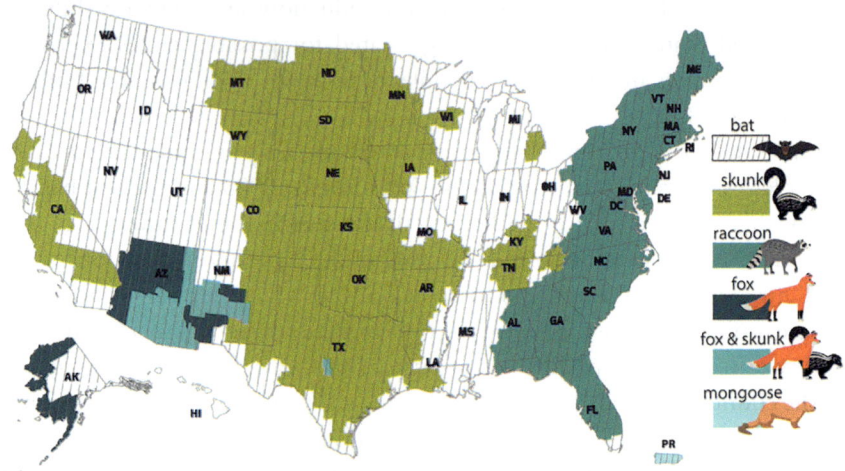

Fig. 18.3 *Geographical distribution of rabies virus variants and species reservoirs in the United States. From United States Centers for Disease Control and Prevention. Data Visualization: Rabies—a Forgotten Killer, US National Rabies Surveillance System, 2019. Public-use data file and documentation. https://www.cdc.gov/vitalsigns/rabies/data-visualization.html.*

and threat of rabies persisting in the North American psyche. In literature and in the cinema, rabies is often part of the narrative. The 1957 Disney film *Old Yeller* comes to mind. The movie is about a family and a dog, Old Yeller, that they come to love. However, toward the end of the movie, Travis, the teenage caretaker of Old Yeller, has to shoot his faithful companion. After protecting the family from a rabid wolf attack, Old Yeller gets infected and becomes raving mad, fierce, and aggressive—the furious form of rabies—trying to rip through the shed in which he is confined to get to the youngest child, Arliss (Fig. 18.4). In the 1981 horror novel and 1983 film *Cujo*, a gentle, friendly St. Bernard dog turns into an aggressive monster rampaging through a neighborhood after getting bitten on the nose by a rabid bat (Fig. 18.4).

Rabies is not just a disease of "mad dogs." Presently in the United States, most confirmed rabies viruses are found in wildlife—bats, raccoons, skunks, and foxes. In 2023 in Texas, only the skunk and bat variant of the rabies virus was identified and confirmed in 159 bats, 125 skunks, 20 raccoons, 16 foxes, 13 cats, 9 cattle, 4 dogs, 2 horses, and 1 goat. This is a major difference from prior to the 1960s when most rabies cases were in domestic animals, primarily dogs but also in cats. The Washington DC Public Health lab confirmed that the fox that bit at least nine people in the area around the US Capitol

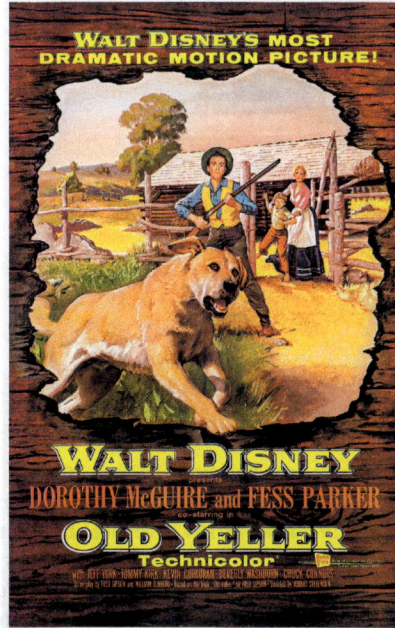

Fig. 18.4 *Movie poster for Walt Disney's* Old Yeller. *Old Yeller gets infected with rabies from a wolf, and because there is no treatment for animals with rabies, Old Yeller is shot.* Courtesy Wikimedia Commons Public Domain.

building in 2022 was captured, humanely euthanized, and tested positive for the rabies virus (Fig. 18.5).

The virus

Rabies virus is found globally, present on all continents except Antarctica. Different variants of the lyssavirus, the causative agent of rabies, are classified by the particular mammalian reservoir host, for example, canine, bat, raccoon, skunk, fox, and mongoose. In nature, only mammals can become infected with rabies, but only a few species are reservoir hosts. In North America, the major reservoir species and the rabies variant types have unique geographical distributions. In Canada, skunk rabies is centered around southern Saskatchewan and Manitoba; fox rabies is in the northern regions; raccoon rabies is in the lower island area of New Brunswick; and a combination of fox rabies and raccoon rabies is found in southern Ontario (Fig. 20.1). In the United States, bat rabies is found in every state except for Hawaii; skunk rabies is found throughout the Midwest from North Dakota

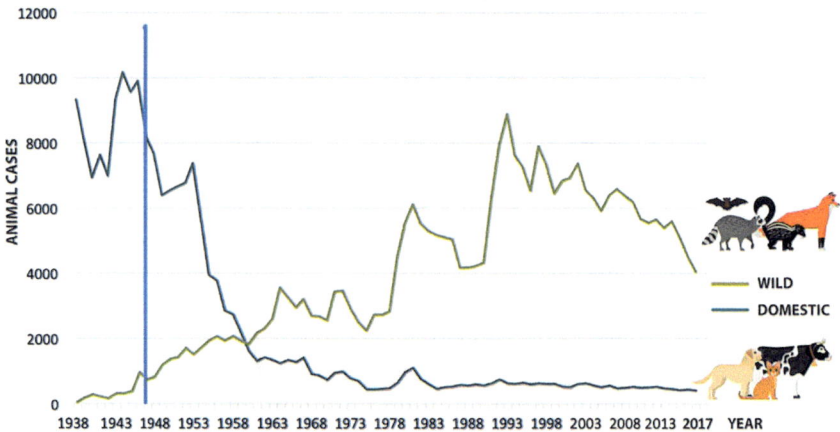

Fig. 18.5 *Cases of rabid animals reported from 1938 to 2017. In 1938, there were 9321 cases of rabies in domestic animals and only 44 cases in wild animals. With the intro-duction of vaccinating dogs for rabies in 1947 (vertical blue line), the distribution of rabies cases has reversed so that in 2017 there were 399 cases of rabies in domestic animals and 4055 cases of rabies in wild animals. Concurrently during that time period, cases of human rabies dramatically dropped from approximately 40 cases annually prior to 1947 to 1 to 2 cases annually from 1960 to the present (not depicted in the graph). From United States Centers for Disease Control and Prevention. Data Visualiza-tion: Rabies—a Forgotten Killer, 2019. Public-use data file and documentation. https://www.cdc.gov/vitalsigns/rabies/data-visualization.html.*

to Texas and also in California and parts of Kentucky, Tennessee, Virginia, and North Carolina; raccoon rabies is found all along the eastern United States from Maine to Florida; fox rabies is found in the southwestern states and Alaska; and mongoose rabies is located in Puerto Rico (Fig. 20.2). It should be stressed that these geographical distributions are not exclusive [3].

There are two major environmental cycles of rabies: the urban cycle, where dogs act as the major reservoir host, and the sylvatic cycle, where wildlife species are the predominant animals with rabies. These cycles are not completely independent as presently in North American cities on the East Coast, raccoons are the predominant infected species and are the reser-voir of the rabies virus in the urban, suburban, and rural areas. Rabid rac-coons may potentially attack domestic pets.

The disease

Rabies virus is a neurotropic virus, which means it lives and replicates in nervous tissue. Thus the clinical presentation is reflective of neurological problems. Initially, there are nonspecific signs such as fever and pain, along with other vague nonspecific symptoms. In humans, there is also an unusual or unexplained tingling, prickling, or burning sensation at the site of the wound.

There are generally two recognized forms of clinical presentation with rabies: furious and paralytic (also referred to as dumb rabies). Despite this dichotomy, clinical signs of rabies are highly variable and don't fulfill all criteria for either form.

Furious rabies is the classical and best-recognized form of rabies. Clinical signs include restlessness, wandering, howling, drooling, and excitable aggressive behavior with unprovoked attacks on other animals, people, or objects. Animals frequently lose their natural wariness and fear of humans, advancing upon and attacking people.

The paralytic form, which is less frequent (<20% of cases), is characterized by progressive paralysis with the animal becoming somnolent or depressed. Changes in vocalization, such as abnormal bellowing in cattle or a hoarse howling in dogs, have been reported. Facial paralysis with drooping of the lower jaw also occurs. Unlike in the furious form, aggressiveness and biting are uncommon.

When the virus infects the cells of the brain and spinal cord, inflammation of the central nervous system occurs. It is the severe and progressive inflammation of the brain that results in death. In people who have managed to survive clinical rabies (<20 people as of this printing), most end up with severe neurological deficits.

Historically, dogs that bit were considered rabid, but diagnosis of rabies was only based upon clinical signs of the furious form. Today, as in the past, in any instance where an animal has unexplained behavioral changes or paralysis, rabies should always be a consideration, especially in endemic areas or when there is exposure to bats and other potential reservoir animals (Box 18.1).

Transmission

The incubation period of rabies ranges from 1 week to a year but is typically 1 to 3 months in both humans and animals. This range is due to factors such as the location of viral introduction into the body, the viral load of the inoculum, and the animal's immune system.

After infecting and replicating in the brain, the virus travels along the facial nerve to the cluster of neurons in the salivary glands. The virus

BOX 18.1 What killed Edgar Allan Poe?

The cause of death of the great American writer Edgar Allan Poe remains an interesting mystery. On October 3, 1849, he was found unconscious outside a bar in Baltimore. He was assumed to be in an alcoholic stupor. He was in such a state that he was admitted to hospital. He died on October 7 as a result of "congestion of the brain." Although most considered Poe to have been drunk, the attending physician did not think it was alcohol poisoning. The symptoms were not consistent with alcohol withdrawal. It has been suggested that Poe's symptoms were, in fact, consistent with clinical rabies. He drifted in and out of consciousness and hallucinated when he was awake. He also showed symptoms of extreme aggression at times. In addition, he was reported to have great difficulty in drinking water—hydrophobia. He eventually slipped into a coma and died. This pattern is characteristic of rabies. However, there were no obvious bite wounds on the body!

replicates in the neurons of the salivary glands and thus is excreted in large amounts in the saliva; the saliva of infected animals contains high concentrations of the rabies virus. This makes the salivary glands an important site for viral replication and an efficient avenue of exit for the virus. Infected animals can only spread rabies once the virus is replicating in the salivary glands, but it should be pointed out that dogs, and most likely other reservoir species, can excrete the virus in the saliva up to 14 days before the appearance of clinical signs.

The virus cannot penetrate intact skin, so usually there is an accompanying penetrating wound—a deep bite or scratch from an infected animal—for transmission to occur. This is the predominant manner in which humans and other animals become infected with rabies virus. Infrequently, transmission can occur if the saliva of infected animals comes into direct contact with the mucosa of the mouth or eyes. Rabid animals lack their natural wariness and develop aggressive and biting behavior, advancing on objects and people, and repeatedly biting them. Thus the virus in the saliva can be introduced into the bitten animal, move to adjacent nerves, replicate, and migrate up nerves to the spinal cord and the brain. Interestingly, in the late stages of rabies in humans, as in other animals, when the virus is replicating in the salivary glands, behavioral changes occur, and the individual develops a propensity to bite and gnaw on objects.

Human-to-human transmission through bites or saliva is theoretically possible but has never been confirmed. In California in the 1950s, a person dying of rabies escaped from a hospital. While being subdued, nurses, doctors, and policemen were bitten, thus necessitating the bitten individuals to take postexposure prophylactic vaccines. In addition, it has been theorized that the vampire legend may have originated in a rabies epidemic in the Balkans during the late 1600s and early 1700s, where men were reported to be running naked with dogs and wolves, aggressively attacking others, and biting them. During sexual intercourse, vampires (i.e., people with rabies) had a propensity of biting their partners, predominantly in the neck, thus reinforcing the illusion that after being bit or having your blood sucked by a vampire, a person turns into a vampire—they could have been infected by the rabid person's saliva during the bite.

Aerosol transmission of rabies through inhalation of virus-containing bat guano has been described but is extremely rare. In fact, only two cases of rabies have been attributed to probable aerosol exposure to bat guano in caves, but an unnoticed or undetected bite from a bat may actually have been the route of transmission.

In North America, human deaths due to dog-mediated transmission have been very rare since the advent of mandatory dog and cat rabies vaccination. Because the reservoir for the rabies virus is now wildlife, these animals are the predominant means of human exposure. Deaths following exposure to raccoons, skunks, foxes, and other wild animal host species are also very rare. In North America, the main source of transmission and human deaths is from hematophagous bats (blood consumers, i.e., vampire bats) and other bat species. This is in sharp contrast with the less developed countries, mainly in Africa and Asia, where dog attacks and bites are responsible for 99% of the over 60,000 global deaths annually [4].

Treatment

Once clinical symptoms appear, rabies is virtually 100% fatal. Therefore any bite by a dog was considered to be from a rabid dog. Prior to the 20th century, before the germ theory and the advent of vaccines, treatment consisted of washing the wound, cauterization, and possible excision.

Another therapy that was in vogue over the centuries was the use of a "madstone" to draw the poison out of bites. As reported in *Medical Repository* (c. 1801), a child was bitten by a dog suspected to have rabies. The "snakestone [madstone] was applied by which every particle of poison was extracted, and no ill consequences resulted from the accident." The most common madstone was a bezoar, a round stony concretion formed in the stomach of ruminants or horses. The madstone was applied to the bite wound, and if it stuck to the wound, poison was present. The madstone would suck the poison out. When it finally fell off, the madstone was placed in boiling milk so that the poison was released into the milk. Sometimes this would result in a green discoloration of the milk, which indicated that the poison was released into the milk. This continued until the madstone no longer adhered to the wound, and then the individual was free of the poison. Because not all dog bites came from rabid dogs, most people did not come down with rabies, and thus, these treatments were considered useful.

A very lucky young lady

As stated previously, rabies is virtually 100% fatal once a person begins to show signs and symptoms of the disease. However, if a postexposure prophylaxis (PEP) regimen is started within 1 week of exposure to a rabid animal, that is, bitten and before clinical symptoms occur, there is a 99.99% chance of survival. PEP consists of wound washing, administering human rabies immune globulin (HRIG), and a series of rabies vaccinations. Wound washing should be done as soon as possible. The wound should be gently but thoroughly cleaned with water or a dilute water–iodine solution. Wound cleaning reduces the amount of virus in and around the bite area. This markedly reduces the possibility of infection. HRIG contains preformed antibodies directed against the rabies virus. The HRIG is injected into and around the wound, thus immediately providing antibodies to attack the virus until the body starts to produce its own. The rabies vaccine stimulates the immune system to produce antibodies against the rabies virus (for more information on the immune system and vaccines, refer to Chapters 2 and 3).

In 2004, 15-year-old Jeanna Giese found a bat in the back of her church in Fond du Lac, Wisconsin. Upon picking the bat up and attempting to release it outdoors, the bat bit her left index finger. Her mother subsequently washed the bite wound; no other medical attention was sought, and PEP was not administered. Jeanna remained healthy, with no symptoms, and actively engaged in numerous sports—she was an active athlete. One month after the bat bite while playing volleyball, she seemed to lose her coordination and experienced generalized fatigue and paresthesia of the left hand. She was seen by a local physician. When she told the physician about the incident with the bat, the physician examined her finger and stated that rabies was of no concern because the wound had healed well. Over the next

5 days, Jeanna's condition progressively worsened with symptoms including double vision, blurred vision, unsteadiness, nausea and vomiting, weakness of her left leg, tremors of her left arm, slurred speech, and a fever. Jeanna was admitted to the Medical College of Wisconsin at Milwaukee and diagnosed with rabies [5].

Knowing that rabies is fatal in individuals displaying clinical symptoms, Jeanna's parents permitted an aggressive novel approach to treat rabies, which included a chemical-induced coma, antiviral therapy, and necessary supportive care. She was put into an induced coma for 14 days, and after 76 days she was discharged to her home. Intensive rehabilitation and physical therapy followed, and over the following months, Jeanna regained cognitive functions and other skills. Though she continues to have some neurologic deficits, Jeanna earned her driver's license, graduated high school, earned a bachelor's degree in 2011, married, and had a son in 2018.

Jeanna Giese became the first known unimmunized survivor of rabies. Since that initial treatment, which became known as the Milwaukee Protocol for treating rabies, approximately 20 people have survived clinical rabies. Unfortunately, neither these expensive protocols nor the PEP regimen is readily available to the hundreds of thousands of people exposed to rabies around the world. Because dogs are the predominant cause of rabies exposure globally, a concerted worldwide canine vaccination program is the only economical and feasible tactic to reduce the burden of rabies in humans. However, though advocated for over many decades, the commitment by world and national health agencies is lacking; thus, rabies will continue to be a threat to human health.

Vaccines

In the United States, wild mammals currently account for over 90% of human rabies exposures. Thus wild animals must be vaccinated if rabies is to be controlled [6]. This requires the oral administration of vaccines. Several different vaccines have been used in oral rabies vaccination programs. Most have used live, attenuated rabies virus. More recently, wildlife vaccinators have switched to recombinant vaccines using either vaccinia or adenovirus vectors. This vaccine is distributed in a small plastic sachet and either surrounded by solid fishmeal polymer (FP bait) or coated with wax and fishmeal crumbs. It is licensed for use in the United States in coyotes, raccoons, and gray foxes. These baited vaccines can be distributed over a very wide area by dropping from aircraft. They have been used in Ontario to prevent the spread of raccoon and red fox rabies in the Eastern United States to establish a barrier along the Appalachian Ridge to prevent the westward spread of raccoon rabies, and in Texas to block the northward spread of coyote rabies from Mexico and to eliminate rabies in gray foxes in west-central Texas.

In Ontario, oral rabies virus baits were distributed by aircraft or hand. The baiting started near the center of the outbreak and was then extended during successive years. Within 5 years, the reported cases of rabid foxes dropped from 203 cases/year to 4 cases. Spillover cases in skunks and livestock dropped from 36 and 42 cases/year to 0, 5 years later.

Oral baiting for rabies in South Texas was initiated in 1995 in the form of a northern barrier. In subsequent years, the baited area was extended south toward the border with Mexico. Since 1995, 17.5 million doses of vaccinia-vectored rabies vaccine have been airdropped over 255,500 square miles (661,745 km^2) of Texas. The number of rabies cases dropped from 122 cases in 1995 to 0 in 2004. Currently, Texas continues to drop the baits in a zone 30 to 65 kilometers wide along the international border to provide a barrier of immune animals. Thus these vaccines are effective against rabies in coyotes, raccoons, and red foxes.

Oral rabies vaccines have been extensively tested in several less-developed countries in an effort to prevent the spread of rabies by pet and feral dogs. Pet dogs may simply be provided a vaccine sachet by the owner, "the hand-out model." Dogs that cannot be handled or dogs encountered in the street are simply offered bait. Feral dogs may simply have access to scattered baited vaccines in suitable locations such as garbage dumps or public markets. Baits that have been used include chicken heads, minced meat with breadcrumbs, boiled chicken intestines, or dog biscuits. Interestingly, dogs in different countries have different bait preferences. This strategy can complement traditional dog vaccination programs to achieve the desired level of herd immunity.

References

[1] Wan J. Rabies, medicine, and culture: dogs, disease, and urban life in the United States, 1840–1920. In: Rupprecht CE, editor. History of rabies in the Americas: from the pre-Columbian to the present, vol. 1. Switzerland: Springer International Publishing; 2023.
[2] Rosatte RC. Rabies in Canada: history, epidemiology and control. Can Vet J 1988;29:362–5.
[3] Baer GM, editor. The natural history of rabies, vols. 1 and 2. New York: Academic Press; 1975.
[4] Willoughby RE, Tieves KS, Hoffman GM, et al. Survival after treatment of rabies with induction of coma. N Engl J Med 2005;352:2508–14.

[5] Willoughby RE. Rabies rare human infection—common questions. Infect Dis Clin N Am 2015;29:637—50.

[6] Finley D. Mad dogs: the new rabies plague. College Station: Texas A&M University Press; 1998.

COVID-19—2020

Contents

In November 2002, an outbreak of "atypical pneumonia" broke out in Guandong province of southern China. Dozens of people experienced headaches, muscle pains, and a severe cough. This disease was found to be caused by a unique coronavirus, and the disease was named severe acute respiratory syndrome, or SARS. The coronavirus probably originated in bats. The disease spread worldwide but was rapidly brought under control and largely forgotten. In 2019, history repeated itself.

Great American Diseases
ISBN: 978-0-443-31404-9
https://doi.org/10.1016/B978-0-443-31404-9.00019-7

Origins

The first cases of severe respiratory disease in Wuhan likely occurred between early October and late November 2019 (perhaps as early as October 4). The first "official" case of respiratory infection, the index case, developed in mid-November 2019, in Wuhan, Hubei province, China. The disease was "officially "recognized in early December 2019 [1].

On December 1 a patient was admitted to a Wuhan hospital with severe breathing difficulty caused by a previously unknown coronavirus. By December 17, 27 similar cases had been diagnosed. By December 27, there were more than 180 cases. Some of these early cases were reportedly linked to the Huanan Seafood Wholesale Market in Wuhan. These so-called "wet" markets sell fresh fish and meat as well as assorted reptiles and wild mammals and house large numbers in close proximity. Mammals such as bats, pangolins, and civets may be sold as pets or for food. Viruses can spread readily within such dense populations of stressed animals. From the animals, they can easily jump to humans. This is called zoonotic spillover [2]. The COVID-19 coronavirus most likely originated in bats. Although some of the original cases were associated with the "seafood" market, it was soon obvious that the virus was also readily transmitted among humans.

Although the disease most likely originated from human–animal contact in a wet market, it is also possible that the virus may have accidentally escaped from a laboratory at either the Wuhan Institute of Virology or The Wuhan Center for Disease Control and Prevention where research was being undertaken on bat coronaviruses. Three researchers from the Virology Institute became sick in November 2019 and had to seek hospital treatment. However, the cause of their illness is unclear. The WHO investigated this possibility, accepted Chinese assurances, and reported that laboratory escape was unlikely, though Chinese transparency and thus the relevance of the WHO's findings are still in question. Once it infected humans, the virus spread rapidly through the population.

Within a few weeks of the onset of the outbreak, the offending virus was identified, isolated, and its genome sequenced, something that could not have happened in prior pandemics. On January 7, 2020 the Chinese authorities announced that the disease was caused by a novel alphacoronavirus. Its complete genome was published on January 10, and development of the first diagnostic polymerase chain reaction (PCR) assay was reported on January 15. This was an incredibly rapid scientific response.

On January 15, 2020 a 35-year-old man returned to Washington state after visiting his family in Wuhan, China. On January 19, 2020 he presented at a clinic in Snohomish County, Washington with a 4-day history of cough and fever; he sat in the waiting room for 20 minutes, without any mask or face covering, before being seen by a physician. Oral and nasal pharyngeal swabs were tested by PCR assays at the CDC and found to be positive for SARS-CoV-2 [1]. The patient developed severe pneumonia but eventually recovered. He was probably not the first case of the infection in the United States. Analysis of donated blood samples looking for specific antibodies subsequently indicated that a slightly different coronavirus was probably circulating in California prior to January 19. The first confirmed SARS-2 death in that state occurred on February 6. It too appears to have been of Chinese origin. Subsequent analyses indicate that the virus also reached New York City sometime in February by way of tourists returning from California, Iran, or Paris. A retrospective analysis of donated blood samples found antibodies to SARS-CoV-2 in blood from Illinois donated on January 7, 2020. Because it takes about 2 weeks for these antibodies to develop, those infections were probably acquired as early as mid-December 2019. The virus was eventually named SARS-CoV-2. The disease that it causes was named coronavirus disease (COVID)-19 by the World Health Organization [3].

Coronaviruses

Coronaviruses are enveloped, nonsegmented positive-sense RNA viruses. They derive their name from the crown-like halo of spikes around each spherical virion. They have the largest genomes known among the RNA viruses. The *Coronavirinae* are subdivided into four groups, the alpha-, beta-, gamma-, and deltacoronaviruses. Alpha- and betacoronaviruses typically cause disease in mammals. Gamma- and deltacoronaviruses mainly infect birds. The first coronavirus, identified in 1937, was the virus that causes infectious bronchitis in chickens. As its name implies, this is a respiratory disease, but in cattle, pigs, dogs, and cats, other coronaviruses usually cause severe gastroenteritis. Coronaviruses readily adapt to changes in their hosts. For example, it is likely that bats have provided the gene pools for the alpha- and betacoronaviruses, whereas birds are the most likely sources of the gamma- and deltacoronaviruses [3]. SARS-CoV-2 grows readily in ferrets, mink, and cats. Cats (and tigers) and dogs are susceptible to airborne infection [4,5].

Coronaviruses cause respiratory diseases ranging from mild, such as some forms of the common cold, to seriously lethal, such as the diseases caused by SARS-CoV, MERS-CoV, and SARS-CoV-2. Four human coronavirus (HCoV) cause about a third of common cold cases—a runny nose.

The spikes that cover coronaviruses are used by the virus to invade cells. Thus the spike protein of SARS-CoV-2 binds strongly to a human cell surface protein, called angiotensin-converting enzyme-2 (ACE2). ACE2 is found on the cells that line our airways. As a result, inhaled viruses can bind, enter, and kill these cells.

Clinical disease

Initial studies in China soon determined that the incubation period of COVID ranged from 2 to 14 days. Most symptoms appeared in 4 to 5 days. It appeared to be especially lethal for the elderly (the median patient age was 59 years) and those with underlying diseases. The number of cases in the early epidemic in China doubled every 7.2 days. The hospitalization rate was in the region of 20%. SARS-CoV-2 killed between 1% and 3.4% of these hospitalized patients.

Most respiratory viruses infect either the upper or lower respiratory tracts, but SARS-CoV-2 infects both simultaneously. After it kills the cells in the airways, the dead cells and fluids are shed into the bronchi, where they are drawn into the lungs and clog the airways, resulting in severe breathing difficulty. Because it infects the upper airways, COVID-19 is readily shed in the airborne droplets generated by coughing, sneezing, singing, or even speaking loudly. Some of these droplets may be very small and remain suspended in the air for a long time.

SARS-CoV-2 causes a spectrum of illnesses ranging from asymptomatic to lethal pneumonia. It is estimated that about 40% to 45% of infected individuals never show symptoms. As a result, there is an "iceberg" effect where many infections are inapparent and thus not detected. On the other hand, in US cases, it has been estimated that about 10 people in every 1000 will die from the disease. This is about 50 to 100 times more lethal than seasonal influenza.

However, these numbers are affected by the structure of the population as COVID-19 is especially lethal for older individuals with preexisting medical conditions. These medical conditions may result in a two- to threefold increase in disease severity. They include chronic kidney disease, chronic obstructive pulmonary disease, immunosuppression, obesity, heart disease, sickle cell disease, and type 2 diabetes mellitus. Persons over 50 are about 2.5 times more likely to develop severe disease than the general population. The case fatality rate in the United States for persons over 75 is 13.83%. Smoking makes it worse.

Although the prime targets of the virus are the cells of the respiratory tract, the heart, kidneys, intestine, and brain may also be affected. Once inhaled, the virus gains access to the nose and throat, where many cells carry ACE2. The disease commonly starts as a flu-like illness. The victim develops a fever, sore throat, loss of smell and taste, as well as headaches, body and muscle aches, intense fatigue, diarrhea, nausea, and vomiting. Some may develop skin rashes and "COVID toes," a discoloration and swelling to the toes. Young people especially develop chronic fatigue. Patients often have great difficulty in breathing. About 20% of patients suffer heart damage; some develop clots in their bloodstream that can block vital arteries. About 25% of patients develop kidney failure. Five percent to ten percent develop brain damage and encephalitis. About half of the affected patients develop diarrhea.

Many infected individuals continue to experience multiple symptoms for 6 months or longer. This so-called "long COVID" includes extreme fatigue, "brain fog" (an inability to think clearly), shortness of breath, muscle aches, and chest pain. Its mechanisms of long COVID are unclear at this time but appear to result from the production of a protein called interferon-gamma.

Many people infected with SARS-CoV-2 show no symptoms. This appears to be due to a preexisting T cell—mediated immunity. It is inherited due to a common allele present in their major histocompatibility complex, HLA-B [6].

Prevention and control

Given that SARS-2 has a lethality of about 1% and that initially nobody had any immunity to it, it was easy to calculate that it could potentially infect all 340 million people in the United States and kill 3.4 million of them! This did not happen because as the virus spread through the population, individuals who recovered developed immunity.

The basic principle of infectious disease control is to ensure that the R_0 drops below 1. In order to do this, the opportunities for viral spread must be minimized. This can be done by reducing the exposure of susceptible people by the use of isolation and quarantine, by wearing properly fitting masks, by avoiding indoor crowds, and by social distancing (staying 2—3 meters

(6—10 feet) apart) so that an infected victim cannot transmit the virus to other susceptible individuals. Additionally, if everyone is vaccinated and hence immune, then the virus will be unable to encounter susceptible victims. Thus logical control procedures consist of three steps, quarantine, social distancing and vaccination.

The events of 2020

As described, the first definite US case of COVID-19 was identified on January 19, 2020 in Washington state. It is therefore probable that the coronavirus had reached most major US cities by early February and had spread widely before the first wave of cases appeared in late February and early March. Restrictions on flights from China to the United States were put in place on January 31. On March 13, travel restrictions were placed on 26 European countries. Unfortunately, these restrictions were easily circumvented, so they could only delay but not stop the spread [7].

The COVID death rate in the United States climbed rapidly, as expected with a pathogenic virus in a population with low immunity. It reached a peak of almost 20,000 weekly deaths by mid-April before dropping to under 5000 by mid-June. It peaked again in late July but subsequently climbed significantly to over 25,000 cases/week in January 2021 before dropping rapidly (Fig. 19.1).

It also peaked at about 21,000 deaths in February 2022. This reflects the indoor winter season as well as school activities season and its spread by the respiratory rate. As the population developed immunity, either by infection or vaccination, the number of deaths subsequently dropped with a small peak of about 3000/week in January 2023 and another peak of about 2000 deaths in January 2024. As of February 2024, the total number of COVID-19 deaths in the United States had reached 1,174,626 persons (Fig. 19.2).

The first wave of the pandemic centered primarily on the Northeast, especially New York City, while much of the rest of the country remained relatively unaffected. New York City was severely affected. Its hospitals were almost overwhelmed by the number of cases. To assist, the US Navy sent the hospital ship *Comfort*, with a 1000 bed capacity and 12 operating rooms, "to provide relief to frontline health care providers" and to increase local hospital capacity; the ship arrived in New York Harbor on March 30. By April 8 there had been 799 COVID-related deaths. By June, New York City, with a population of 8.6 million, had more than

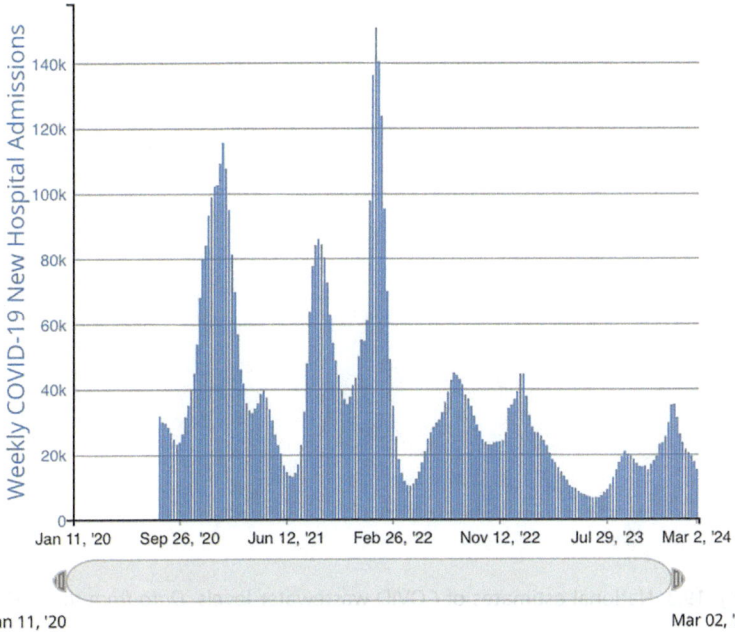

Fig. 19.1 COVID-19 hospital admissions by week in the United States, reported to the CDC.

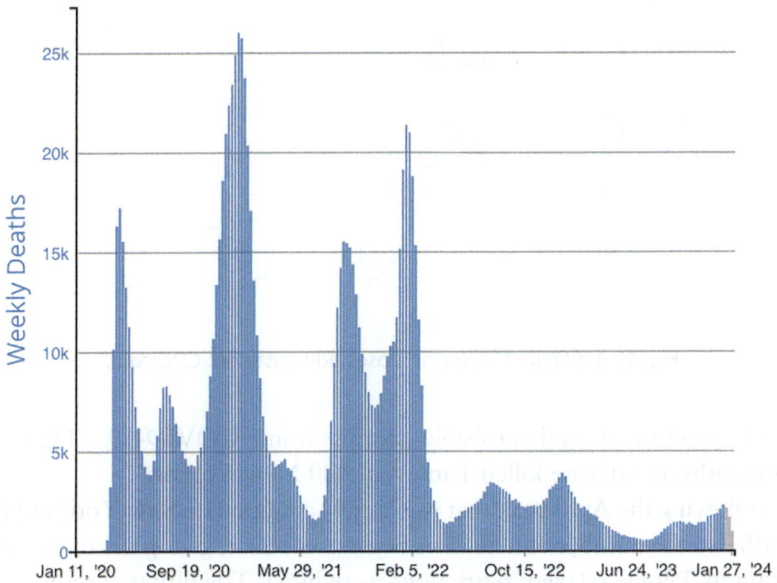

Fig. 19.2 COVID-19 deaths by week in the United States, reported to the CDC.

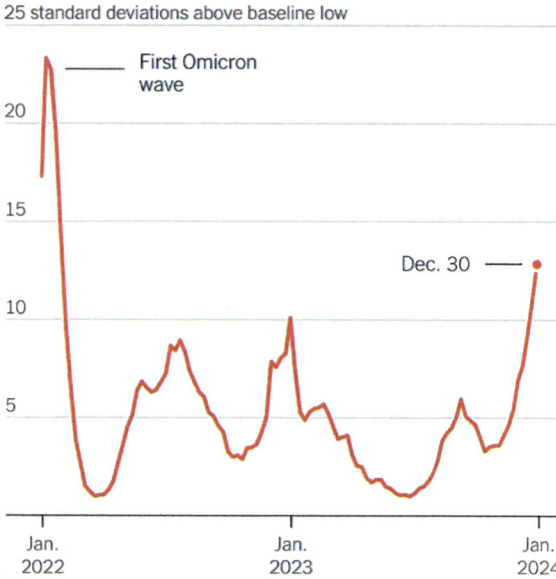

Fig. 19.3 National estimates of COVID wastewater levels. *Data from the CDC.*

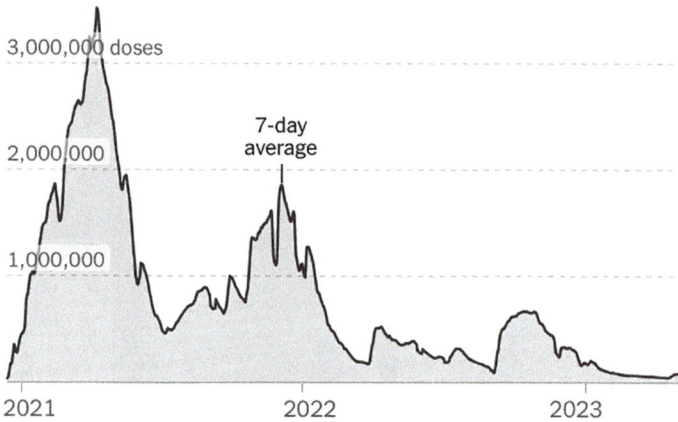

Fig. 19.4 COVID-19 vaccine doses administered. *CDC data.*

23,508 confirmed and probable deaths from COVID-19. Thus, in 3.5 months, the disease killed 1 in every 400 New Yorkers.

Following the April peak that was largely confined to New York and the Northeast, the number of new cases began to decline, giving rise to cautious optimism. People relaxed; restrictions were lifted. The lifting of restrictions was welcomed by many, especially businesses and those who felt the need to gather and celebrate. Thus, beginning at the Memorial Day holiday and recurring on July 4, large crowds of young people gathered, many in the belief that the young had little to fear from COVID. They were wrong. By the second week of June, case numbers began rising again, first in Arizona and subsequently in Texas, Florida, California, and other Sun Belt states. This rise was attributed by many to increased testing, but while the number of tests performed indeed increased, so too did the percentage of positive results. The demand for beds, intensive care units, and respirators began to skyrocket. Hospitals in some cities were overwhelmed by the demand.

Interestingly, while cases and the demand for hospital beds exploded, the rise in deaths was not as great, reflecting increased experience in treating patients, a decline in the average age of patients, and the availability of drugs to treat the most severe cases. As a result of this explosive surge of cases, many states began to reimpose restrictions on businesses and mandating the wearing of face masks. Once again, case numbers declined.

However, beginning in mid-October, the numbers of COVID cases nationally exploded and climbed steadily to peak in early February 2021. This was a sharp peak that declined rapidly. The reasons for the decline were a result of a combination of social distancing, enforcement of mask mandates, and the rapid and effective introduction of new vaccines.

Issues arising

Unfortunately, the United States was unprepared for the COVID pandemic. As a result, whereas the United States has just 4% of the world's population, it has had 25% of its reported COVID cases.

Masks

There was considerable confusion at the beginning of the pandemic. Not knowing that it was spread by airborne droplets, experts first suggested that masks did not help prevent transmission. There were no scientific data on the efficacy of face masks. It took until the summer of 2020 to generate the data that confirmed that masks were indeed effective. The "experts" were forced to reverse themselves, but confidence had been eroded significantly. They also underestimated the virus' ability to change its structure and hence its antigenicity. There was a loss of trust in initial pronouncements. One feature of American society and culture is that it prioritizes individual freedoms over government regulations. There is thus a strong aversion to mandating collective behavior. This aversion was reflected in the decision to wear face masks. Some states required it; others resisted.

Mask wearing became a partisan issue and, as a result, inconsistent. In a Gallup poll on July 13, 2020, 34% of men and 54% of women reported that they always wore a mask outside the home; 20% of men and 8% of women said that they "never" wore a mask! Face masks also appear to be much more effective than first thought at reducing transmission. These masks, when properly fitted, helped reduce transmission by a person shedding the virus from their nose and mouth.

Movement restrictions

In the absence of herd immunity, the simplest method of dropping the R_0 below 1 is social distancing. Many cities and states also introduced social distancing measures in their efforts to fight COVID-19, but this was largely uncoordinated and adjacent jurisdictions may have had very different rules.

Because the obvious method of preventing the disease spread was social distancing, this principle was expanded to require an economic shutdown in many communities. Theaters, restaurants, bars, and gyms were closed in many states at a time when the disease numbers were low. The early closure of events where crowds gather as well as schools and universities contributed to reducing disease spread and its associated mortality.

In a country such as the United States where each state was responsible for controlling COVID-19, there was a patchwork of responses. New York and adjacent states that suffered initially were hesitant to reopen, and their cases continued to decline. In contrast, the early opening states in the South saw explosive outbreaks. Thus the country was faced with the task of opening up one region while suppressing the pandemic in another.

By mid-June 2020, the number of new cases had declined to such an extent that people began to gather in groups again and restaurants to reopen. Unfortunately, an economic disconnect also developed in relatively unaffected states. Few people knew a COVID patient, but many were adversely affected by the economic results of the shutdown. Political and economic pressure to reopen grew rapidly, and by early May, many state governors, especially in Arizona, Texas, and Florida, lifted stay-at-home orders and ended many restrictions on business activities. Some declined to require people to wear face masks and prevented local governments from imposing such requirements. They ignored warnings regarding premature reopening.

People progressively lost their fear of the disease while the economic shutdown took its toll. As a result, the strict isolation and social distancing progressively lapsed. States with relatively few cases opened up first, but within weeks the isolation consensus broke down. Many, perhaps most,

individuals remained cautious and practiced social isolation and mask wearing. A growing minority, however, dismissed the threat. States felt
compelled to reopen even if prevalence was high and tests were unavailable.
Businesses reopened. Life began to return slowly to normal. The virus also
returned.

The consequences of the COVID shutdown were uneven. Some large
organizations adapted to the new opportunities. Conversely, many small
businesses such as restaurants, shops, and bars were forced to close with no
alternative income sources. Small stores and malls had been suffering from
a shrinking market share for years and were unable to compete, so bankrupt-
cies climbed. Shopping malls were deserted, and much retail space was
closed. Thus the shutdown accelerated long-term trends so that big companies increased their market share.

Lack of equipment

A severe shortage of personal protective equipment developed in the early
months of the pandemic. Hospitals, like many modern businesses, operate
on a "just-in-time" basis and do not stockpile large quantities of equipment.
In an interconnected world, complex supply chains often originate in distant
countries with cheap labor. For example, half the world's face masks were
made in China. As a result, a serious shortage of masks and other protective
equipment developed early. The federal authorities decided that solving the
problem was a role for the States and individual hospitals. In order to obtain
access to limited supplies, hospitals were forced to depend on capitalism and
connections. Inevitably, rich hospitals were more successful.

CDC problems

2020 was a disaster for the US Public Health Agencies, and they lost a lot of
credibility. Though the disease had reached epidemic proportions in China by
the end of 2019, the World Health Organization did not declare a Public
Health Emergency of International Concern until January 30, 2020. On
January 31, Health and Human Services Secretary Alex Azar declared a Public
Health Emergency. This declaration provided funds and federal assistance for
state, local, and tribal health departments in their COVID-19 responses, and
as noted above, a limited travel ban was instituted. However, federal and state
governments and emergency preparedness programs did little to prepare the
country, accumulate supplies, or establish an effective surveillance system.
The threat was underestimated. No diagnostic tests were available, nor
were any in development; thus testing and tracing capabilities were lacking.

Ongoing problems included not only a lack of protective equipment but also testing failures. There was, at times, a shortage of the nasopharyngeal swabs used to swab the airways; a shortage of extraction kits for obtaining the purified RNA from the swabs; a shortage of the required chemical reagents; and a shortage of individuals trained to perform the tests. Supply was unable to keep up with demand. Unfortunately, the first diagnostic tests were produced slowly by the CDC, and they were much more complicated than that employed in other countries. The use of contaminated chemicals resulted in false-positive results. Thus, in the early weeks of the pandemic, effective diagnostic tests were in short supply. By the end of February, tens of thousands of Americans were infected, but only several hundred had been tested.

By May 6, 2021 the CDC estimate of the number of COVID deaths in the United States had reached 574,403. However, detailed analysis has suggested that the actual number of deaths on that date, based on excess mortality estimates, was 905,000. The official underestimates are attributed to local health systems being overwhelmed and to insufficient testing. More recent estimates in late 2023 have suggested that the "official" estimates of COVID deaths are underestimated by 16%. Again, this is a result of low levels of testing and a lack of awareness. Lesson: we need to prepare for new diseases.

Global significance

According to the World Health Organization, as of April 2021, 148 million people were known to have been infected, and more than 3 million had died of COVID-19. Of these, at least 17,000 health workers had died from the disease. Economically, the world lost US$10 trillion in output by the end of 2021, and it was estimated that this would compound to US$22 trillion over the 2020 to 2025 period. At the peak of the pandemic in 2020, 90% of schoolchildren were unable to attend school. Ten million more girls were at risk of early marriage, gender-based violence increased fivefold, and 115 to 125 million people were pushed into extreme poverty.

As of early 2024, excess deaths between 2020 and November 2023 globally totaled 27 million. This estimate is based on what would have been expected based on previous years. The official death toll based on individuals actually tested for COVID-19 infection is about 17 million.

The victims

As an airborne virus, it is much easier to become infected by the still air within a building than in the breezy, sunny outdoors. It was estimated that the coronavirus was 19 times easier to catch indoors than outdoors. As a result, outbreaks occurred predominantly in dense indoor populations. These included cruise ships, church halls, prisons, and especially long-term care and nursing homes for the aged. Even politicians could become infected, especially if they resisted mask wearing and social distancing. President Trump and Prime Minister Boris Johnson of the United Kingdom were both hospitalized for a short time as a result of severe respiratory difficulty caused by COVID-19 [8].

The aged

During December 2020, an average of 1.7 individuals died from COVID-19 every minute in the United States. Through December 2020, it is estimated collectively Americans lost about 4.5 million years of life. It was estimated that US life expectancy fell by 1 year, to 77.8 years in the first half of 2020. This was the largest drop since WWII.

The effectiveness of the immune system declines with age, and as a result, the elderly become increasingly susceptible to infectious diseases. Given both their aged immune system as well as the relatively crowded conditions under which the institutionalized elderly live, it is unsurprising that a very high proportion of lethal COVID cases occurred in those over 70. Although nursing homes and care facilities hold less than 1% of the US population, by mid-June 2020, they had accounted for 40% of COVID deaths. In New Jersey, the early wave of disease killed one in every eight nursing home residents. As with healthcare workers, the elderly living in care homes and assisted living were among the first to receive the new vaccines that became available in December and January 2020-2021.

Healthcare workers

Among the most severely affected groups of individuals were physicians, nurses, and caregivers exposed to the virus on a daily basis. If they sickened or were forced into quarantine, this also increased the stress and the burden on their colleagues. This was also a consistent feature of historical plagues where those ministering to the sick were seriously overworked and contracted the disease themselves. Healthcare workers were seriously affected during the COVID-19 outbreak, especially when appropriate protective equipment was not always available. However, once the new vaccines became available these workers were prioritized and became the first to receive their shots.

Minorities

In the United States, minorities have been disproportionally affected. Thus the age-adjusted hospitalization rates per 100,000 population were 261 for Native Americans, 213 for African Americans, 205 for Hispanics, 58 for Asians, and 46 for Whites. These disparities are mainly due to factors such as inequalities in healthcare access, uninsured rates, preexisting and predisposing health conditions, work demands, housing quality, and language barriers [9].

Life expectancy for the Black population dropped by 2.7 years over the same period. The death rate for Black Americans was twice as high as for White Americans. About 1 in 800 Black Americans had died and 1 in 1325 White Americans [10].

COVID vaccines

The first two COVID vaccines, one produced by Pfizer-BioNTech and one by Moderna, were approved by the FDA for emergency use in November 2020 and began to be administered to those in the highest priority groups, healthcare workers and the elderly living in institutions, by mid-December. These two vaccines contained messenger RNA (mRNA) encoding the SARS-2 coronavirus spike protein incorporated in lipid nanoparticles. Once injected, these lipid particles are readily taken up by cells, and the mRNA is then translated into the viral spike protein. This foreign protein is presented on the cell surface and recognized by the recipient's immune system. As a result, it triggers a protective immune response involving both T and B cells. The initial trials showed, and follow-up studies confirmed, that both these vaccines were 94% to 95% effective after two doses in preventing disease and death and had remarkably few adverse side effects. As manufacturing ramped up, initial vaccine shortages were overcome, and the vaccine became available to a second priority group, those over 65 and those with a predisposing medical condition as described above.

A third vaccine received conditional approval by the FDA. This was a single-dose vaccine manufactured by Johnson and Johnson. It differed from the previous two in that it used a slightly different technology. It employed a harmless adenovirus into which the genetic code for the coronavirus spike protein had been inserted. When this "recombinant" adenovirus entered cells, the infected cells used this code to make coronavirus spike protein that in turn triggered an immune response. The vaccine was reported to be 85% effective against severe COVID.

Vaccine hesitancy

There are three prime reasons for a failure to vaccinate. First, there is a belief that these vaccines are unnecessary: the disease is not a serious threat to me or my family. Second, there is concern about adverse effects and its safety. Third, there is lack of knowledge regarding both the disease and the vaccine. The relative importance of these three factors has varied over the course of American history [12], [13].

Attempts to mandate COVID vaccination met with vigorous opposition. Likewise, attempts to regulate disinformation spread on social media platforms have proven to be largely inadequate. As a result, public confidence in COVID-19 vaccines was eroded before the first ones were even made available to the public [14].

As a result of failure to vaccinate, there was a higher rate of COVID-19 deaths among unvaccinated individuals. It became a pandemic of the unvaccinated. A Brown University study [14] showed that between January 1, 2021 and April 30, 2022, there were 641,305 deaths attributed to COVID-19 in the United States. These investigators calculated that had each state maintained its peak vaccination rate at 100%, there would only have been 322,0324 deaths nationally. Even if they had only maintained it at 85% of their peak rate, there would only have been 463,305 deaths. They estimated that if vaccination coverage across the country had been maintained at 90% of people over 18, at least 224,427 deaths would have been averted [15].

The end result

By November 30, 2022, the United States had administered more than 665 million doses of COVID-19 vaccines. As a result, 80% of the population has received at least one dose.

Analysis of numbers indicates that this vaccination has saved the United States more than 1 billion dollars in medical costs. Vaccination has also prevented more than 18 million hospitalizations and over 3 million additional deaths. Current vaccines are 95% to 97% effective—a clear victory for vaccines [16]. It prevented between 112,698,238 and 127,129,565 additional hospitalizations. In addition, it is also apparent that mRNA vaccine technology has great potential to revolutionize the development of vaccines against other infectious diseases [17].

New viral variants

Coronaviruses, like most RNA viruses such as influenza, make no effort to ensure that they replicate faithfully. Thus, as these viruses grow within cells and their RNA is copied, gradual errors and mistakes creep in. This means that over time, their RNA sequences change, as do the viral proteins encoded by this RNA. If these changes are great enough, then it is possible that antibodies directed against the parent strain may not be able to bind and neutralize the proteins in the new variants. This is a similar situation to that in influenza and is the reason why flu vaccines have to be made and administered annually. It is also why the mild coronaviruses that cause the common cold can recur after several years. It is possible that a similar situation may occur with the COVID-19 coronavirus vaccines and that we may require annual booster vaccinations to maintain immunity. It is still too early to know if this will be the case.

RNA viruses such as SARS-2 and influenza lack any mechanisms for ensuring that they correctly replicate their nucleic acids as they divide. As a result, mistakes creep in and changes accumulate. Thus SARs-2 has progressively changed since the first outbreaks in China. This presents the immune system with problems because immunity to one strain may not be protective against a different strain. By the end of 2020 the initial isolate from China had gone through several changes. The World Health Organization then began to give a Greek letter to subsequent variants.

The first major variant, Alpha (B.1.1.7), was reported from the United Kingdom in September 2020. Shortly thereafter the Beta strain (B.1.351) was reported from South Africa. It spread rapidly. The Gamma variant was reported in Brazil in November 2020 but was not widespread. Epsilon was identified in California in July 2020. The Delta variant appeared to originate in India and rapidly spread worldwide. It caused more disease and hospitalizations than other variants. Omicron was identified in late November 2021 and rapidly overtook Delta. Omicron is more transmissible than the original variant and the Delta variant. However, Omicron caused less severe disease. Its common symptoms were cough, fatigue, congestion, and a fever. Omicron has given rise to minor variants called subvariants.

Subvariants such as BA.1 and BA.2 have emerged from Omicron. Other variants soon followed, BA.4 and BA.5. Then XBB variants emerged. In December 2023 a new variant, JN.1 related to BA2.86, emerged. At the beginning of December 2023, JN.1 accounted for 15% to 30% of COVID cases in the United States. The predominant circulating sublineage at the beginning of 2024 was JN-1.

Predicted outcomes

COVID-19 caused a typical, "virgin soil" pandemic. Because no human had encountered the virus previously, then nobody had any immunity, either antibody mediated or cell mediated. Thus, unless controlled, SARS-2 will spreads from individual to individual until a majority of the world's people are infected and subsequently either dead or immune. The final proportion of those who become infected will depend ultimately on the development of herd immunity, but it is estimated that this will likely need to reach at least 60% to 70% of the population to be effective in preventing significant disease outbreaks.

Box 19.1 The reverse transcriptase polymerase chain reaction (PCR) test.

The standard test used for COVID-19 testing in the United States is the reverse transcriptase PCR. This test is designed to detect the presence of coronaviral RNA in a nasal or mouth swab. Because we know the complete gene sequence of the coronaviral RNA, we can select a short sequence that is unique to the coronavirus. We can then construct an artificial sequence that is complementary to this viral RNA and will bind only to that sequence—this is called the primer sequence. To perform the test, a nasopharyngeal swab is extracted in saline, and any viral RNA is converted to DNA through the use of a reverse transcriptase enzyme (this is the same type of enzyme used by retroviruses; Chapter 17). A primer sequence is then mixed with the newly formed DNA. If the coronavirus-specific DNA is present, it will react very specifically with the primer sequence. An enzyme called DNA polymerase is then added to the mixture and the conditions manipulated so that the selected DNA sequence is massively increased through many cycles of a polymerase chain reaction (PCR). Any newly generated DNA, if present, can then be detected by the process of electrophoresis. A positive result can be confirmed later by sequencing the newly formed DNA and confirming that it is indeed complementary to the coronavirus RNA.

Much depends on the duration of immunity as well as the availability and effectiveness of vaccines. With a good vaccine and reasonable immunity, COVID-19 will likely become another seasonal respiratory virus. Models also suggest that the infection will persist globally and flare up at intervals in different countries. In the most pessimistic models, if SARS-CoV-2 infects 60% of the world's population and its case fatality rate is 1%, then 40 million will die, almost matching the number of deaths from the 1918 pandemic.

Perhaps the world will also recognize that epidemic diseases have not gone away and prepare for the inevitable next one. Perhaps the US healthcare system will be restructured to prevent sickness rather than profit from it.

Box 19.2 Viruses in wastewater.

People infected with SARS-CoV-2 shed the virus in their feces whether or not they are showing symptoms of the disease. As a result, the virus is also present in sewage-derived wastewater. This virus can be detected and measured using very sensitive techniques. As a result, wastewater surveillance can show where the virus is present and if it is spreading. It can also identify long-term trends and so act as an early warning system. It can warn communities and permit them to take preventive measures in advance. The system has worked so well that the CDC has established a National Wastewater Surveillance System (NWSS) across the United States (Fig. 19.3).

Wastewater samples are collected as they flow into a treatment plant. These are then sent to a public health laboratory for testing. The results are then sent to the NWSS at the CDC for analysis. If virus levels are high, the wastewater may be tested directly. If low, then the sample must first be concentrated by the process of ultrafiltration. Although wastewater contains a very complex mixture of organisms, any RNA present can be chemically extracted from the sample and subjected to a reverse transcriptase polymerase chain reaction assay (RT-PCR). This involves specifically transcribing only selected genes from COVID viral RNA (because they have their own unique sequence). This involves the initial conversion of the viral RNA to DNA using a reverse transcriptase enzyme before beginning the PCR cycle. The PCR cycle then greatly increases the quantity of this DNA by using the enzyme reverse transcriptase and then measuring the DNA generated. This is a measure of the amount of the virus present in the wastewater sample. Wastewater PCR assays are not only useful in detecting the presence of trace amounts of COVID RNA in tissues; they are currently being employed in Pakistan to follow the presence of wild poliovirus in Pakistani cities and hopefully assist the polio eradication campaign (Chapter 13).

References

[1] Roberts DL, Rossman JS, Jaric I. Dating the first cases of Covid 19. PlosOne 2021. https://doi.org/10.1371/journal.ppat.1009620.

[2] Morens DM, Fauci AS. Emerging pandemic diseases: how we got to COVID-19. Cell 2020. https://doi.org/10.1016/j.cell.2020.08.021.

[3] Zhang Y-Z, Holmes EC. A genomic perspective on the origin and emergence of SARS-CoV-2. Cell 2020;181:223—7.

[4] Menachery VD, Yount BL, Debbink K, Agnihothram S, et al. A SARS-like cluster of circulating bat coronaviruses shows potential for human emergence. Nat Med 2015; 21(12):1508—13.

[5] Lytras S, Xia W, Hughes J, Jiang X, Robertson DL. The animal origin of SARS-CoV-2. Science 2021. https://doi.org/10.1186/science.abh0117.

[6] Augusto DG, Murdolo LG, Chatzileontiadou M, Sabatino JJ, et al. A common allele of HLA is associated with asymptomatic SARS-CoV-2 infection. Nature 2023;620: 128—36.

[7] Anderson CG, Rambaut A, Lipkin WI, Holmes EC, Garry RF. The proximal origin of SRS-CoV-2. Nat Med 2020;26:450—2.

[8] Yong E. How the pandemic defeated America. The Atlantic 2020.

[9] Wade E. An unequal blow: in past pandemics people on the margins suffered the most. Science 2020;368:700—3.

[10] Fauci AS. The story behind COVID-19 vaccines. Science 2021;372(6538):109.

[11] Barro RJ. Vaccination rates and COVID outcomes across the US states. National Bureau of Economic Research working; 2022. https://doi.org/10.3386/w29884. *Paper* 29884.

[12] Wilson SL, Wiysonge C. Social media and vaccine hesitancy. BMJ Global Health 2020;5. https://doi.org/10.1136/bmjgh-2020-004206.

[13] Sun R, Budhwani H. Negative sentiments toward novel coronavirus (COVID-19) vaccines. Vaccine 2022. https://doi.org/10.1016/j.vaccine.2022.10.037.

[14] Zong M, Kshirsagar M, Johnston R, Dodhia R, Glazer T, et al. Estimating vaccine-preventable COVID-19 deaths under counterfactual vaccination scenarios in the United States. MedRxiv 2022. https://doi.org/10.1101/2022.05.19.22275310.

[15] Butt AA, Omer SB, Yan PShaikh OS, Mayr FB. SARS-CoV-2 vaccine: effectiveness in a high-risk national population in a real-world setting. Annals Intern Med 2021. https://doi.org/10.7236/M21-1577.

[16] Fitzpatrick MC, Moghadas SM, Pandey A, Galvani AP. Two years of US COVID-19 vaccines have prevented millions of hospitalizations and deaths. To the Point. Commonwealth Fund; 2022. https://doi.org/10.26099/whsf-tp90.

[17] Yap C, Ali A, Prabhakar A, Pal A, et al. Comprehensive literature review on COVID-19 vaccines and the role of SARS-CoV-2 variants in the pandemic. Ther Adv Vaccines Immunother 2021;9:1—21.

Glossary

Aerobic: Requiring air (oxygen) in order to grow.

Anaerobic: An organism that only grows in the absence of oxygen.

Anemia: A lack of red blood cells.

Antibiotic: A chemical produced by bacteria or fungi that kills other microorganisms.

Antibody: A protein produced during an immune response that binds to foreign invaders and marks them for destruction.

Antigen: A general term for any foreign molecule that triggers an immune response.

Antiserum: Serum from an immunized animal that contains high levels of specific antibodies.

Asymptomatic: Showing no symptoms of an infection.

Atrophy: Decreasing in size or wasting away.

Attenuation: A procedure that causes an organism to lose its virulence and hence the ability to cause disease.

Bacteremia: Bacteria in the bloodstream.

Bacterium: A single-celled organism with a cell wall but no organized nucleus.

Capsid: The protein structure that encloses and protects the nucleic acids of a virus.

Carrier: An individual persistently infected with a pathogen but suffers no ill effects and is a potential source of infection for others.

Commensal: An organism that lives within the body and shares resources such as food with its host.

Contagious: A disease that can spread between individuals by direct or indirect contact.

Cytokine: A protein that carries signals between nearby cells.

Dysbiosis: An imbalance in the normal composition of a mixed bacterial population.

Ectoparasite: A parasite that attacks the outside of the body, such as a flea or mosquito.

Embolus: A particle such as a blood clot or an air bubble that develops within the bloodstream and can become lodged in a blood vessel, thus blocking it.

Encephalitis: Inflammation of the brain.

Endemic: An infectious agent that can be regularly isolated from a specific population.

Endoparasite: A parasite that lives within the body. Examples include gastrointestinal worms and protozoa in the bloodstream.

Endotoxin: Toxic bacterial products released when the organism dies, which can cause nonspecific signs and symptoms of sickness.

Enteric disease: Disease involving the gastrointestinal tract.

Enzootic: An infectious agent that persists indefinitely within an animal population.

Epidemic: An outbreak of infectious disease that affects large numbers of individuals in a region over a short period of time.

Eukaryote: An organism in which the genetic material (DNA) is contained within a nucleus.

Exotoxin: A toxic protein secreted by certain bacteria such as the tetanus bacillus and *Vibrio cholerae*.

Gene sequencing: Determining the precise order of nucleotides in a DNA molecule or genome.

Heirloom disease: An infectious disease that likely affected the anthropoid ancestors of humans.

Immune globulin: A preparation made from serum containing a high level of purified antibodies, which can be administered to a patient to treat an infection.

Incidence: The rate at which a disease is contracted over a specific period of time.

Interferon: A cytokine with potent antiviral activity.

Latent disease: A hidden or inactive disease.

Lymphocyte: A form of white blood cell that mediates both antibody- and cell-mediated immune responses.

Macrophage: Large white cell found within tissues whose function is to capture and eat (phagocytose) invading microbes.

Microbiota: The population of microorganisms living within a specific habitat, such as on the skin or within the intestine.

Morbidity: The rate of disease in a population caused by a specific infectious agent.

Mortality: The rate of deaths in a population caused by a specific infectious agent.

Necropsy: A postmortem examination of an animal.

Opportunistic organism: An organism that only causes disease when the immunity of its victims is somehow diminished.

Pandemic: An outbreak of disease affecting a very large area or multiple countries and continents.

Pathogen: An organism that can cause disease and death.

Prevalence: The proportion of a population infected with a specific infectious agent at any specific time.

Prognosis: A determination of the likely outcome of a disease or situation.

Prokaryote: A cell in which the genetic material is contained within the cytoplasm because it lacks a nuclear membrane.

Prophylaxis: The prevention of disease, usually through the use of a vaccine that induces protective immunity or a drug that prevents infection.

Protozoa: A single-celled eukaryotic organism.

Reservoir: A population of humans or other animals that is chronically infected with an agent and so serves as a source of infections for others.

Serotype: A strain of a bacterium that can be distinguished by its reaction with a specific antiserum.

Signs: The obvious external signs of a disease determined by clinical examination.

Symptoms: The effects of a disease as reported by a patient.

Syndrome: A group of clinical signs and symptoms that consistently occur together.

Vector: An organism such as an insect that serves as a means of spread of an infectious agent from one individual to another.

Viremia: Viruses in the bloodstream.

Virulence: A measure of the degree of pathogenicity of an infectious agent.

Virus: A molecular construct consisting of genetic material (nucleic acid) enclosed within a protein coat. Some viruses may also be enclosed in a lipid envelope. They can only grow within living cells.

Zoonosis: An infectious disease acquired from an animal source.

Further reading

[1] Abutaleb Y, Paletta D. Nightmare scenario. Inside the Trump administration's response to the pandemic that changed history. New York, NY: Harper Collins; 2021.

[2] Adams GW. Doctors in blue. Baton Rouge, LA: Louisiana State University Press; 1996.

[3] Andam CP, Worby CJ, Chang Q, Campana MG. Microbial genomics of ancient plagues and outbreaks. Trends Microbiol 2016;24:978—90.

[4] Anderson RM, May RM. Population biology of infectious diseases. Parts 1 and 2. Nature 1799;280; 361—7; 455—61.

[5] Barry JM. The great influenza: the epic story of the deadliest plague in history. New York, NY: Viking, Penguin Group; 2004.

[6] Blevins SM, Bronze MS. Robert Koch and the golden age of microbiology. Int J Infect Dis 2010;14:e744—51.

[7] Bollet AJ. Plagues and poxes; the impact of human history on epidemic diseases. New York, NY: Demos Medical Publishing; 2004.

[8] Brandt AM. No magic bullet. A social history of venereal disease in the United States since 1880. Cambridge, MA: Oxford University Press; 1985.

[9] Bridges EL. Uttermost part of the earth. Indians of Tierra del Fuego. New York, NY: Dover Publications Inc.; 1949.

[10] Cantor NF. In the wake of the plague: the black death and the world it made. New York, NY: Perennial: Harper Collins; 2001.

[11] Carrell JL. The speckled monster. A historical tale of battling smallpox. New York, NY: Dutton: Penguin Group; 2003.

[12] Carter R, Mendis KN. Evolutionary and historical aspects of the burden of malaria. Clin Microbiol Rev 2002;15(4):564—94.

[13] Cartwright FF, Biddiss M. Disease and history. 2nd ed. Guilford: Sutton Publishing; 1972.

[14] Chase M. The barbary plague. The Black Death in Victorian San Francisco. New York, NY: Random House; 2003.

[15] Coker RJ. From chaos to coercion: detention and the control of tuberculosis. New York, NY: St Martin's Press; 2000.

[16] Cox FEG. History of the discovery of the malaria parasites and their vectors. Parasites Vectors 2010;3:5—13.

[17] Cunningham HH. Doctors in gray. Baton Rouge, LA: Louisiana State University Press; 1960.

[18] Daniel TM. The impact of tuberculosis on civilization. Infect Dis Clin North Am 2004;18:157—65.

[19] Defoe D. A journal of the plague year. Mineola, NY: Dover Publications; 2001.

[20] Diamond J. Guns, germs and steel. New York, NY: Norton; 1997.

[21] Dubos R, Dubos J. The white plague. New Brunswick, NJ: Rutgers University Press; 1952.

[22] El-Najjar MY. Human treponematosis and tuberculosis: evidence from the new world. Am J Phys Anthropol 1979;51:599—618.

[23] Esposito E. The side-effects of immunity: Malaria and African Slavery in the United States. Am Econ J Appl Econ 2022;14(3):290—328.

[24] Fen EA. Pox Americana: the great smallpox epidemic of 1775—82. New York, NY: Hill and Wang; 2001.

[25] Fried S. Rush: revolution, madness and the visionary doctor who became a founding father. New York, NY: Penguin—Random House; 2018.

[26] Frith J. History of tuberculosis. Part 1—phthisis, consumption and the white plague. J Mil Vet Health 2019;22(2):29—35.

[27] Frith J. History of tuberculosis part 2—the sanatoria and the discoveries of the tubercle bacillus. J Mil Vet Health 2019;22(2):35—44.

[28] Grob GN. The deadly truth: a history of disease in America. Cambridge, MA: Harvard University Press; 2002.

[29] Helbert M. Flesh and bones of immunology. Philadelphia, PA: Mosby/Elsevier; 2006.

[30] Hoff B, Smith C. Mapping epidemiology. A historical atlas of disease. New York, NY: Franklin Watts/Grolier Publishing; 2000.

[31] Honigsbaum M. The pandemic century. One hundred years of panic, hysteria and hubris. Cambridge, MA: Penguin; 2020.

[32] Jackson M. The Routledge history of disease. New York, NY: Taylor & Francis Group; 2017.

[33] Jones JH. Bad blood: the Tuskegee syphilis experiment. New York, NY: Free Press; 1981.

[34] Kelly J. The great mortality. New York, NY: Harper Collins; 2005.

[35] Kolata G. Flu: the story of the great influenza pandemic of 1918 and the search for the virus that caused it. New York, NY: Farrar, Straus and Giroux; 1999.

[36] Levy B. Conquistador. Hernan Cortes, King Montezuma, and the last stand of the Aztecs. New York, NY: Bantam Books, Random House; 2008.

[37] MacQuarrie K. The last days of the Incas. New York, NY: Simon and Schuster; 2007.

[38] Mann CC. 1491: new revelations of the Americas before Columbus. New York, NY: Alfred Knopf; 2005.

[39] Mann CC. 1493: uncovering the new world Columbus created. New York, NY: Vintage Books. Random House Inc.; 2011.

[40] Marineli F, Tsoucalas G, Karamanou M, Androutsos G. Mary Mallon (1869—1938) and the history of typhoid fever. Ann Gastroenterol 2013;26:132—4.

[41] McCall S, Vilensky JA, Gilman S, Taubenberger JK. The relationship between encephalitis lethargica and influenza: a critical analysis. J Neurovirol 2008;14(3):177—85. https://doi.org/10.1080/13550280801995445.

[42] Moote AL, Moote DC. The great plague: the story of London's most deadly year. Baltimore, MD: Johns Hopkins University Press; 2004.

[43] Murray JF, Schraufnagel DE, Hopewell PC. Treatment of tuberculosis: a historical perspective. Ann Am Thorac Soc 2015;12(12):1749—59.

[44] Nunn N, Qian N. The Columbian exchange: a history of disease, food and ideas. J Econ Perspect 2010;24(2):163—88.

[45] Oldstone MBA. Viruses, plagues and history. 2nd ed. Oxford: Oxford University Press; 2010.

[46] Oshinsky DM. Polio: an American story. Oxford: Oxford University Press; 2005.

[47] Pepin J. The origins of AIDS. 2nd ed. Cambridge: Cambridge University Press; 2021.

[48] Pierce JR, Writer J. Yellow Jack: how yellow fever ravaged America and Walter Reed discovered its deadly secrets. Hoboken, NJ: John Wiley; 2005.

[49] Quammen D. The chimp and the river. How AIDS emerged from an African forest. New York, NY: WW Norton Co; 2015.

[50] Quammen D. Ebola: the natural and human history of a deadly virus. New York, NY: WW Norton Co; 2014.

[51] Quammen D. Spillover: animal infections and the next human pandemic. New York, NY: WW Norton Co; 2013.

[52] Ryan F. The forgotten plague: how the battle against tuberculosis was won—and lost. Boston, MA: Little, Brown and Company; 1993.

[53] Schilts R. And the band played on. New York, NY: St. Martin's Griffin; 2003.

[54] Shah S. The fever: how malaria has ruled humankind for 500,000 years. New York, NY: Sarah Crichton Books; 2010.

[55] Sherman IW. The power of plagues. Washington, DC: American Society of Microbiology; 2006.

[56] Thomas H. Conquest: Montezuma, Cortes, and the fall of old Mexico. New York, NY: Touchstone Books; 1993.

[57] Tizard I. A history of vaccines and their opponents. San Diego, CA: Academic Press/ Elsevier; 2023.

[58] Tizard I. Veterinary immunology: an introduction. 11th ed. St Louis, MO: Elsevier; 2024.

[59] Tizard I. Immunology: an introduction. 4th ed. Philadelphia, PA: Saunders College Publishing; 1995.

[60] Torrey EF, Yolken RH. Beasts of the earth: animals, humans and disease. New Brunswick, NJ: Rutgers University Press; 2005.

[61] Verghese A. My own country: a doctor's story of a town and its people in the age of AIDS. New York, NY: Vintage Books; 1994.

[62] Walters MJ. Six modern plagues and how we are causing them. Washington, DC: Island Press/Shearwater Books; 2003.

[63] Webster N. A brief history of epidemic and pestilential diseases; with the principal phenomena of the physical world, which precede and accompany them, and the observations deduced from the facts stated. In Two Volumes. Hartford, CT: Hudson and Goodwin; 1799.

[64] Willrich M. Pox: an American history. New York, NY: Penguin Press; 2011.

[65] Woodward SD. The story of smallpox in Massachusetts. In: Massachusetts. Annual oration; 1932. http://www.massmed.org.

[66] Wooten HG. The polio years in Texas. Battling a terrifying unknown. College Station, TX: Texas A&M Press; 2009.

[67] Wright L. The plague year: America in the time of COVID. Alfred A. New York, NY: Knopf; 2021.

[68] Zinsser H. Rats lice and history. Boston, MA: Little Brown Company; 1935.

Index

Note: Page numbers followed by b indicate boxes and f indicate figures.